Seawater Reverse Osmosis Desalination

Assessment and Pre-treatment of Fouling and Scaling

Seawater Reverse Osmosis Desalination

Assessment and Pre-treatment of Fouling and Scaling

SERGIO G. SALINAS-RODRÍGUEZ

JAN C. SCHIPPERS

GARY L. AMY

IN S. KIM

MARIA D. KENNEDY

Published by: **IWA Publishing**
Unit 104 – 105, Export Building
1 Clove Crescent
London E14 2BA, UK
Telephone: +44 (0)20 7654 5500
Fax: +44 (0)20 7654 5555
Email: publications@iwap.co.uk
Web: www.iwapublishing.com

First published 2021

British Library Cataloguing in Publication Data
A CIP catalogue record for this book is available from the British Library

Library of Congress Cataloguing in Publication Data
A catalogue record for this book is available from the Library of Congress

Reference:
Salinas Rodriguez, S. G., Schippers, J. C., Amy, G. L., Kim, I. S. & Kennedy, M. D. (eds.) (2021). *Seawater Reverse Osmosis Desalination: Assessment and Pre-treatment of Fouling and Scaling*, London: IWA Publishing. doi: 10.2166/9781780409863

Cover design: Hans Emeis
Graphic design: Hans Emeis

ISBN 9781780409856 (Hardback)
ISBN 9781780409863 (eBook)

Foreword

Seawater Reverse Osmosis Desalination:
Assessment and Pre-treatment of Fouling and Scaling

Editors:

Sergio G. Salinas, Jan C. Schippers, Gary L. Amy, In S. Kim*, Maria D. Kennedy

IHE Delft Institute for Water Education, Delft, The Netherlands

This book is an introduction to desalination and the development of membrane technology around the world in the effort to constantly improve the system and lower the cost of water purification to bring water to a growing thirsty population. The problem of fouling and scaling in sea and brackish water reverse osmosis plants is still a major issue in the desalination process and is thus stressed in the book.

I am often reminded of my visit to Prof. Ronald Probstein at MIT after I founded the journal Desalination in 1966. He asked me what was the most important problem in desalination. The answer was "fouling" which we see is still with us today.

The textbook focuses on theory and practice and is intended for designers, operators, consultants, suppliers and students. The chapters are written by IHE's present and former staff and by former students who are now active professionals in the field of desalination. And essential contributions have been made by scientists from GIST, S. Korea, QEERI, Qatar, KAUST, Saudi Arabia, Synauta and Northwest A&F University, China.

The editors of this book are foremost eminent scientists and teachers in the world of education and practice in research and industry who play a prominent role in the field of desalination and water treatment. Over 23,000 water professionals from more than 190 countries have been educated at IHE and now apply their expertise back home. One of their first graduates has led the effort in the team of editors and authors of this book.

IHE Delft Institute for Water Education in The Netherlands is the largest Institute for Water Education in the world. It is an eminent international graduate water education institute which confers MSc and PhD degrees in collaboration with partner universities. It is under the auspices of UNESCO in keeping with its aims of training students of water technology to help in capacity building of competence of students who will serve as local experts, mainly in the global south to bring clean water to a growing population in a sustainable manner.

---

* Gwangju Institute of Science and Technology, S. Korea

It is heartwarming for me to introduce this book composed by colleagues and friends who are bringing the desalination technology to the laps of eager students and seasoned colleagues in the fascinating world of desalination technology and practice.

Miriam Balaban
Desalination and Water Treatment, Editor in Chief
European Desalination Society, Secretary General

Contributors

	Chapter(s)
Prof. Miriam Balaban *European Desalination Society (EDS), Italy*	*Foreword*
Dr. Abayomi Babatunde Alayande *Gwangju Institute of Science and Technology (GIST), South Korea*	5
Prof. Dr. Gary L. Amy *Clemson University, USA*	10
Prof. Dr. In S. Kim *Gwangju Institute of Science and Technology (GIST), South Korea*	5
Prof. Em. Dr. Jan C. Schippers *IHE Delft Institute for Water Education, Netherlands*	1, 2, 3, 4, 7, 8, 9
Dr. Lijo Francis *QEERI, Qatar, and KAUST WDRC, Saudi Arabia*	10
Dr. Loreen O. Villacorte *Grundfos Holding A/S, Denmark*	5
Prof. Dr. Maria D. Kennedy *IHE Delft Institute for Water Education, Netherlands*	2, 3, 4, 8, 9
Dr. Mike Dixon *Synauta, Canada*	5
Nasir Mangal, MSc *IHE Delft Institute for Water Education, Netherlands*	8
Prof. Dr. Noreddine Ghaffour *King Abdullah University of Science and Technology (KAUST), Saudi Arabia*	10
Dr. Sergio G. Salinas-Rodríguez *IHE Delft Institute for Water Education, Netherlands*	1, 2, 3, 4, 8, 9
Dr. Siobhan F. E. Boerlage *Boerlage Consulting, Australia*	4
Dr. Thanh-Tin Nguyen *Gwangju Institute of Science and Technology (GIST), South Korea*	5
Dr. Victor A. Yangali-Quintanilla *Grundfos Holding A/S, Denmark*	8
Dr. Zhenyu Li *Northwest A&F University., China*	10

About the editors

Sergio G. Salinas-Rodríguez is Associate Professor of Water Supply Engineering at IHE Delft. He is a desalination and water treatment technology professional with experience in Latin America, Middle East, and Europe. Sergio has a PhD in Desalination and Water Treatment from the Technical University of Delft (Netherlands), an MSc in Water Supply Engineering from UNESCO-IHE Institute for Water Education (Netherlands), a Master's in Irrigation and Drainage and a BSc in Civil Engineering from San Simon Major University (Bolivia). He also obtained the University Teaching Qualification, a qualification of pedagogical competences of university teachers. He has over 50 publications in international peer-reviewed journals, book chapters in top publishing houses, and conference proceedings in the areas of seawater and brackish water desalination, water treatment, water reuse, and natural organic matter characterization. Sergio is involved in teaching and curriculum development of the MSc Programme in Urban Water and Sanitation (UWS) at IHE Delft. Furthermore, he has worked in capacity building and in research and innovation projects, and has been project leader in several of them. He has mentored more than 40 MSc students, co-promoted 3 PhD students, and currently supervises 3 PhD students.

Jan C. Schippers is Professor Emeritus in Water Supply Technology at IHE Delft Institute for Water Education, and consultant. He has extensive professional experience in drinking and industrial water supply projects in Morocco, Qatar, United Emirates, Gabon, Cab Verde, Namibia, Uzbekistan, France, Chile and The Netherlands. Jan advised the World Bank and Ministries in the Netherlands.

His specializations are: consultancy, research, training and education, in the field of integral drinking and industrial water production and desalination/membrane related technologies.

Jan gave courses on membrane technology, fouling, scaling and pre-treatment in membrane technology, aquatic chemistry, conventional filtration techniques and membrane bio-reactors in Cyprus, Morocco, China, Jordan, Yemen, Bahrain, Iran, Oman, Saudi Arabia, Chile, Italy and The Netherlands.

Gary L. Amy is Dean Distinguished Professor at Clemson University. Dr. Amy's main areas of expertise are drinking water treatment and wastewater reclamation/reuse, with specific expertise in membrane rejection and fouling, selective adsorption, natural organic matter characterization, disinfection by-product formation and control, and natural systems. More recently, his major research emphasis has been on low-energy seawater desalination technologies, energy-harvesting wastewater treatment processes, and managed aquifer recharge for wastewater reuse. Over a career of 35 years, he has published over almost 400 articles in refereed publications, and supervised almost 50 PhD students. He serves on the editorial boards of Water Supply and Technology (IWA), International Ozone Association (IOA), and the Journal of Drinking Water Engineering (on-line). He has received best paper awards from the Journal AWWA and the Journal of the Water Environment Federation. His PhD students have received best dissertation awards from the AWWA and the IOA. He was recipient of a Fulbright Award for Germany (2003-2004), was invited as Distinguished Lecturer, Korea Brain 21 Program (2002), and was appointed Visiting Scholar at Kyoto University, Japan (2001).

In S. Kim is a Professor in the School of Earth Sciences and Environmental Engineering. He is Vice President for Research (VPR) at Gwangju Institute of Science & Technology (GIST) in Korea. His specific research area is membrane-biotechnology for water reuse, drinking water desalination and renewable energy. He has been working as a member of Presidential Advisory Council on Science and Technology (PACST) in Korea.

He is leading the Global Desalination Research Center (GDRC) as a director (follow-up of SeaHERO Program). He has served president of Korean Desalination Plant Association (KDPA). He is a member of Korean Academy of Science and Technology and a fellow of International Water Association (IWA). He has served the chair of water reuse specialist group in IWA. He has been working as chief review board for environment & renewable energy field in National Research Foundation (NRF).

Prof. Kim was awarded outstanding university service award by GIST (1995), outstanding paper by Korean Federation of Science & Technology Association (2001), Prime Minister award (2008), National Order of Merit (2014) and others. Prof. Kim is serving as editor and associate editor in several international technical journals.

Maria D. Kennedy is Professor of Water Treatment Technology at IHE Delft. She has over 28 years of experience in education, research, consulting and capacity development in water treatment. During the last 28 years, she has been involved in the supervision of over 200 MSc participants and 22 PhD research fellows in the areas of water quality, groundwater treatment, disinfection, advanced oxidation, surface water treatment, desalination and membrane related technology, natural treatment systems, water reuse, water transport and distribution and biological stability. She has over 150 publications in peer reviewed journals (h index > 34), and she has edited several books/book chapters on various aspects of water treatment. Professor Maria Kennedy has organized numerous international short courses in the field of desalination & membrane related technology e.g., in Jordan, Palestine, Oman, Bahrain, Israel, St. Maarten, Iran, Yemen, South Africa, Korea and Chile. She is or has been the director of several (large) capacity development projects in the Middle East region e.g., in Jordan, Palestine, Yemen and Iran. She was also involved in several large EU/Horizon 2020 research projects such as EU MEDINA, EU TECHNEAU, EUROMBRA, H2020 MIDES and EU India H20.

Professor Maria Kennedy is a past president of the European Desalination Society (EDS), and Chairman of the board of directors. She was also a member of the Science and Technology Board of the EU Joint Programming Initiative (JPI), and she is a jury member of several prestigious international technology events such as the USAID Water for Food Desalination Prize (2014 – 2015), the Oman Humanitarian Desalination Challenge and the Aquatech Innovation Award.

Contents

Chapter 1

Introduction to desalination

Sergio G. Salinas-Rodríguez, Jan C. Schippers

The main learning objectives of this chapter are the following:

- Discuss the main drivers and applications for desalination
- Present and discuss the world desalination capacity
- Identify the main desalination technologies
- Present and discuss the energy consumption and costs
- Discuss the environmental concerns and solutions in desalination

1.1 DRIVERS

Desalination capacity of seawater and brackish water has grown rapidly over the last thirty years to reach an existing world capacity of over 100 million cubic meter per year. This growth is driven by the need of alternative water sources to the renewable ones to cope with increasing world population, increasing demand of industry, more water consumption per capita due to an increased economy. By 2050 the world population is expected to reach 9.7 billion (United Nations, *et al.*, 2019).

Despite progress, 2.2 billion people around the world still lack safely managed drinking water, including 785 million without basic drinking water (United Nations and Department of Economic and Social Affairs, 2020).

Cities living along the coast or close to the coast may consider the use of sea water as an alternative source for drinking water production, water for agriculture, or water for industry. Around 680 million people live in low-lying coastal zones - that is expected to increase to a billion by 2050 UN, 2020. Nearly 2.4 billion people live within 100 km of the coast (United Nations, 2017). 65 million live in small island developing states (UN-OHRLLS, 2015). In total, approximately 44 percent of the world's population lives within 150 km of the ocean (UN Atlas of the Oceans).

Editors: Sergio Salinas, Jan Schippers, Gary Amy, In Kim, Maria Kennedy
doi: 10.2166/9781780409863_0001

Besides availability of water, independently of its use, the quality is also important. In this regard, desalination technologies are considered robust technologies capable of removing most contaminants and emerging compounds.

Water use has been growing at more than twice the rate of population increase in the last century (FAO, 2013). Combined with a more erratic and uncertain supply, this will aggravate the situation of currently water-stressed regions, and generate water stress in regions with currently abundant water resources.

The economic and demographic growths are two main drivers for over-abstraction of conventional freshwater resources in various parts of the world, which leads to the situation of water scarcity. Water scarcity is normally considered when the total annual runoff available for human use is less than 1,000 m^3/capita/year (Brown and Matlock, 2011). The rapid increase in the population growth and the trend of rural-urban migration will intensify the issue of water shortage in these countries mainly due to the withdrawal of fresh water to satisfy the demand for municipal and agricultural use (Bremere, *et al.*, 2001).

Water stress already affects every continent (Figure 1). About four billion people live under conditions of severe physical water scarcity for at least one month per year (Mekonnen and Hoekstra, 2016). Around 1.6 billion people, or almost a quarter of the world's population, face economic water shortage, which means they lack the necessary infrastructure to access water (UN Water, 2014). By 2050, 40 % of the world's population is projected to live under severe water stress, including almost the entire population of the Middle East and South Asia, plus significant parts of China and North Africa (UNESCO World Water Assessment Programme, 2020). The main drivers being, population growth, urbanization, and climate change.

Climate change will affect the availability, quality and quantity of water for basic human needs, threatening the effective enjoyment of the human rights to water and sanitation for potentially billions of people (UNESCO World Water Assessment Programme, 2020). The alteration of the water cycle will also pose risks for energy production, food security, human health, economic development and poverty reduction, thus seriously jeopardizing the achievement of the Sustainable Development Goals (UNESCO World Water Assessment Programme, 2020).

There are several technical solutions that can help to solve water scarcity all over the world:

1. **Saving water**	Increasing productivity in agriculture & industry Reducing leakages in public water supply Implementing progressive tariffs for consumption
2. **Water transport**	Normally requires transport over long distances with potential high energy costs
3. **Aquifer storage**	River water during high flow
4. **Water reuse**	Increasing reuse/recycling in industry & domestic wastewater in agriculture
5. **Desalination**	Brackish water, wastewater, seawater

Among the different alternative solutions to solve the issues of water scarcity, desalination is usually only implemented as a last resort where conventional freshwater resources have been stretched to the limit. Yet, desalination can be considered as a drought-proof water source, which does not depend on river flows, reservoir levels or climate change. Desalination may be an option to alleviate scarcity in the industry and coastal cities.

Desalination, or desalting of water, consists of a water treatment process by which sea or brackish water is converted into potable water for supplying communities that have the most difficulty accessing freshwater.

Although the most well-known application of desalination (and related membrane technology) is to produce freshwater from seawater, it can also be used to treat slightly salty (brackish) water, low-grade surface, and groundwater, and treated effluent resources. The current global trend shows that desalination technology is finding new outlets as an alternative source for supplying water to meet growing water demand in most of the water-scarce countries (Bremere, *et al.*, 2001). However, there have been barriers to its widespread adoption of technology mainly due to its cost, energy demand, lack of expertise, and the footprint.

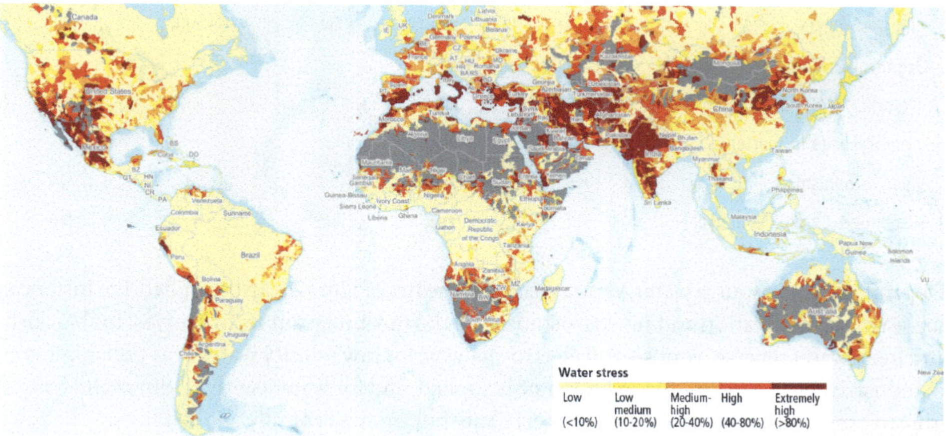

Figure 1 Map of water stress in 2020 (Water resources institute, 2020)

1.2 DESALINATION TECHNOLOGIES

There are several desalination desalination technologies (thermal-based and membrane-based processes) currently employed that have been developed over the years. Six different membrane technologies are applied for the production of drinking and industrial water, namely: microfiltration (MF), ultrafiltration (UF), nanofiltration (NF), reverse osmosis (RO), electro-dialysis (ED), and electro-deionization (EDI).

1.2.1 Reverse osmosis

Reverse osmosis has main applications in seawater and brackish water desalination. Electrodialysis is applied in desalination of brackish water. Nanofiltration is mainly applied for removing of sulphate, hardness and natural organic matter. Ultra- and micro-filtration are applied for removing suspended and colloidal matter and for disinfection of drinking water. Table 1 summarizes a comparison of the removal capacities of various membrane technologies.

Ultra- and micro-filtration are applied: *i*) in drinking water production (for removal of micro-organisms such as viruses, giardia, and cryptosporidium, for removal of suspended and colloidal matter, and for algae removal); *ii*) as pre-treatment for RO and NF (for removal of suspended & colloidal matter e.g., turbidity, SDI & MFI, organic polymers e.g., transparent exo-polymer particles); *iii*) in wastewater treatment as membrane bio-reactors (MBR) or in water reuse (for removal of suspended and colloidal matter, removal of bacteria, cysts, and viruses).

Table 1 Comparison of removal of inorganic and organic compounds, micro-organisms, and suspended and colloidal matter by different membrane technologies

Removal	RO	NF	UF	MF	ED
Inorganic compounds					
mono-valent: Na^+, Cl^-	+	+/-	No	No	+
di-valent: SO_4^{2-}, Ca^{2+}	++	+	No	No	+
Organic compounds					
synthetic organic compounds	+	+	-	-	-
natural organic matter	+	+	-	--	
Micro-organisms	+	+	+	+	No
Suspended / colloidal matter	+	+	+	+/-	No

Depending on the source water, various technologies (see Figure 2) can be applied. For instance, for seawater, distillation and reverse osmosis are the most relevant technologies, for brackish and fresh water reverse osmosis and electro-dialysis; for low salinity water or as polishing step in industrial water treatment ion exchange is applied, and for waters with hardness and colour (due to presence of natural organic matter) nano-filtration is typically applied.

Electrodialysis is a separation process based on the transport of ions through membranes as a result of an electrical current.

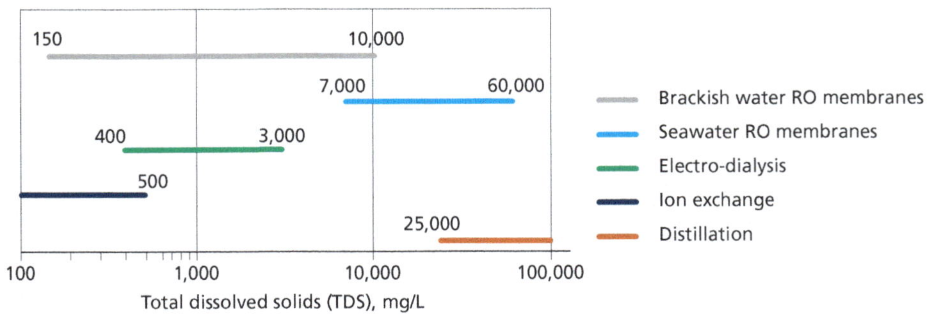

Figure 2 Normal operation range of desalting technologies based on salinity of water

Membranes consist mainly of capillaries or flat sheets having a thin membrane layer. They are usually made of organic polymers, with very small pores. Membranes are assembled in membrane elements. Whether particles can pass a membrane or (partially) not depends mainly on the size of the particles and the size of the pores in the membranes. The mechanism of sieving is governing for an important part the process. Next to sieving based on size, rejection can also be caused by other characteristics, such as the electrical charge of membrane pores, the nature of the membrane material, the electrical charge of particles (in particular charge of the ions), the valence of the ions, the diffusion coefficient of particles (ions), and the process conditions e.g., temperature, salinity, filtration rate (flux e.g., L/m^2/h) play an important role as well.

Example 1– Rejection of ions

The size of inorganic ions (including attached water molecules) is the following:
Positive ions: H^+ 0.053 nm, K^+ 0.25 nm, Na^+ 0.37 nm, Ca^{2+} 0.62 nm, Mg^{2+} 0.70 nm.
Water molecule: H_2O 0.33 nm.
Negative ions: Cl^- 0.24 nm, NO_3^- 0.26 nm, HCO_3^- 0.42 nm, SO_4^{2-} 0.46 nm.
Which ions are better rejected by RO membranes: Na^+ or Ca^{2+}? What about Cl^- and SO_4^{2-}?
Why is H_2O passing membranes better than Cl^-?

Answers:

Ca^{2+} is larger in size than Na^+ (0.62 nm > 0.37 nm). In addition, Ca^{2+} is divalent where Na^+ is mono-valent.
Water is neutrally charged in comparison with Cl^-.

Reverse osmosis makes use of membranes with small pores (< 1 nm pore size) e.g., flat sheets, capillaries. Water is forced to flow through these pores with the help of (high) pressure to overcome the osmotic pressure and the hydraulic resistance of the membrane. Salts cannot pass the small pores (are rejected due to slow diffusion and sieving mechanism).

The salinity of seawater, brackish water or fresh water is the result of presence of cations and anions. The most important combination of these ions is sodium / chloride. Several other cations and anions are usually present as well e.g., calcium, magnesium, potassium, ammonium, sulphate, hydrogen carbonate, nitrate, fluoride, boron.

Figure 3 illustrates the various components of a RO desalination plant, including the pre-treatment, the high pressure pump units, the assembly of RO elements in pressure vessels, and the post-treatment required to re-mineralize the RO permeate water. With the help of energy recovery devices, the pressure of the RO concentrate after leaving the pressure vessel is transferred hydraulically to the feed water. Pre-treatment needs to guarantee that the RO feedwater has a value of a silt density index (SDI) less than 5 but preferably less than 3. Post-treatment will introduce back minerals in the RO permeate and will make sure the final water is fit for purpose.

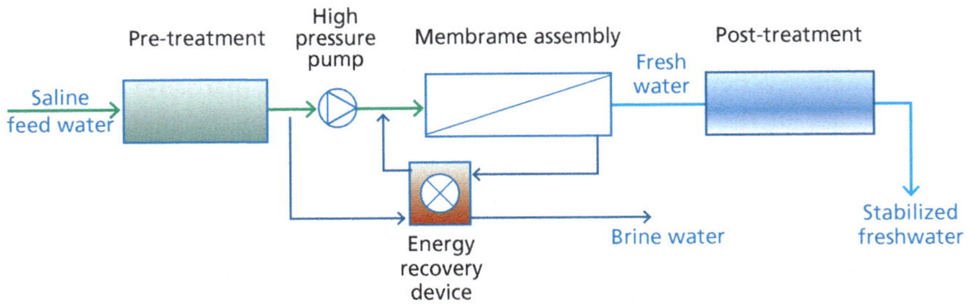

Figure 3 Schematic of a RO system including pre-treatment and post-treatment (Adapted from Buros, 1980)

Figure 4 illustrates the placement of the RO elements inside a RO pressure vessel. O-rings and brine seals make sure that there is no mix between the various water streams. Typically, in seawater RO, the recovery ranges 40 to 50 % with 6 to 8 elements placed in series in one stage.

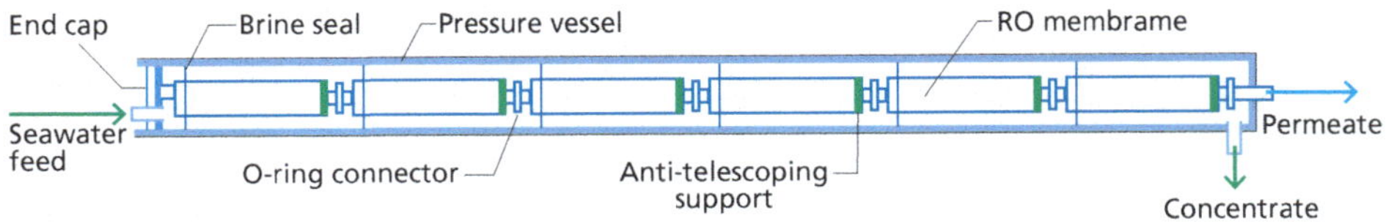

Figure 4 Schematic of a RO pressure vessel containing 6 RO elements (Adapted from Buros, 1980)

1.2.2 Distillation

The theory of distillation, obtaining clean water out of steam, is not new. It has been employed by alchemists, chemists for the separation of e.g., alcohol from water. Distillation of saline water for potable use was of early interest to sailors on long sea voyages. Patents were issued in the 17th century in England for commercial units. Distillation is the oldest known process for producing fresh water from seawater.

When salt water is boiled, the salt ions remain behind as freshwater vapor is boiled away. In the distillation process water is first boiled and then the steam (water vapor) is cooled in a clean vessel. This cooling condenses the steam (water vapor) to water again.
The energy required to evaporate 1 kg water with temperature of 25 °C at 100 °C and 1 bar (1 atm) amounts:

- The specific heat capacity of water is 4,200 joules per kilogram per degree Celsius (J/kg °C). This means that it takes 4,200 J to raise the temperature of 1 kg of water by 1 °C (BBC, 2021)
- increase temp 25 °C → 100 °C: (100 °C – 25 °C) × 4.2 kJ/kg °C = 315 kJ/kg
- heat of vaporization at 100 °C and 1 bar = 2256 kJ/kg
- making a total = 2571 kJ/kg

In order for water vapor to condense to a liquid, it is necessary that the heat of condensation is removed.

The heat of condensation is equal to the heat of vaporization. A simple distillation unit is presented in Figure 5.

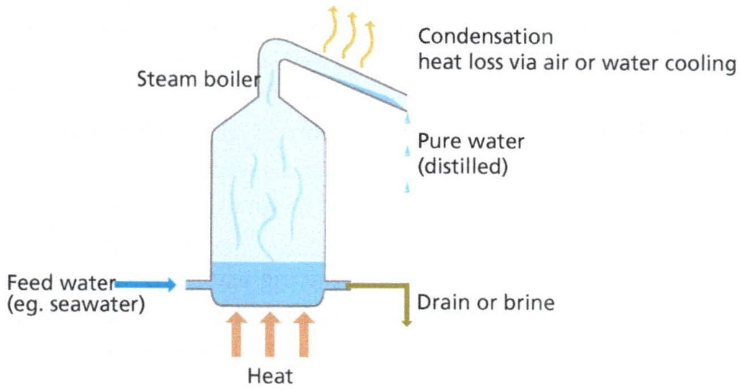

Figure 5 Schematic of a simple distillation unit

This unit has a very high energy consumption as for the production of 1 kg fresh water about 2600 kJ is needed. Moreover, the efficiency of the addition of the heat is poor.

Table 2 Minimum energy requirements for seawater desalination at 25 °C (Spiegler and El-Sayed, 2001)

Conversion R	Theoretical separation energy	
0%	0.71 kWh/m^3	2.6 MJ/m^3
25%	0.82 kWh/m^3	3.0 MJ/m^3
50%	0.99 kWh/m^3	3.6 MJ/m^3
75%	1.35 kWh/m^3	4.9 MJ/m^3
100%	3.10 kWh/m^3	11.2 MJ/m^3

Energy consumption for distillation is much higher than for membrane-based desalination with RO. For instance, to raise 1 kg water 10 °C in temperature, 4.2 kJ/kg energy is needed. The heat of vaporization at 100 °C equals 2256 kJ/kg. Consequently, the heat required for evaporation of 1 m^3 water of 25 °C amounts about 2600 MJ. The heat of combustion of oil is about 40 MJ/kg. The current world market price of crude oil amounts about 0.27 $/kg (at 40 $ per crude oil barrel (July 2020), 1 barrel = 160 L, density about 0.9 kg/L).

The evaporation costs for 1 m^3 water (* 100% combustion efficiency assumed), amounts (2,600 MJ/m^3 /40 MJ/kg) × 0.27 $/kg (July 2020) = 17.6 $/m^3, which is much too high to be payable, compared to desalination with RO (< 1$/m^3).

Finally, the boiling during the vaporization process is violent and salt water droplets are entrained in the vapor produced. These droplets must be removed to keep the salt content of the condensate low.

There are three main methods that have been used for implementing vaporization in distillation units in the past 25 years. These are: submerged tube, flash; thin film.

Submerged tube: Water is brought to the boiling point by the addition of heat in tubes which are submerged in a pool of water. Configurations that have been used include helical, curve, and straight tube bundles, with steam being condensed on the inside of the tubes to supply the heat. When submerged in saline solutions, these tubes are subject to the formation of scale on the outside of the tubes since calcium carbonate and calcium sulphate precipitate out of the solution at high temperatures. This scale can severely reduce the heat transfer of the tubes. Submerged tube design is frequently used in small single stage units.

Flash: Vigorous boiling can be promoted by introducing water into a chamber through an orifice, thereby reducing the water's pressure below that of the equilibrium vapor pressure required for boiling. This causes the water to immediately begin to boil vigorously when introduced into the chamber. This method is used in the majority of plants built in the past 50 years. One advantage to flash distillation is that once the flashing process begins the saline water does not come in contact with hotter heat transfer surfaces. Consequently, the chance of scaling (precipitation of calcium carbonate and calcium sulphate) is limited.

Thin film: In this process saline water is applied as a film on the inside or outside of the tubes which are being heated by condensation of vapor (steam) on the opposite side. Two methods are applied: falling film, and spray film. The film is usually applied on the inside of vertical tubes and the outside of horizontal tubes. One disadvantage to thin film vaporization is that heat-induced scaling can occur on the surfaces of the heated tubes.

Condensation of steam (vapor) takes place on the inside or outside of tubes. These tubes are made of material which is capable of a high degree of heat transfer (e.g., copper alloys). For efficient condensation to occur the surface must rapidly remove the heat and allow the condensed liquid to flow to a collection point.

The heat input into the distillation unit must be balanced by the heat output of the unit. The heat input considers solar energy, condensing steam and hot water. The heat output considers: radiation and general heat loss (usually of minor importance), distillate, brine, cooling water. The temperature of the distillate and brine steams is elevated above the ambient feedwater temperature.

There are three major distillation processes being used in the industry today:

- Multi-effect evaporation / distillation (MED)
  - submerged tube evaporation (ST),
  - vertical tube evaporation (VTE),
  - horizontal tube evaporation (HTE).
- Multi-stage flash (MSF).
- Vapor compression (VC).

In **Multi effect evaporators** each effect steam (vapor) is condensed on one side of a tube and the heat of condensation derived from this is utilized to evaporate saline water on the other side of the tube wall (Figure 6).

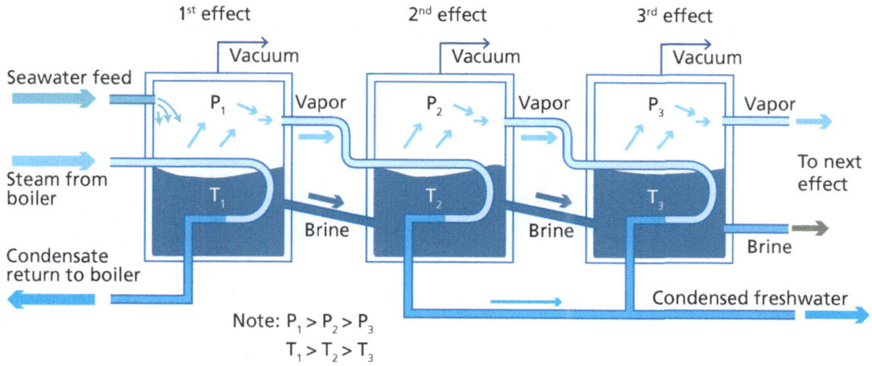

Figure 6 Example of a multi effect distillation unit with 3 effects (Adapted from Buros, 1980)

The subsequent use and reuse of the heats of vaporization and condensation reduces the heat consumption significantly. In theory each effect produces about: 1 ton fresh water per ton of steam (supplied by a boiler). Consequently, when 3 effects are applied 3 tons of fresh water per ton of steam are produced. In practice, however, the steam economy in each effect is not 1.0 but 0.7 to 0.85, which means that overall "steam economy" is lower than the theoretical value. Steam economy is defined as the number of tons water produced for each ton of steam utilized. A "steam economy" (for the whole plant) amounts in practice about 10, which means that more than 10 effects are applied to achieve the economy. The energy costs will be reduced when more effects are applied. However, the investment costs are higher when more effects are installed.

In Figure 7, the principle of the **submerged tube multiple effects distillation process** is shown. One of the last major municipal multiple-effect submerged-tube distillation plants was built in 1958. It was a 10,000 m^3/day facility consisting of 5 units (2,000 m^3/day) each having 6 effects. These units were operated for 22 years before being taken out of service in 1980.

The greatest problems with the submerged-tube units are: the brine pool cannot be vaporized as efficiently as in other configurations, because of the smaller relative surface area exposed; scale often forms on the hot submerged tubes and produces a coating which reduces the heat transfer. Submerged-tube plants utilizing waste heat for industrial and marine installations are still manufactured.

The **vertical-tube evaporator** configuration was intended to resolve some of the problems of the submerged tube configuration. Compared to the submerged-tube configuration the vertical-tube units have the potential for increased thermal efficiency and reduced scaling. The vertical-tube plants are more complex and require more external piping and pumps.

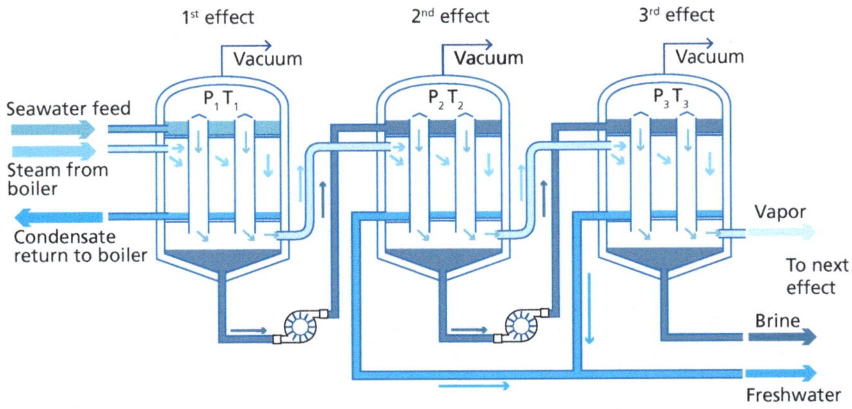

Figure 7 The principle of multi effect vertical tube evaporation process (Adapted from Buros, 1980)

The principle of the **horizontal tube multi effect distillation process** is shown in Figure 8. The principle of the operation is the same as for the vertical-tube evaporator. However, the brine and steam are applied on the opposite sides of the tubes in both systems. Scale formation and removal is significantly less problematic in horizontal-tube than in vertical-tube units.

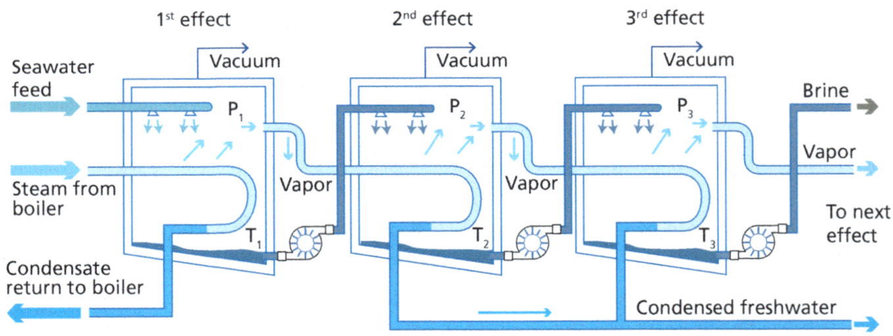

Figure 8 The principle of multi effect horizontal tube evaporation process (Adapted from Buros, 1980)

The principle of the **multi stage flash distillation** process is shown in Figure 9. In this process the incoming seawater is first heated by the condensing vapor and before entering the first stage the feedwater is further heated by externally supplied steam. This raised the feedwater to its top temperature after which it is passed through the various stages where flashing takes place.

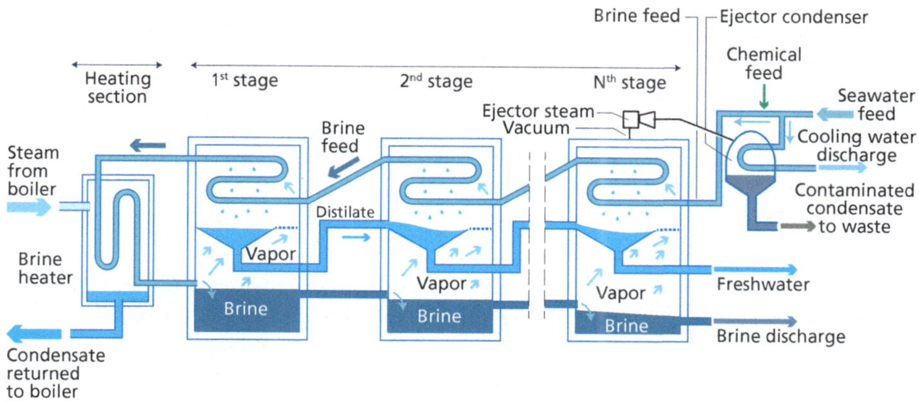

Figure 9 The principle of multi stage flash (MSF) process (Adapted from Buros, 1980)

The number of stages in a MSF-plant varies depending on the application, efficiency desired etc. The number usually ranges from 20 to 50. The number of stages is in general increased to improve the efficiency of recovery heat. The "steam economy" amounts about 6 to 12 depending on the design of the plant.

The **vapor compression** process differs from the other distillation processes in that it does not utilize an external heat source. It makes use of the compression of water vapor (by e.g., a compressor) to increase the vapor's pressure and condensation temperature (See Figure 10). The compressor serves a dual purpose: it compresses the vapor raising its condensation temperature, and it lowers the pressure on the feedwater brine and reduces its boiling temperature. There are two methods used to compress the water vapor: mechanical compressor, and steam ejector.

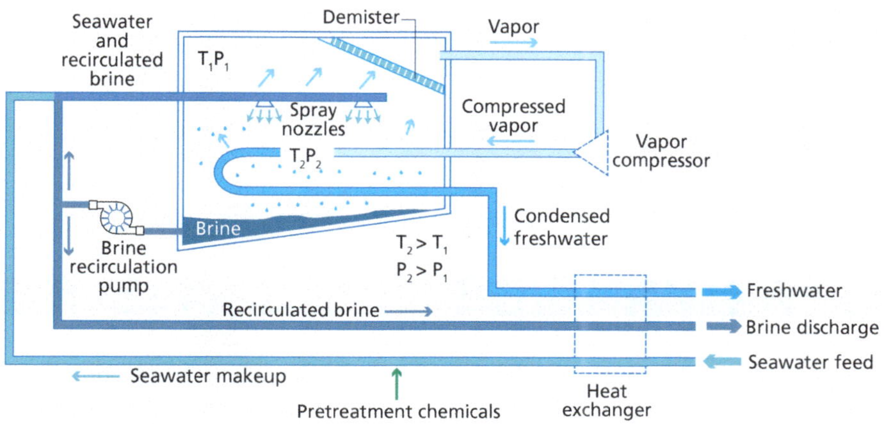

Figure 10 The principle of vapor compression (VC) process (Adapted from Buros, 1980)

The major problem in operating seawater distillation plants is the formation of scale caused by precipitation of: calcium carbonate, calcium sulphate and magnesium hydroxide. This phenomenon occurs due to the increase of the brine temperature and the increase of the concentration due to evaporation. Scale formation can be prevented in three ways: controlling the temperature, controlling the pH (for calcium carbonate and magnesium hydroxide), or introducing additives (for calcium sulphate) e.g., sodium-hexa-meta-phosphate (SHMP), poly acrylic acids, etc.

1.2.3 Energy consumption and cost

From the start of all six membrane technologies, energy was a major issue. Electrodialysis makes use of an electrical current, where the energy consumption is proportional with the amount of removed salts (ions). Reverse osmosis, nanofiltration and ultra- and microfiltration are pressure driven membrane techniques, where water is forced to flow through small pores in RO, NF, UF and MF membranes.

Electrical power is traditionally generated with: *i*) diesels, using diesel as a energy source, *ii*) steam /turbines using oil, coal and gas as an energy source, iii) natural gas turbines using natural gas. The result of this approach is the large amounts of carbon dioxide produced, which is responsible for global warming. Renewable energy is available from various sources, including: *i*) hydropower stations which are commonly applied when available, *ii*) wind farms which are gradually implemented, and *iii*) photo voltaic generation through solar photo voltaic farms.

Table 3 Energy consumption and pressure for various treatment technologies

Technology	Pressure, bar	Energy consumption, kWh/m³	Heat	Cost, euro or $ per m³
Conventional drinking water	0.1 - 0.2	-		
Electro-dialysis				0.25 - 0.50
Ultra- and micro-filtration	0.5 - 2	0.1 - 0.2	-	0.05 - 0.10
Nano-filtration	5 - 10	0.3 - 0.5	-	0.15 - 0.25
Brackish RO	10 - 20	0.5 - 1.0	-	0.25 - 0.50
Seawater RO	50 - 90	3 - 4	-	0.50 - 1.00
Distillation	-	1 - 4	160 MJ/m³	
Cost of energy		**0.05-0.1 $/kWh**	**5-15 $/GJ**	

The ranges of energy consumption and pressure, including a reference production cost for various technologies are presented in Table 3. The treatment of freshwater by conventional water treatment is the less energy demanding in comparison with the other technologies. The energy consumption for MF/UF is also comparable with the one of conventional drinking water treatment. As the pore sizes of the membranes decreases, more pressure needs to be applied and thus, the energy consumption also increases. In membrane-based sea water desalination, the energy consumption is in average 3-4 kWh/m^3 with pressure range between 50 and 90 bar.

Example 2– Energy consumption and cost in a sea water RO plant
What is the power required for a seawater RO plant with a capacity of 40,000 m^3/day (14.6×10^6 m^3/year)? And what is the power cost?

Answer:
Considering an average energy consumption of 3 kWh/m^3.
40,000 m^3/day × 365 days/year × 3 kWh/m^3 = 43,800,000 kWh/year
or equivalent to: 43,800,000 kWh/y / (365 d/y × 24 h/d) = 5,000,000 kW = 5 MW
In case of using renewable energy, a wind turbine of 5 MW generates on average 20 % power = 1 MW. Consequently, 5 wind turbines are needed.
Considering an energy cost of 0.10 $/kWh, the power cost per year will be:
3 kWh/m^3 × 40,000 m^3/d × 365 d/year × 0.10 $/kWh = 4.4×10^6 $/year.

The production cost in sea water reverse osmosis plants can be divided in the following categories, as presented in Figure 11, energy consumption represents about 40 % of the total production cost, amortization also amounts for about 40 %, staff costs amounts 4-11 %, consumption of chemicals during treatment 2-6.5 %, costs of RO membranes 2-5%, plant maintenance 3.5-4.5 %, and cleaning of the RO membranes about 0.2-0.3 %. Any optimization in energy consumption will decrease the production cost. It is expected that by using renewables energies, the energy costs will decrease while at the same time minimizing effects on environment.

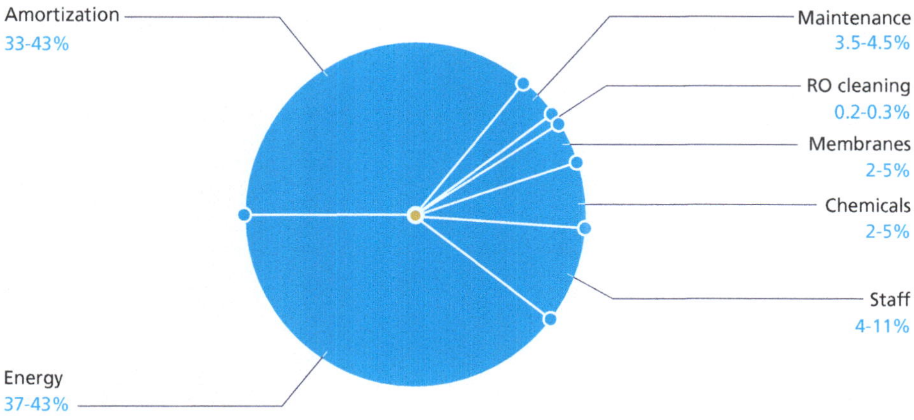

Figure 11 Production costs in sea water reverse osmosis plants (Sanz, 2020)

1.3 GLOBAL DESALINATION CAPACITY

Currently, about 21,000 desalination plants are operational with a production capacity larger than 100 Mm^3/d located all over the world in about 180 countries. Although brackish water and waste water treatment methods offer a great future potential, desalination of seawater will remain the dominant desalination process for years to come. Table 4 presents a summary of the existing number of plants, their status, and plant capacity as reported in 2020. It is remarkable to point out that there are in 2020 about 275 plants under construction with a capacity of about 11 Mm^3/d.

Table 4 Summary of the world desalination capacity in 2020 (Global Water Intelligence, 2020)

Nr. Plants	Desalination plants status	Capacity, m^3/d
20,957	Total plants	115,625,178
3,823	Off-line	7,193,546
16,860	In operation	97,305,664
274	Under construction	11,125,968
17,134	In operation + under construction	108,431,632

Figure 12 presents the global historical cumulative production capacity of desalination plants for all raw water sources, including: seawater, brackish water, fresh water, wastewater, pure water. Over two-thirds of the current total capacity is produced by membrane-based desalination technology (reverse osmosis) and less than one-third is produced by thermal processes (multi-stage flash distillation, and multi-effect distillation). One of the reasons why sea water reverse osmosis production capacity grows faster than thermal processes is the lower investment costs and the lower energy consumption (3-4 kWh/m^3). In the last thirty years, the online production capacity has increased from 13.7 Mm^3/d to the current 101.6 Mm^3/d, which is about 7.5x more capacity. In the last 10 years, the growth in desalination capacity has been about 41 % and mostly related to the new plants making use of reverse osmosis as main desalination technology. It is expected that by 2030 the world desalination capacity will double (Sanz, 2020).

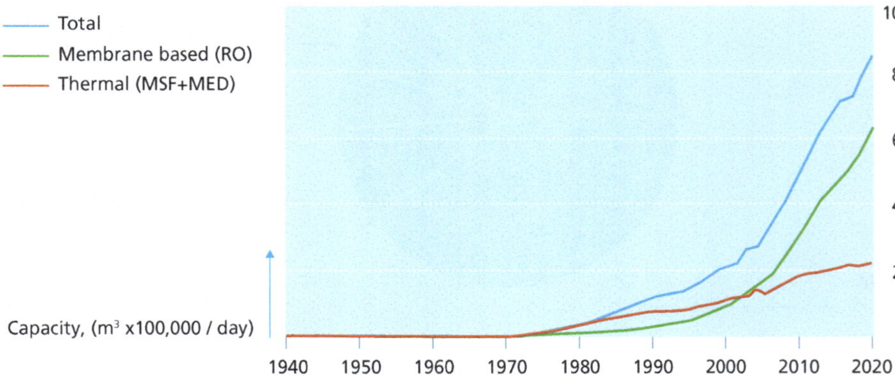

Figure 12 Total desalination capacity in the world (seawater, brackish, wastewater, and fresh water) (Global Water Intelligence, 2020)

The implementation of desalination plants has increased in many parts of the world. Much of the growth of the desalination capacity takes place in the sea water desalination industry, although wastewater desalination and brackish water desalination is becoming more relevant. Besides the number of desalination plants increasing, also the capacity of the plants has significantly increased over time, as presented in Figure 13, illustrating the preference for XL plants (>50,000 m^3/d) over the large size (10,000-50,000 m^3/d), medium size (1,000-10,000 m^3/d) and small capacity ones (<1,000 m^3/d). More XL sea water RO plants are expected in the future and thus reliable pre-treatment systems are mandatory for these XL plants as frequent cleaning-in-place (CIP) is difficult (>1/year).

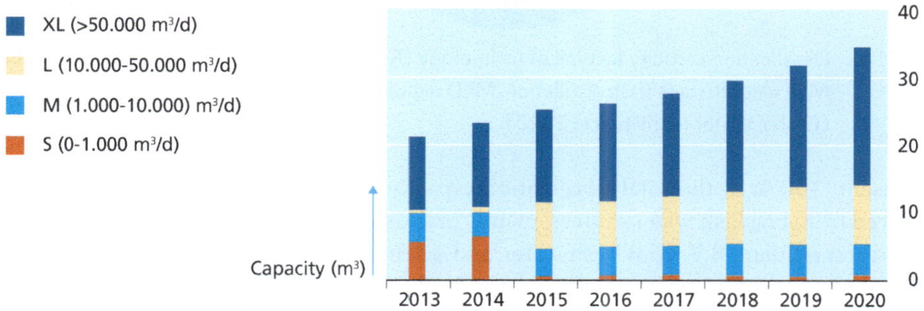

Figure 13 Plant size of seawater reverse osmosis over time (Global Water Intelligence, 2020)

In many countries, like in the Netherlands, conventional ground water treatment is being upgraded by treatment with reverse osmosis, due to the robust RO treatment approach to also remove micropollutants (endocrine disruptors, medicines, personal care products, micro-plastics, etc.) that could be present in the raw water sources.

1.3.1 Desalination capacity by technology and source water type

For all source water types reverse osmosis (RO) is the preferred desalination technology. It accounts for 69.2 % (67 Mm^3/d) of the global capacity (Figure 14); 24 % or 23.2 Mm^3/d of the global capacity is produced by distillation plants, either multi-stage flash (MSF) or multi-effect distillation (MED) plants, with relative market shares of 17 % (16.6 Mm^3/d) and 7 % (6.6 Mm^3/d), respectively. Electrodialysis (ED) process with about 2 % market share (1.97 Mm^3/d), and other processes, such as electro-de-ionization (EDI) account for 0.3 % (0.3 Mm^3/d), nano-filtration (NF) accounts for another ~2 % (1.8 Mm^3/d) of the world desalination capacity.

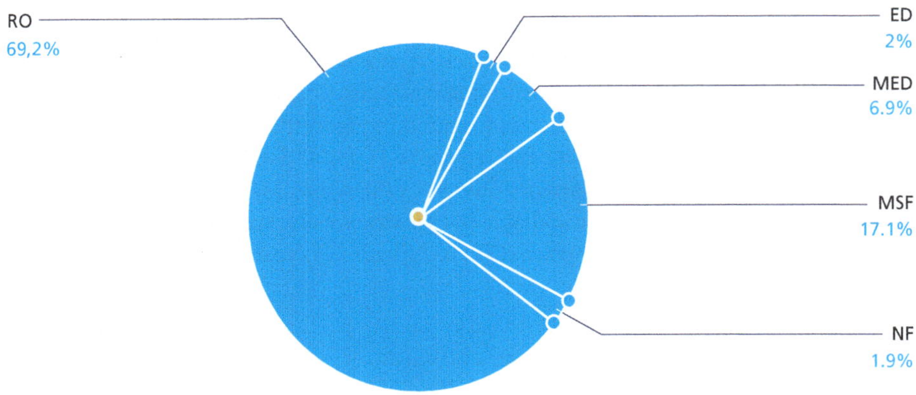

Figure 14 Desalination capacity by type of technology (RO = reverse osmosis, NF = nano-filtration, MSF = multi-stage flash distillation, MED = multi-effect distillation, ED = electro-dialysis). (Global Water Intelligence, 2020)

At present, ~60 % of the total desalination capacity is produced from seawater, 20 % is produced from brackish water sources, mainly brackish groundwater, 8 % is produced from waste water effluent, 8 % from fresh water, and 4 % from pure water. Seawater is hence the predominant source water for desalination and accounts for a worldwide water production of ~60 Mm^3/d.

Figure 15 distinguishes between the different source-water types and the technologies that are applied. For seawater, RO and thermal processes dominate the global sea water desalination production (34.4 Mm^3/d and 25.7 Mm^3/d). MSF is the main thermal process, accounting for 31 % of the global seawater desalination production. RO is the dominant process for brackish water (90 %, 17.8 Mm^3/d) and for waste water (91 %, 6.9 Mm^3/d) desalination.

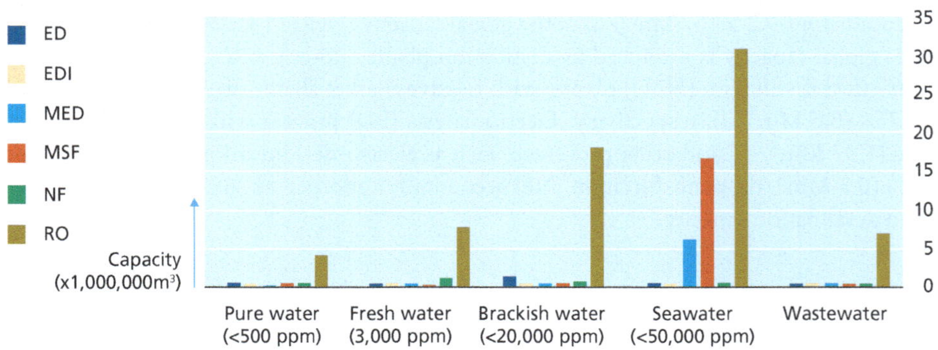

Figure 15 Desalination production capacity per raw water source and per technology for plants online and presumed online (Global Water Intelligence, 2020)

1.3.2 Desalination capacity by region

Globally, 53 % (54 Mm^3/d) of desalination capacity is sited in the countries of the Middle East and North Africa, 16 % (16 Mm^3/d) in East Asia and Pacific countries, 10 % (9.9 Mm^3/d) in North America, 8 % (8 Mm^3/d) in Western Europe, 6 % (5.7 Mm^3/d) in Latin America and Caribbean countries, 3 % (3.7 Mm^3/d) in Southern Asia, 2 % (1.8 Mm^3/d) Sub-Saharan Africa, and 2 % (2.2 Mm^3/d) in Eastern Europe and Central Asia. The global desalination capacity per region is presented in Figure 16 distinguishing between three water sources (seawater, brackish, and wastewater).

In all the regions, seawater is the main water source for desalination with exception of North America where brackish water desalination accounts for 73 % (7.3 Mm^3/d) of the regional capacity followed by 19 % wastewater (1.9 Mm^3/d).

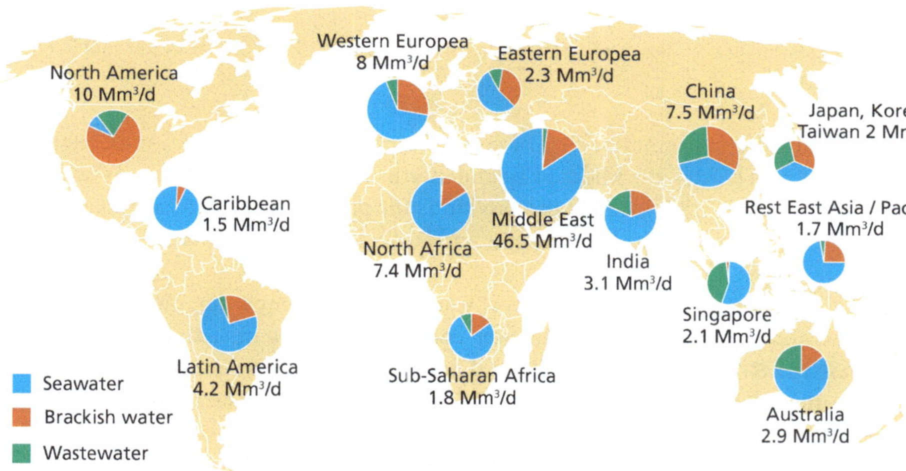

Figure 16 Desalination capacity in different regions of the world per percentage capacity production of various water sources (seawater, brackish water, wastewater effluent) (Information from Global water intelligence, 2020). For example: North America desalinates water with a total capacity of 10 Mm^3/d of which 73 % is produced from brackish water, 19 % from waste water effluent and 8 % from seawater.

Japan, Korea, and Taiwan combined desalination production of about 1.95 Mm^3/d is distributed from seawater (35 %), brackish water (29 %) and waste water (36 %). In the case of Singapore, the production capacity is about 2 Mm^3/d produced from seawater (55 %), brackish water (2 %) and waste water (43 %). Australia with a production capacity of 2.9 Mm^3/d from seawater (63 %), brackish water (15 %) and wastewater effluent (22 %).

The sea water desalination capacity per region is presented in Figure 17. Middle East and North Africa accounts for about 70 % of the world seawater desal capacity of which 55 % is thermal-based produced.

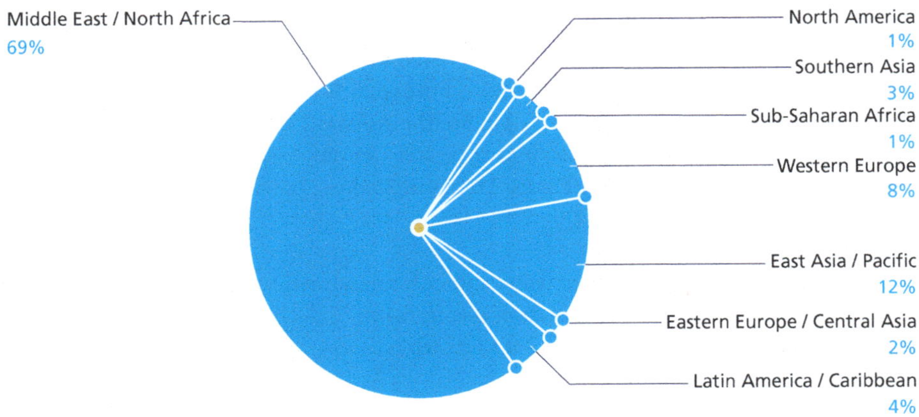

Figure 17 Desalination of seawater reverse osmosis location in 2020 (Global Water Intelligence, 2020)

Seawater desalination accounts for 79 % of the total capacity in Latin America and Caribbean, sub-Saharan Africa (77 %), the Middle East and North Africa (84 %), Southern Asia (61 %), and Western Europe (65 %), and is the predominant process in most remaining regions except for North America; Northern Europe; and Japan, Korea, and Taiwan.

Distillation is only of relevance for sea water. When comparing membrane-based desalination and thermal-based desalination per region (see Figure 18), it is only in the Middle East and North Africa where thermal desalination has more capacity than reverse osmosis plants (57 % vs. 43 %). In the rest of the world reverse osmosis is (83 % vs. 17 %) the dominant technology. Altogether, the world average production capacity is 45 % for thermal processes vs. 55 % for reverse osmosis.

1.3.3 Desalination capacity per type of customer

The existing production of drinking water based on RO from various raw water sources is as follows: from seawater 24 Mm^3/d, from brackish water 9 Mm^3/d and from fresh water about 3.2 Mm^3/d. This production of drinking water can be translated into 330 million inhabitants in the world receiving drinking water supplied by desalination plants (estimated at 120 L/p/d), which is a great contribution to the sustainable development goals, in particular to the SDG6.

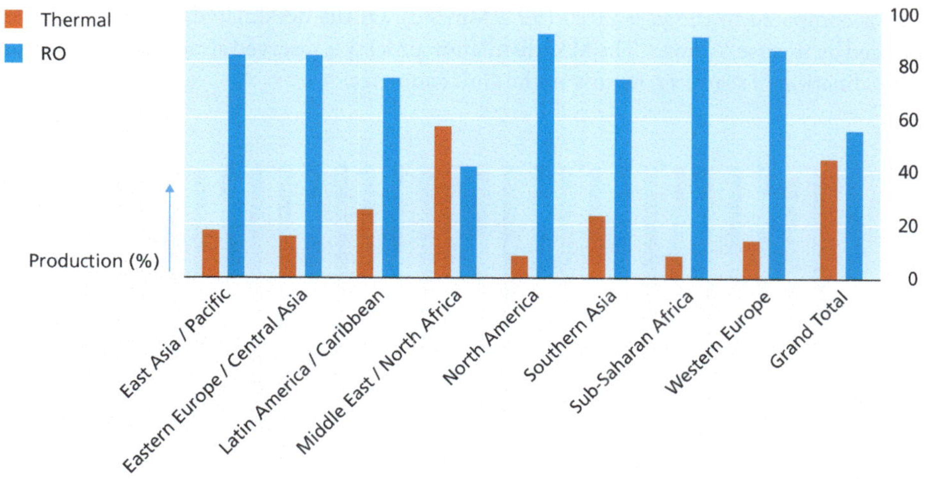

Figure 18 Thermal vs. Membrane-based seawater desalination in 2020 (Global Water Intelligence, 2020)

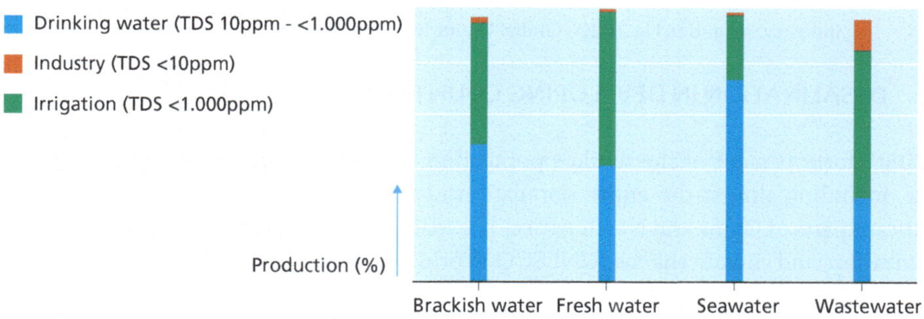

Figure 19 Desalination end-user type per raw water type produced by RO in 2020 (Global Water Intelligence, 2020)

The majority of seawater desalination is used as municipal water (73 %), followed by 25 % use in industry and about 1 % for irrigation. Waste water desalination is used as indirect water reuse (30%), in industry 55 %, and for irrigation 11 % (see Figure 19).

The countries with desalination capacities larger than 650,000 m^3/d are presented in Figure 20. The use of desalination for irrigation is relevant in three countries, namely, Spain, Kuwait, and Morocco. China, India, South Korea, Brazil, Japan, Taiwan, Indonesia rely on desalination for industry applications. Saudi Arabia, USA, UAE, Spain, Kuwait, Algeria, Oman, Israel, Singapore, Bahrain, Libya, Morocco rely on desalination for municipal use.
In conclusion, about 68 % or 68.5 Mm^3/d of the worldwide desalination capacity was produced from seawater sources in 2020. The global desalination capacity increased

by 41 % compared to the year 2010 (59.2 Mm^3/d). Of the desalinated seawater, 57 % is produced by reverse osmosis. The MSF distillation process is reserved almost exclusively for the desalination of seawater, mainly in the Gulf countries.

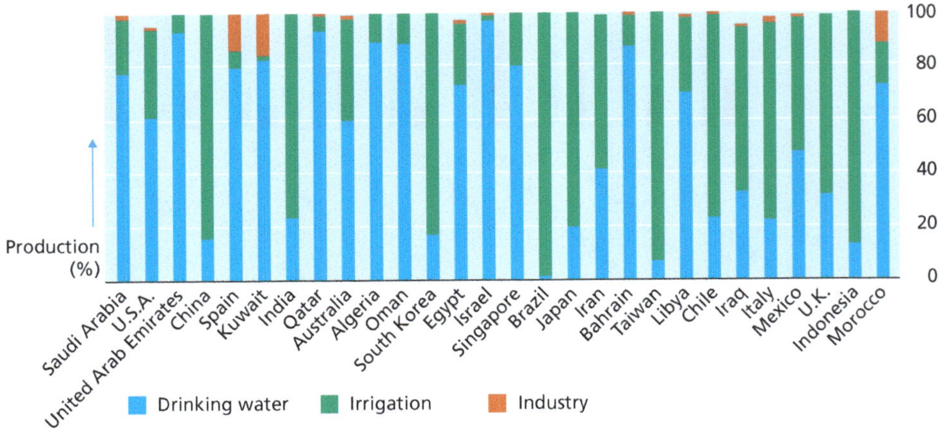

Figure 20 Highest capacity desalination countries and main use of desalination (drinking water, industry, irrigation) in 2020 (Global Water Intelligence, 2020)

1.4 DESALINATION IN DEVELOPING COUNTRIES

By 2050, forty percent of the world's population is projected to live under severe water stress, including almost the entire population of the Middle East and South Asia, plus significant parts of China and North Africa. The main drivers being the population growth, urbanization, and climate change (UNESCO World Water Assessment Programme, 2020).

Considering that about 785 million people still lacked even a basic drinking water service in 2019 (UN, SDG progress, 2019), that nearly 2.4 billion people live within 100 km of the coast (UN, Ocean Conference 2017) and the challenges with increased water stress – less renewable water and decreased water quality with more challenging emerging compounds; desalination is already an alternative that many countries all over the world are relying upon. For instance, in Kenya, in Likoni in Mombasa County, are planning the construction of a desalination plant with capacity of 100,000 m^3/day (Construction review online, 2019). In Mexico, the government considered in its water and sanitation investment plan for the coming five years, the construction of 4 desalination plants.

Many countries with economies in transition are already implementing desalination plants, and thus, the need for research and capacity development in these regions is very urgent for achieving a sustainable implementation of desalination projects.

Africa can be divided in North African countries and sub-Saharan countries. The current desalination capacity in North Africa is about 7.4 Mm^3/d while in sub-saharan countries the capacity is about 1.8 Mm^3/d. In North Africa, 87% of the desalination is from seawater and 12 % from brackish water, while in sub-Saharan Africa 66% is from seawater, 21% from brackish water and 13 % from wastewater effluent. This is illustrated per country in Figure 21.

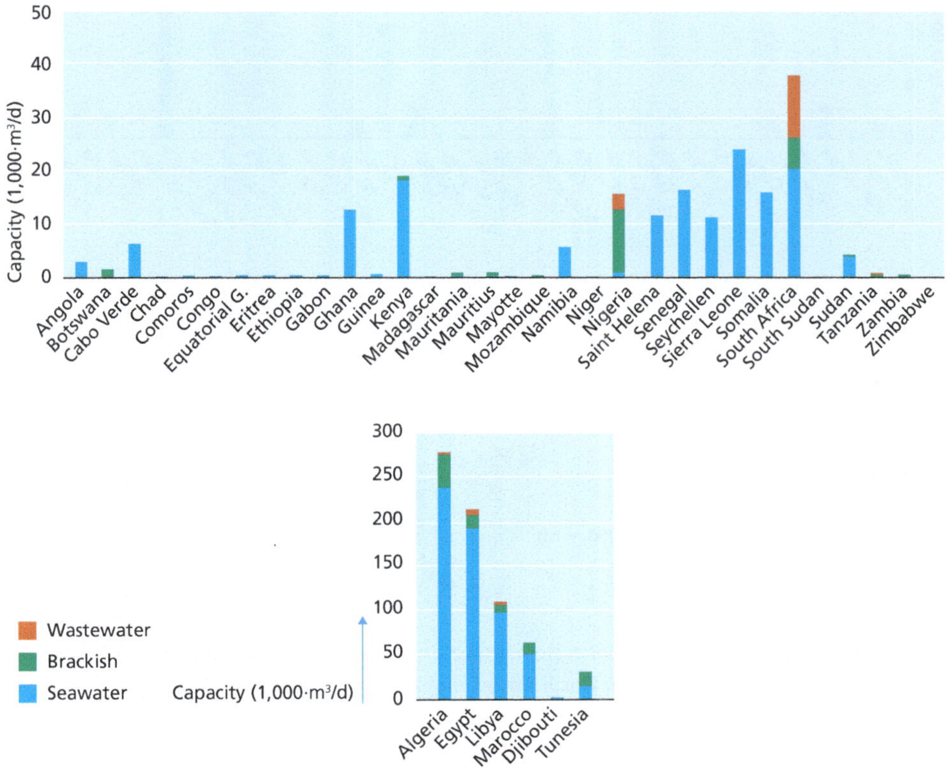

Figure 21 Desalination in Africa per feedwater sources (seawater, brackish, wastewater effluent), top: Sub-Saharan Africa, bottom: North Africa (Global Water Intelligence, 2020)

In North Africa, 81 % of the desalinated water is used for provision of drinking water and 17 % for industry, while in sub-Saharan Africa 47 % of the desalinated water is used for drinking water while 52 % is used for industry. Figure 20 shows per country the customer type of the desalination plants.

Energy is in many cases the limiting factor for implementing desalination plants. Taking sub-Saharan Africa as an example, how much energy is required to desalinate water today? Considering that the current power consumption per capita is about 500 kWh in sub-Saharan Africa (World Bank, 2020), and the population in sub-Saharan Africa is about 1.1 billion inhabitants (World Bank, 2020). Then, the total power consumption is about = 5.53×10^{11} kWh.

The current desalination production in sub-Saharan Africa is about 1.8 Mm^3/day. Assuming that the total installed capacity is realized with SWRO, then the energy demand ≈3 kWh/m^3.

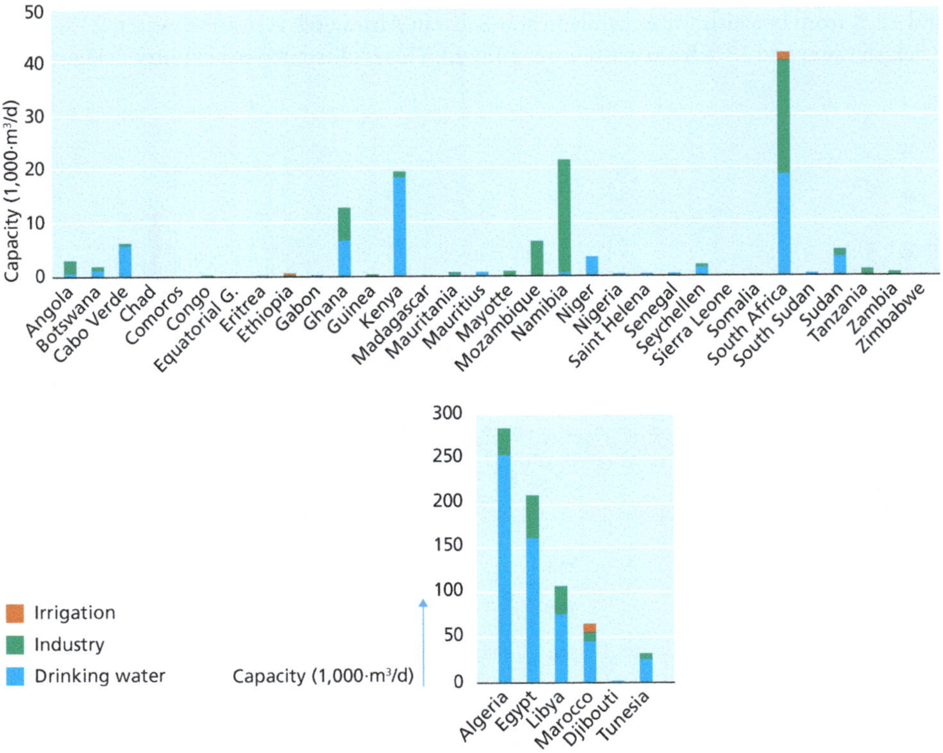

Figure 22 Desalination in Africa per customer type (irrigation, industry, drinking water), top: Sub-Saharan Africa, bottom: North Africa (Global Water Intelligence, 2020)

The total energy requirement for desalination in Africa today equals the capacity per year × 3 kWh/m^3 = 1.8×10^6 m^3/day × 360 d/year × 3 kWh/m^3 = 1.94×10^9 kWh. This power demand equals to about 0.35 % of the electrical power consumption in sub-Saharan Africa in 2020!

Industry has already turned to desalination to meet their water needs (India, China, Brazil, & Chile) – this strategy may be applied to other developed and developing countries. Energy is a key issue and will remain a challenge because of the "high" cost of renewable energy.

1.5 ENVIRONMENTAL CONCERNS

Desalination is a water treatment method that is "often chemically, energetically and operationally intensive, focused on large systems, and thus requires considerable infusion of capital, engineering expertise and infrastructure..." (Shannon, *et al.*, 2008). Like all human activities, desalination plants have also environmental impact. Despite many efforts, there are still some environmental concerns (Lattemann and Höpner, 2008), such as:

- Disposal of material use
- Land use

- Energy use to desalinate water and greenhouse gas (GHGs) emission
- Discharge of concentrate
- High volume of chemical use
- Loss of aquatic organism from marine pollution and open seawater intake

The use of fossil fuels to desalinate the water emits the greenhouse gas, which includes carbon monoxide (CO), nitric oxide (NO, nitrogen dioxide (NO_2) and sulphur dioxide (SO_2). The recent technological advanced helped to decrease the emission of GHGs and depends upon if oil is used instead of natural gas (Dawoud and Al Mulla, 2012). Likewise, the use of the high volume of chemicals during pre- and post-treatment of seawater is another environmental concern. The main concern is the discharge of chemical into the natural water, which affects the ecological imbalance (Lattemann and Höpner, 2008). Furthermore, the design of open seawater intake has a potential role in the loss of aquatic organisms, as these organisms collide with the intake screen are sometimes drawn into the plant (Dawoud and Al Mulla, 2012).

The summary of environmental challenges and possible sustainable solutions is illustrated in Figure 23.

Some of the possibilities for the sustainable solutions to prevent/minimize the issue listed above are (Lattemann *et al.*, 2012):

- Surface and ground water pollution (concentrate and residual chemicals): minimize chemical use by using best available techniques, treatment of all backwashing and cleaning solutions, use and design diffusers to disperse the concentrate in order to meet mixing regulations.
- Sediments and soil impacts (pollution of sediments, changed erosion, and the deposition processes): place intake and outfall pipelines below ground to minimize the disturbance of coastal and marine sediment
- Land use & landscape impacts: identify suitable sites through EIA process, aesthetic design of facilities, green building and landscaping, noise reduction and shielding measures, minimize land use and compensate habitat loss if necessary.
- Air quality and climate (greenhouse gas and other air pollutant emissions): compensate the remaining energy demand if necessary, e.g., by renewable energy or reforestation projects.
- Resource consumption (energy, water, materials, chemicals, land): minimize energy use by using best available techniques such as pressure exchangers, conduct MCA&LCA studies to identify processes and modes of operation that reduce resource consumption, improve recyclability or identify options for beneficial reuse.
- Ecosystem impacts (effluent toxicity, construction impacts, habitat loss, intake effects): conduct EIA studies including: field monitoring studies, whole effluent toxicity studies, hydrodynamic modelling studies; establish mixing zone regulations; use tunnelling for intake and outfall pipelines to minimize disturbance of sensitive benthic ecosystems.; use subsurface or offshore submerged intakes to lower chemical use in pre-treatment and to minimize impingement and entrainment (with low intake velocity for submerged intakes).

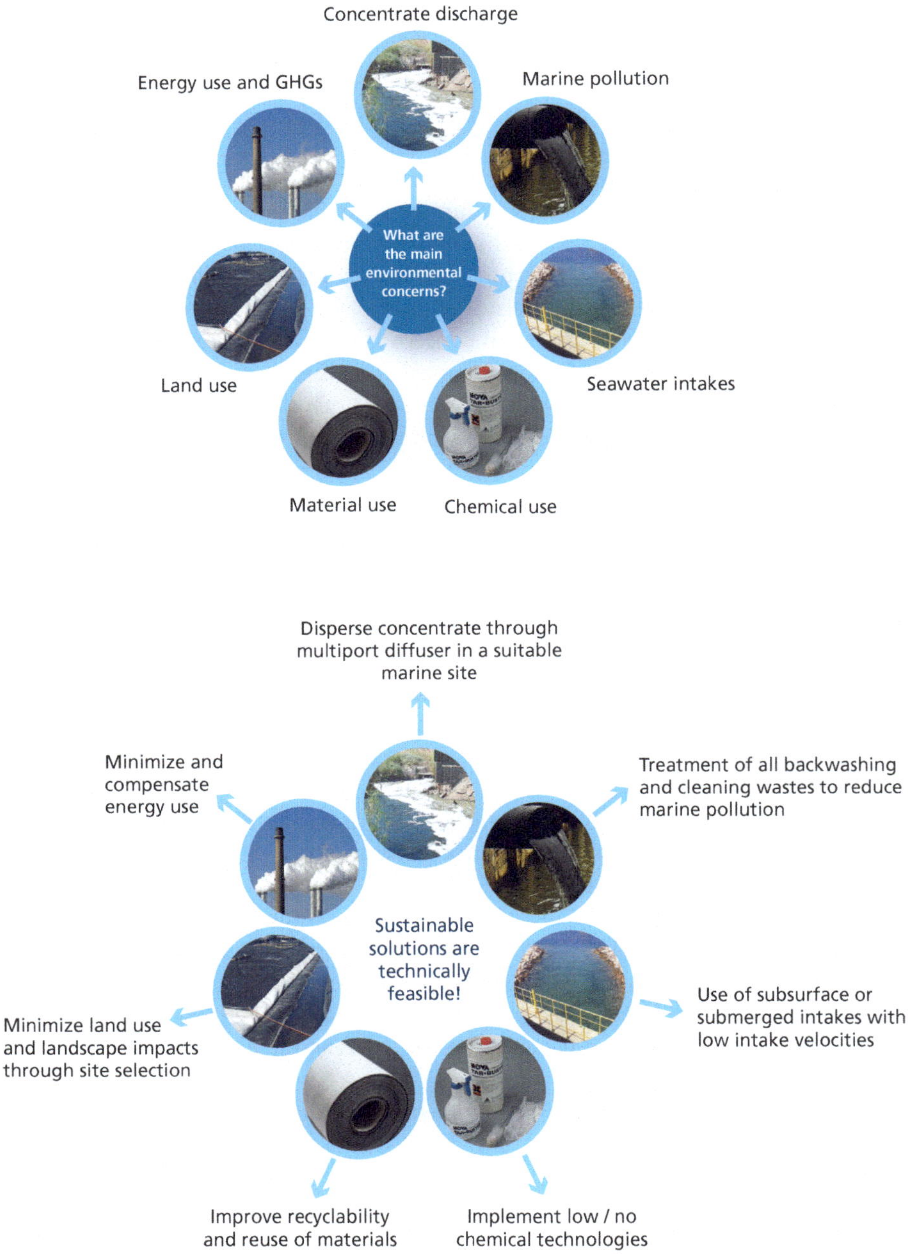

Figure 23 Environmental concerns (top) and sustainable and technical solutions(bottom) in membrane-based desalination plants. ((Adapted from Lattemann, *et al.*, 2012)

1.6 MEMBRANE FOULING

Membrane fouling is still the main "Achilles heel" for the cost-effective application of reverse osmosis (Flemming, *et al.*, 1997).
The types of fouling are categorized into *i*) particulate/colloidal fouling, *ii*) inorganic fouling, *iii*) organic, *iv*) biofouling, and *v*) scaling. Moreover, the particulate and colloidal fouling are mostly controlled with this improvement in the pre-treatment; but the occurrence of organic and biofouling is still a major issue in SWRO membranes.

To prevent the occurrence of membrane fouling, pre-treatment in RO plants is essential. Pre-treatment can take place in the form of media filters with or without coagulation, MF/UF, etc.

The consequences of fouling in RO membrane systems are:

- Increase in head loss across the feed spacer of spiral wound elements
- Higher energy consumption to maintain the constant flux operation
- Higher chemical cleaning frequency
- Increase the replacement of membrane due to irreversible membrane fouling
- Decrease the rate of water production due to longer downtime during chemical cleaning and membrane replacement
- Increase salt passage and thus deteriorate the permeate quality

Reliable methods to monitor the membrane fouling potential of raw and pre-treated water is important in preventing and diagnosing fouling and to develop the effective fouling control strategies for the cost-effective operation of SWRO membranes. The most relevant and important parameters/indicators/methods are presented in Table 5. The details these indicators are described in the following chapters of this book.

Table 5 Relevant indicators/parameters to monitor the membrane fouling in SWRO membranes

Particulate matter and fouling	Organic fouling	Biofouling	Others
Turbidity	Total organic/ dissolved organic carbon (TOC/ DOC)	Transparent exopolymer particles (TEP)	Algal cell concentration
Particle counters	Liquid chromatography organic carbon detection (LC-OCD)	Assimilable organic (carbon (AOC)	Chlorophyll-a concentration
Silt density index (SDI)	UV_{254}	Bacterial growth potential (BGP) based on flow cytometry or based on adenosine tri-phosphate	
Modified fouling index ($MFI_{0.45}$)	Fluorescence excitation and emission matrix (FEEM)	Membrane fouling simulator (MFS)	
Modified fouling index ultrafiltration (MFI-UF)	Fourier transform infrared spectroscopy (FTIR)		

1.7 CONCLUDING REMARKS

- By 2050, about 40 % of world population will be strongly hit by water scarcity, and about 2 billion of these people may live in developing countries.
- The experience of some countries (e.g., India, China, South Korea, Brazil, etc) in using desalination water to industrial users may be adopted in other nations to solve the issue of water scarcity by 2050.
- The current global trend showed that the desalination technology is finding new outlets as an alternative source for supplying water to meet growing water demand in most of the water-scarce countries. However, there have been barriers to its widespread adoption of technology mainly due to its cost, energy, lack of expertise, and the footprint.
- Desalination cannot deliver the promise of improved water supply (in developing countries) unless underlying weaknesses are addressed: reduction of non-revenue water, appropriate cost recovery, environmental impact assessments, capacity building and training, integrated water resources management.

1.8 REFERENCES

BBC. (2021). Temperature changes and energy [Online]. Available: https://www.bbc.co.uk/bitesize/guides/zpjpgdm/revision/4 [Accessed 6 April 2021]

Bremere I, Kennedy M, Stikker A, Schippers J (2001) How water scarcity will effect the growth in the desalination market in the coming 25 years. Desalination 138: 7-15 DOI https://doi.org/10.1016/S0011-9164(01)00239-9

Brown A, Matlock MD (2011) A review of water scarcity indices and methodologies. The sustainability consortium Whitepaper 106

Buros, O.K. (1980). The USAID desalination manual. Englewood, NJ, USA, IDEA Publications

Kenya to commence construction of US $154m desalination plant in Likoni (2019) https://constructionreviewonline.com/news/kenya/kenya-to-commence-construction-of-us-154m-desalination-plant-in-mombasa/ Cited 23 December 2020

Dawoud MA, Al Mulla MM (2012) Environmental Impacts of Seawater Desalination: Arabian Gulf Case Study. International journal of environmental and sustainability 1: 22-37

FAO (2013) Coping with Water Scarcity: An Action Framework for Agriculture and Food Security. FAO Water Reports No. 38. FAO, Rome

Flemming H-C, Schaule G, Griebe T, Schmitt J, Tamachkiarowa A (1997) Biofouling—the Achilles heel of membrane processes. Desalination 113: 215-225

Global Water Intelligence (2020) 32nd Worldwide Desalting Plant Inventory Media Analytics Ltd.

Lattemann S, Höpner T (2008) Environmental impact and impact assessment of seawater desalination. Desalination 220: 1-15 DOI https://doi.org/10.1016/j.desal.2007.03.009

Lattemann S, Salinas Rodríguez SG, Kennedy MD, Schippers JC, Amy GL (2012) Environmental and Performance Aspects of Pretreatment and Desalination Technologies. In: Lior N (ed) Advances in Water Desalination:79-195

Mekonnen MM, Hoekstra AY (2016) Four billion people facing severe water scarcity. Science Advances 2 DOI doi.org/10.1126/sciadv.1500323

Sanz MA (2020) Disruptive water and desalination. Paper presented at the Technology Innovation Pioneers (TIP) 2020, Abu Dhabi, UAE, 4-5 February 2020

Shannon MA, Bohn PW, Elimelech M, Georgiadis JG, Mariñas BJ, Mayes AM (2008) Science and technology for water purification in the coming decades. Nature 452: 301-310 DOI 10.1038/nature06599

Small island developing states in numbers - Climate change edition 2015 (2015) https://sustainabledevelopment.un.org/content/documents/2189SIDS-IN-NUMBERS-CLIMATE-CHANGE-EDITION_2015.pdf. Cited 04.08.2020

Spiegler, K. S. & El-Sayed, Y. M. (2001). The energetics of desalination processes. Desalination, 134, 109-128

UN Ocean conference. Facts-figures: People and Oceans (2020) https://www.un.org/en/conferences/ocean2020/facts-figures. Cited 04.08.2020

UN Water, International Decade for Action, Water for Life 2005–2015 (2014) www.un.org/waterforlifedecade/scarcity.shtml.

UNESCO World Water Assessment Programme (2020) The United Nations World water development report 2020: water and climate change

United Nations (2017) Factsheet: People and Oceans. In: UN (ed) The Ocean Conference, United Nations, New York.

United Nations, Department of Economic and Social Affairs (2020) The Sustainable Development Goals Report 2020 United Nations Publications,, New York, NY, 10017, United States of America

United Nations, Department of Economic and Social Affairs, Population Division (2019) World Population Prospects 2019: Highlights (ST/ESA/SER.A/423). In: United Nations (ed) United Nations,, pp. 46.

World Bank (2020) Data Population. In: World Bank (ed).

World Bank (2020) Data Power consumption. In: World Bank (ed)

Chapter 2

Basic principles of reverse osmosis

Sergio G. Salinas-Rodríguez, Jan C. Schippers, Maria D. Kennedy

The main learning objectives of this chapter are the following:

- Understand and be able to apply the basic principles of reverse osmosis, such as: recovery, salt passage, salt rejection, concentration polarization, effect of temperature, energy consumption.

2.1 INTRODUCTION

Reverse osmosis (RO) systems are capable of separating dissolved ions from a feed stream based on salt diffusion mechanism. In RO systems, feed water is split into two streams: one with a (very) low salinity and one with a high salinity. The low salinity stream is known as permeate or product water while the high salinity stream is known as concentrate, brine, or reject.

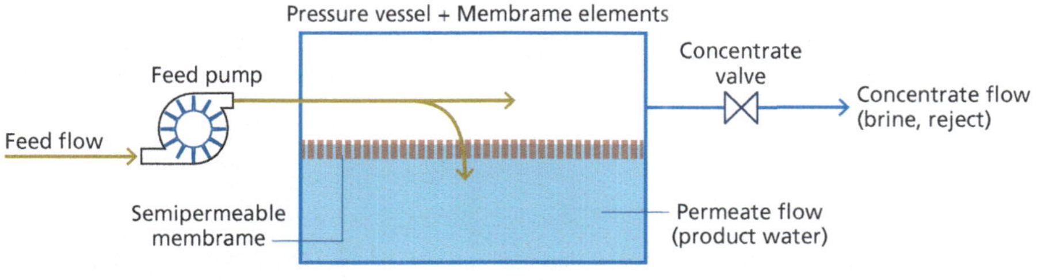

Figure 1 Basic schematic of a reverse osmosis system (Adapted from DuPont, 2020)

The quantity of water (Q_w) flowing through a membrane is proportional to the differential pressure feed-permeate (ΔP), membrane surface area (A) and permeability of the membrane (K_w). This relationship is expressed with the following equation:

$$Q_w = \frac{dV}{dt} = (\Delta P - \Delta \pi) \cdot K_w \cdot A \qquad \text{Eq. 2.1}$$

Editors: Sergio Salinas, Jan Schippers, Gary Amy, In Kim, Maria Kennedy
doi: 10.2166/9781780409863_0029

where:

Q_w	permeate flow (m^3/h)
V	total filtered volume water (permeate) (L or m^3)
t	time (h, min, s)
ΔP	differential pressure (pressure feed – pressure permeate) (bar)
$\Delta\pi$	differential osmotic pressure (bar)
	(osmotic pressure feed – osmotic pressure permeate)
K_w	permeability constant for water (m^3/m^2.s.bar)
A	surface area of the membrane(s) (m^2)
$(\Delta P - \Delta\pi)$	net driving pressure (NDP) (bar)

In membrane technology, flux is defined as the ratio between the permeate flow and surface area of the membrane. It is expressed as:

$$J = \frac{Q_W}{A} = \frac{1}{A} \cdot \frac{dV}{dt} \qquad \text{Eq. 2.2}$$

Flux (J) is the permeate flow through a membrane surface area (Q_w/A) (m^3/m^2.h or L/m^2.h). Then,

$$J = \frac{1}{A} \cdot \frac{dV}{dt} = (\Delta P - \Delta\pi) \cdot K_w \qquad \text{Eq. 2.3}$$

Example 1– Flux and permeate flow of spiral wound membrane elements

Assuming: Pressure is 10 bar; permeability of membrane is

a) 1 L/m^2.h.bar and

b) 5 L/m^2.h.bar; membrane surface area is 35 m^2; and no osmotic pressure.

Question: Calculate the flux (J) and permeate flow (Q_w) under the given conditions.

Answer:

a) 10 bar · 1 L/m^2.h.bar = 10 L/m^2.h then 10 L/m^2.h · 35 m^2 = 350 L/h

b) 10 bar · 5 L/m^2.h.bar = 50 L/m^2.h then 50 L/m^2.h · 35 m^2 = 1,750 L/h

2.2 OSMOTIC PRESSURE

In reverse osmosis systems the osmotic pressure is governed by the salinity of the feed water and the recovery at which the RO system operates.

Osmotic pressure is the pressure which needs to be applied to a solution to prevent the inward flow of water across a semipermeable membrane (Voet, *et al.*, 2001). Osmotic pressure is also defined as the minimum pressure needed to cancel out osmosis. Figure 2.2 illustrates the process of reverse osmosis in which the applied pressure needs to overcome the osmotic pressure head to force water to pass through the semipermeable membrane.

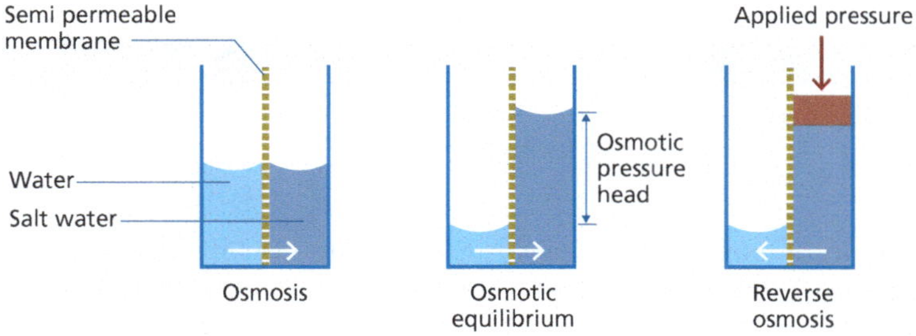

Figure 2 Illustration of reverse osmosis principle. (Adopted from Salinas Rodriguez *et al.*, 2015)

The osmotic pressure reduces the effect of hydraulic pressure; as a consequence, the effective pressure or net driving pressure (NDP) is equal to the hydraulic pressure minus the osmotic pressure.

$$NDP = \Delta P - \Delta\pi \qquad \text{Eq. 2.4}$$

where:

ΔP differential hydraulic pressure (pressure feed – pressure permeate) (bar)

$\Delta\pi$ differential osmotic pressure (osmotic pressure feed – osmotic pressure permeate) (bar)

In membrane filtration, the osmotic pressure hinders the water flow as illustrated in Figure 3.

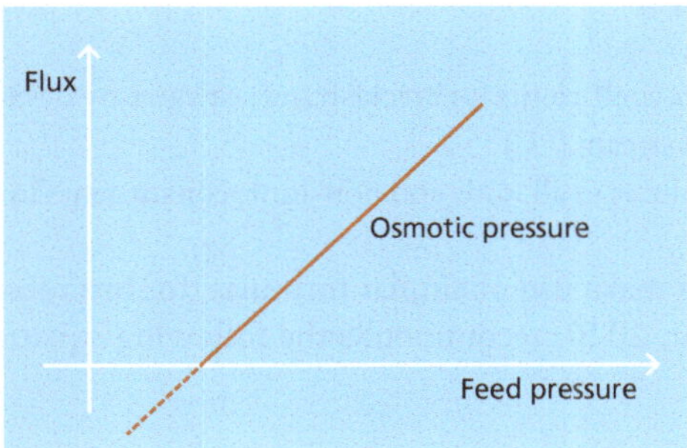

Figure 3 Schematic of osmotic pressure hindering flux

2.2.1 Calculation of osmotic pressure

In practice, feed water can be classified according to the amount of salts it contains as follows: brackish water with salts up to 10,000 mg/L, and seawater with salts content higher than 30,000 mg/L. In Table 1 the osmotic pressure is presented for low salinity brackish water and for seawater.

Table 1 Osmotic pressure for seawater and brackish water

Water type	Salinity, mg/L	Osmotic pressure, bar
Brackish water	1,000	~0.7
Seawater	35,000	~25

In practice a salinity of about 1,300 to 1,400 mg/L equals to 1 bar of osmotic pressure. From this number, the osmotic pressure (in bar) can be estimated with a rule of thumb, as follows:

$$\pi \approx 0.7 \cdot 10^{-3} \cdot C \qquad \text{Eq. 2.5}$$

Where the salt concentration (C) is in milligrams per liter (mg/L). Hydranautics suggests that an approximation for osmotic pressure may be made by assuming that 1,000 mg/L of total dissolved solids (TDS) equals about 11 psi (0.76 bar) of osmotic pressure (Hydranautics, 2001).

On the other hand, the feed osmotic pressure can be calculated more accurately by using the equation provided by ASTM 2000 based on the van't Hoff equation:

$$\pi_f = 8.308 \cdot \phi \cdot (T_f + 273.15) \sum m_i \qquad \text{Eq. 2.6}$$

Where:

π_f osmotic pressure (kPa)
ϕ Osmotic coefficient
Estimates of osmotic coefficients for brackish and seawater are 0.93 and 0.90, respectively.
T_f temperature of feed stream (°C)
Σm_i summation of molalities of all ionic and non-ionic constituents in the water

Membrane manufacturers make use of similar formulas, for instance DOW in its Filmtec technical manual (Dupont, 2020) recommends the following equation for the feed water osmotic pressure:

$$\pi_f = 1.12 \cdot (T + 273) \sum m_i \qquad \text{Eq. 2.7}$$

Where:

π_f osmotic pressure (psi). 1 psi = 6.8948 kPa = 0.068948 bar
T temperature of water (°C)
Σm_i summation of molalities of all ionic and non-ionic constituents in the water

Dupont uses the following simplified equations when calculating the average osmotic pressure in the feed-concentrate stream as a function of salinity. For $C_{fc} < 20{,}000$ mg/L, the osmotic pressure in bars:

$$\pi_{fc} = \frac{C_{fc} \cdot (T + 320)}{491{,}000} \qquad \text{Eq. 2.8}$$

For $C_{fc} > 20,000$ mg/L, the osmotic pressure in bars:

$$\pi_{fc} = \frac{0.0117 \cdot C_{fc} - 34}{14.23} \cdot \frac{(T + 320)}{345} \qquad \text{Eq. 2.9}$$

Where:

T temperature (°C)
C_{fc} salt feed-concentrate concentration (mg/L)

Hydranautics recommends the following equation for the feedwater osmotic pressure (Hydranautics, 2001):

$$\pi_{fc} = 1.19 \cdot (T + 273) \cdot \sum m_i \qquad \text{Eq. 2.10}$$

Where:

π_{fc} osmotic pressure (psi). 1 psi = 6.8948 kPa
T temperature of water (°C)
Σm_i summation of molalities of all ionic and non-ionic constituents in the water

2.3 WATER FLOW

Theory suggests that the chemical nature of the membrane is such that it will absorb and pass water preferentially to dissolved salts at the solid/liquid interface. This may occur by weak chemical bonding of the water to the membrane surface or by dissolution of the water within the membrane structure (Solution Diffusion Theory). The chemical and physical nature of the membrane (e.g., surface charge and pore size) determines its ability to allow the preferential transport of water over salt ions.

Under specific reference conditions, flux and rejection are intrinsic properties of membrane performance (Dupont, 2020).

In constant flux systems, the flux in a RO system can be expressed as:

$$J = (Net\ driving\ Pressure) \cdot K_w = (\Delta P - \Delta\pi) \cdot K_w \qquad \text{Eq. 2.11}$$

With:

$$\Delta P = P_f - P_p \qquad \text{Eq. 2.12}$$

And

$$\Delta\pi = \pi_f - \pi_p \qquad \text{Eq. 2.13}$$

Where:

J_w	water flux ($L/m^2.h$)
ΔP	hydraulic differential pressure (pressure feed - pressure permeate) (bar)
$\Delta\pi$	osmotic pressure difference between the feed water and product water (permeate) (bar)
K_w	permeability constant for water ($m^3/m^2.s.bar$) ($L/m^2.h.bar$)

Example 2 – Net driving pressure

In a seawater RO plant, the feed pressure equals 25 bar. Assume that the salinity of the feed water is 35,000 mg/L and the RO recovery is 50 % under normal conditions.
Question: What is the production at 25 bar? Explain your answer.

Answer:

Feed osmotic pressure = $35 \times 0.7 = 24.5$ bar
NDP = $25 - 24.5 = 0.5$ bar ≈ 0.
Concentrate osmotic pressure is much higher than feed osmotic pressure.
Therefore, no production is expected.

Example 3 – Net driving pressure, flux and permeate production

A sea water RO element operates at a pressure of 60 bar.
Questions: What is the flux and permeate production?
Assume: Salinity = 35,000 mg/L; membrane permeability $K_w = 0.8$ $L/m^2.h.bar$; membrane surface area per element = 35 m^2 and Recovery very low (less than 5 %).

Answers:

Flux: $(60 - 25)$ bar $\cdot$ 0.8 $L/m^2.h.bar$ = 28 $L/m^2.h$
Permeate production: 28 $L/m^2/h \cdot 35$ m^2 = 980 L/h

Example 4 – Feed water pressure for a seawater RO element

Assuming: average flux 25 $L/m^2/h$; membrane surface area 35 m^2; recovery = 12 %; feed water concentration 44,000 mg/L; $K_w = 1.0$ $L/m^2/h/bar$

Answer:

The concentrate concentration is $44{,}000 \times 1/(1\text{-}R) = 50{,}000$ mg/L.
Now, we can calculate the average feed-concentrate concentration = 47,000 mg/L.
$\pi_{avg} = 36$ bar (using the Dupont formula, Eq. 2.8)
The feed pressure will be: $P_f = J / K_w + \pi_{avg}$
$P_f = 25 / 1.0 + 36 = 25 \text{ bar} + 36 \text{ bar} = 61 \text{ bar}$

2.3.1 Salt rejection

The salt rejection (SR) is by definition the ratio of the salt concentration in the feed water minus the salt concentration in the product water over the salt concentration in the feed water and it is expressed as percentage, as follows:

$$SR = \frac{C_f - C_p}{C_f} \cdot 100\% \qquad \text{Eq. 2.14}$$

$$SR = \left(1 - \frac{C_p}{C_f}\right) \cdot 100\% \qquad \text{Eq. 2.15}$$

Where C_f is the salt concentration in the feed water and C_p is the salt concentration in the product water.

2.3.2 Salt passage

The salt passage (SP) is by definition the ratio of the salt concentration in the product water to the salt concentration in the feed water expressed as percentage, as follows:

$$SP = \frac{C_p}{C_f} \cdot 100\% \qquad \text{Eq. 2.16}$$

Salt passage is the opposite of salt rejection.

$$SP = 100\% - SR \qquad \text{Eq. 2.17}$$

Example 5 – Salt rejection and salt passage

A sea water RO plant is processing water with a salinity of 45,000 mg/L. The product water has a salinity of 500 mg/L.

Questions: Calculate the salt rRejection of the plant and calculate the salt passage of the plant.

Answers:

$SP = (C_p / C_f) \times 100\% = (500 / 45{,}000) \times 100\% = 1.1\%$

$SR = 100\% - SP = 98.9\%$

2.4 SALT FLOW

Water can pass a reverse osmosis membrane; salts as well, however, at a much lower rate. The transport of salts through RO membranes is due to diffusion. Diffusion is a result of the motion of ions in water and the tendency of salts to move from high concentration to low concentration. Diffusion is a slow process, but cannot be neglected.

The salinity of the product water (C_p) depends on the relative rates of water and salt transport through a membrane. This relationship is expressed by the following equation:

$$C_p = \frac{Q_s}{Q_w} \qquad \text{Eq. 2.18}$$

Where Q_s is defined by the following equation:

$$Q_s = \Delta C \cdot K_s \cdot A \qquad \text{Eq. 2.19}$$

With

$$\Delta C = C_f - C_p \qquad \text{Eq. 2.20}$$

Where:

Q_s	Flow rate of salt through membrane (kg/s)
ΔC	Salt concentration differential across membrane (kg/m^3) = $C_f - C_p$
K_s	Membrane permeability coefficient for salt (m^3/m^2.s)
A	Membrane area (m^2)
C_f	Feed concentration (mg/L)
C_p	Permeate concentration (mg/L)

Replacing the formula of Q_s in the formula of C_p, we have:

$$C_p = \frac{\Delta C \cdot K_s \cdot A}{(\Delta P - \pi) \cdot K_w \cdot A} \qquad \text{Eq. 2.21}$$

Replacing terms,

$$C_p = \frac{(C_f - C_p) \cdot K_s}{(P_f - P_p - \Delta\pi) \cdot K_w} \qquad \text{Eq. 2.22}$$

Dividing the whole equation by C_f, and rearranging the equation, we have the salt passage:

$$\frac{C_p}{C_f} = \frac{\left(C_f - \frac{C_p}{C_f}\right) \cdot K_s}{(P_f - P_p - \Delta\pi) \cdot K_w} \qquad \text{Eq. 2.23}$$

Then,

$$Salt\ passage\ (SP) = \left(SP\right) = \frac{C_p}{C_f} \cdot 100\% \qquad \text{Eq. 2.24}$$

Looking at the right side of the equation 2.24, since C_p is small compared to C_f, the ratio C_p / C_f is much smaller than 1, therefore the salt transport (Q_s) is constant at a certain C_p and is independent of the pressure. As a consequence, the salt passage ($SP = C_p / C_f$) is lower at high pressure (P_f) and vice versa. This is because the same quantity of salt (Q_s) will be diluted by a larger volume of (product) water and vice versa.

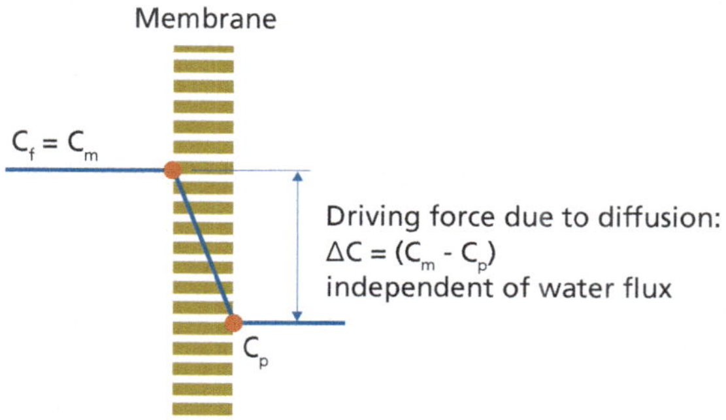

Figure 4 Schematic representation of the salt flux due to diffusion

The salt flux (J_s) is defined as the salt transport per membrane area per hour ($mg/m^2.h$) (mg per m^2 per hour). The salt flux is proportional with the concentration difference between membrane surface at: feed water side (C_f) and product water side (C_p), and the permeability coefficient of membrane for salts (ions) K_s.

$$J_s \approx (C_f - C_p) \cdot K_s \qquad \text{Eq. 2.25}$$

Remark: The larger the pores, the larger the permeability for salt and water. In general, C_p is low in comparison with C_f, so C_p can be neglected in this formula.

$$J_s \approx C_f \cdot K_s \qquad \text{Eq. 2.26}$$

Where:

J_s salt flux ($mg/m^2.h$)
C_f concentration at feed side membrane (mg/L)
C_p concentration at product side membrane (mg/L)
K_s Permeability for salt [($mg/m^2.h$) / (mg/L)]

Example 6 – Salt passage and salt concentration permeate

The flux in an RO membrane is 10 L/m²/h. Salt concentration in feed water is 41,000 mg/L and K_s value is 0.08 [(mg/m².h)/(mg/L)] = 0.08 L/m²/h.

Questions: What is the salt concentration in the permeate? What is the salt passage?

Answer:

In 1 hour, 10 L water is produced by 1 m² membrane surface area. In 1 hour, the following amount of salt will pass 1 m² membrane surface area.

$$C_{perm} = \frac{J_s}{J_w} = \frac{(C_f - C_p) \cdot K_s}{J_w} \approx \frac{C_f \cdot K_s}{J_w}$$

J_s = mg/m².h ≈ $C_f \cdot K_s$ = 41,000 mg/L · 0.08 L/m²h = 3,280 mg/m²h

This amount of salt (namely 3,280 mg) arrives in 10 L.

Resulting in (3,280 mg salt / 10 L water) = 328 mg/L salt in the product water/permeate.

The formula applied is:

$$SP = \frac{C_p}{C_f} \cdot 100\% = \frac{328}{41{,}000} \cdot 100\% = 0.8\%$$

or

SP = 100 % – SP = 100 % – 0.8 % = 92 %

Example 7 – Salt passage and salt concentration permeate

Considering the previous example, the water flux is increased from 10 L/m²/h to 20 L/m²/h.

Questions: What will be salt concentration in the permeate, salt passage and salt rejection?

Answer:

Salt transport is the same (since the concentration of the feed water does not change) and water transport (flux) is double, so C_p will be two times lower. Consequently, *SP* will be two times lower.

$$C_p = \frac{J_s}{J_w} = \frac{C_f \cdot K_s}{J_w}$$

C_f and K_s are constant. J_w is 20/10 = 2 times higher.

As a consequence, C_p will be 2 times lower, so (328 mg/L / 2) = 164 mg/L

Salt passage will be:	SP = (164 / 41,000) · 100 % = 0.4 %	instead of 0.8 %
Salt rejection will be:	SR = 100 % - 0.4 % = 99.6 %	instead of 99.2 %.

Example 8 – Salinity of RO permeate

A seawater RO plant produces water with a salinity of 500 mg/L. Due to technical problems the capacity has to be reduced by 50 %, while the recovery can be kept constant.

What will happen with the salinity of the product water?

[] same [] decrease [] increase

Make a guess of the salinity of the product water.

Answers:

The flux will be 50% lower, since the capacity is reduced by 50 %.

$C_p = (C_{fc} \times K_s) / J$

When we assume that C_{fc} will not change (because the recovery is constant), then C_p will become two times higher

Two different types of RO membranes are identified, namely:

- brackish water (BWRO) membranes up to about 99.5 % rejection, operating at max. 40 bar.
- Sea water (SWRO) membranes with rejections > 99.8%, operating at max. 80 bar.

In general, high salt rejection (low K_s) combines with a low K_w value (due to smaller pores). For instance, in SWRO, K_s = 0.08 L/m^2/h and K_w = 1 L/m^2.h.bar, while in BWRO, K_s = 1.1 L/m^2/h and K_w = 5 L/m^2/h/bar.

2.4.1 Permeate salinity

The salinity product/permeate of a stage (or unit) follows from:

$$C_p = C_{fc}\,(1\text{-SR})$$

or

$$\text{SR} = 1 - \text{SP}$$

and

$$C_p = C_{fc} \times \text{SP}$$

Or more accurately, because SR depends on flux:

$$C_p = C_{fc}\,(K_s) / J$$

Where: $C_{fc} = (C_{fc} + C_c) / 2$

Since salt rejection (SR) depends on flux, K_s is not directly available from manufacturers information. Thus, K_s has to be calculated from test results under standard conditions.

The permeate salinity can be chosen to a certain extend by the choice of:

- Type of membrane and manufacturer
- Recovery: lower recovery results in lower salinity
- Flux: higher flux gives lower salinity

For drinking water, 500 mg/L is usually the guideline. For industrial waters much lower guidelines are often adopted, e.g., 10 to 50 mg/L. Usually a second pass is installed when lower salinity is required. For instance, the product water seawater RO plants use to be rather high.

2.5 RECOVERY AND CONCENTRATION FACTOR

Recovery (R) is also known as conversion. Recovery is by definition the part of feed water that is converted in product water and is expressed as percentage.

$$R = \frac{Q_p}{Q_f} \cdot 100\% \qquad \text{Eq. 2.27}$$

Where:

Q_p Product water flow rate (m^3/h)
Q_f Feed water flow rate (m^3/h)

Recovery affects salt passage and product flow. As recovery increases, the salt concentration on the feed-concentrate side of the membrane increases, which increases the salt transport and the permeate salinity (C_p).

High salt concentration in the feed-concentrate solution increases the osmotic pressure, which consequently reduces the net driving pressure (*NDP*). As a result, the product water flow rate is reduced and the permeate salinity (C_p) increased.

A balance can be performed in a RO system to define the concentration factor as illustrated in figure 5.

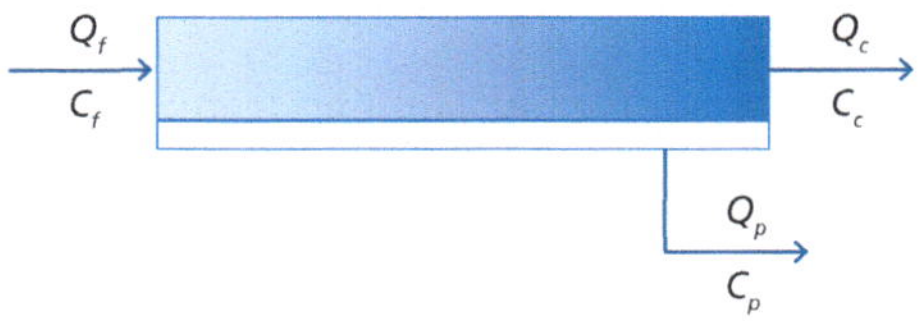

Figure 5 Balance in a RO system

$$Q_f = Q_p + Q_c \qquad \text{Eq. 2.28}$$

Then multiplying by the concentration of salts in each stream, we have:

$$Q_f \cdot C_f = Q_c \cdot C_c + Q_p \cdot C_p \qquad \text{Eq. 2.29}$$

Re-arranging the equation for the feed concentration:

$$C_f = \frac{Q_p}{C_f} \cdot Q_p + \frac{C_c}{C_f} \cdot C_c \qquad \text{Eq. 2.30}$$

From the definition of recovery we can modify the previous equation.

$$\frac{C_c}{C_f} = \frac{Q_f - Q_p}{Q_f} = \frac{Q_f}{Q_f} - \frac{Q_p}{Q_f} = 1 - R \qquad \text{Eq. 2.31}$$

Substituting Eq 2.31 into Eq. 2.30, we have:

$$C_f = R \cdot C_p + (1 - R) \cdot C_c \qquad \text{Eq. 2.32}$$

From the definition of recovery:

$$R = \frac{Q_p}{Q_f} = \frac{Q_f - Q_c}{C_f} = 1 - \frac{Q_c}{C_f} \qquad \text{Eq. 2.33}$$

The concentration factor (CF) in a RO system is by definition:

$$CF = \frac{C_c}{C_f} \qquad \text{Eq. 2.34}$$

From section 2.4.1 we have:

$$C_p = C_f \cdot (1 - SR) \qquad \text{Eq. 2.35}$$

Replacing Eq. 2.35 in Eq. 2.32 we have

$$C_f = R \cdot C_f \cdot (1 - SR) + (1 - R) \cdot C_c \qquad \text{Eq. 2.36}$$

And

$$\frac{C_f}{C_c} = \frac{R \cdot C_f \cdot (1 - SR)}{C_f} + \frac{(1 - R) \cdot C_c}{C_f} \qquad \text{Eq. 2.37}$$

Then

$$(1 - R) \cdot \frac{C_c}{C_f} = 1 - R \cdot (1 - SR) \qquad \text{Eq. 2.38}$$

Then

$$C_c = \frac{C_f \cdot \left[1 - R \cdot (1 - SR)\right]}{1 - R} \qquad \text{Eq. 2.39}$$

Or

$$CF = \frac{\left[1 - R \cdot (1 - SR)\right]}{1 - R} \qquad \text{Eq. 2.40}$$

Since the salt passage is usually low, we may assume that the total amount of salt entering the plant ($Q_{feed} \times C_{feed}$) will increase in concentration in the concentrate (brine) stream. As a result:

$$CF = \frac{1}{1 - R} \qquad \text{Eq. 2.41}$$

and

$$C_c = \frac{C_f}{1-R} \qquad \text{Eq. 2.42}$$

In the concentrate, concentrations of salts are increased, including sparingly soluble compounds. In seawater mainly calcium carbonate, while in brackish/fresh e.g., calcium carbonate, calcium sulfate, strontium sulfate, barium sulfate and silica (SiO_2).

As soon as the solubility is exceeded, precipitation / scaling might occur (scaling potential and kinetics play a role). Results in lower membrane permeability (K_w). Precipitation of calcium carbonate can be avoided by acid dosing and/or adding antiscalants. Supersaturation to a certain extend is allowable by antiscalant dosing.

Example 9 – Conversion / Recovery / Concentration

A reverse osmosis plant is treating seawater with a salinity of 40,000 mg/L.
Having a feed flow of 500 m^3/h and permeate flow of 200 m^3/h.
Questions: What is the recovery of the plant? What is the concentrate flow? What is the salinity in the concentrate? What is the concentration factor?

Answers:

Recovery: $Q_p / Q_f = (200 / 500) \times 100\,\% = 40\,\%$

Concentrate flow: $Q_f - Q_p = 500 - 200 = 300\ m^3/h$

Salinity concentrate: $CF \times C_f = 1 / (1 - R) \times C_f = 1 / (1 - 0.4) \times 40{,}000 = 66{,}667$ mg/L

Assuming salt rejection 100 %, the concentration factor for various recoveries is presented in table 2.

Table 2 Concentration factor versus recovery

Recovery, %	Concentration factor
30	1.4
40	1.7
50	2.0
75	4.0
90	10.0

In practice, brackish water plants operate at 75 %, some up to 90 % recovery. In brackish water reverse osmosis systems, the maximum recovery is mainly governed by scaling potential of feed water. Seawater reverse osmosis systems normally operate at 30-50 %, and the maximum recovery is governed by the high osmotic pressure.

Example 10 – Feed flow

Total capacity of an RO plant:	1,000 m^3/h
Number of units:	3
Recovery/conversion:	45 %

Question: What is the feed flow per unit?

Answer:

$R = Q_p / Q_f$ or $Q_f = Q_p / R$

$Q_f = (1{,}000 / 3) / 0.45 = 740\ m^3/h$ per unit.

Total feed flow is 3 × 740 = 2,220 $m^3/h/L$

Example 11 – Recovery, brine concentration, osmotic pressure

A seawater RO plant is producing 200 m^3/h permeate.

Feed flow is 500 m^3/h; feed water has a salinity of 47,000 mg/L

Questions: What is the recovery of the plant? What is the salinity of the concentrate? What is the osmotic pressure in the feed water and in the concentrate? What should be the minimum feed pressure? How many arrays are in the system?

Answers:

In seawater RO, recovery is usually 40 %.

Concentration factor:	CF = 1/ (1-0.4) = 1.667
Salinity concentrate:	C_c = 47,000 × CF = 78,300 mg/L
Feed osmotic pressure:	π_f = 47 × 0.7 = 32.9 bar
Concentrate osmotic pressure:	π_c = 78.3 × 0.7 = 54.8 bar

Pressure higher than 32.9 bar is required to exceed the osmotic pressure.

One stage.

2.6 PRESSURE DROP

Pressure drop in feed – concentrate channel can be calculated with the formula of Schock and Miquel, (1987).

$$\Delta P = \frac{0.5 \cdot \lambda \cdot \rho \cdot v^2 \cdot L}{dh} \qquad \text{Eq. 2.43}$$

$$\lambda = 6.23\ Re^{-0.3} \qquad \text{Eq. 2.44}$$

$$R_e = \frac{(\rho \cdot v \cdot d_h)}{\eta} \qquad \text{Eq. 2.45c}$$

Or

$$\Delta P = \frac{0.5 \cdot \rho^{0.7} \cdot v^{1.7} \cdot \eta^{0.3}}{d_h^{\ 0.7}} \qquad \text{Eq. 2.46}$$

Where:

ΔP = pressure drop across spacer — λ = friction coefficient
ρ = density water — v = velocity
L = length membrane — d_h = hydraulic diameter

The **normalized pressure drop** (NPD) in a feed – concentrate channel can be calculated with the following formula.

$$NPD = \Delta P_{act} \cdot CF(Q) \cdot CF(T) \qquad \text{Eq. 2.47}$$

Where:

ΔP_{act} is actual pressure drop
CF(Q) is correction factor for flow
CF(T) is correction factor for temperature.

The pressure drop in one element is approximately 0.2 bar.
The empirical formula for normalizing pressure drop is according to Schock and Miquel (1987):

$$NPD = \Delta P_{act} \cdot \left(\frac{Q_{fc,ref}}{Q_{fc,act}}\right)^m \cdot \left(\frac{\eta_{\tau,ref}}{\eta_{\tau,act}}\right)^n \qquad \text{Eq. 2.48}$$

Where:

$Q_{fc,ref}$ average reference feed/concentrate flow
$Q_{fc,act}$ average actual feed/concentrate flow
$\eta_{T,ref}$ viscosity at reference temperature
$\eta_{T,act}$ viscosity at actual temperature
m = 1.4 (Hydranautics); or m = 1.7 (Schock and Miquel)
n = 0.34 (Hydranautics); or n = 0.3 (Schock and Miquel)

The procedure to determine the reference pressure drop, which equals normalized pressure drop at startup, is the following:

1. Measure ΔP_{start} at startup, preferably for each stage
2. Measure $Q_{fc,start}$ at startup, preferably per stage. Use this value as reference $Q_{fc,ref}$
3. Choose a reference temperature e.g., 20 °C
4. Calculate ΔP_{ref} with formula and $\eta_{T,ref}$ and $\eta_{T,start}$
5. Use ΔP_{ref} as reference pressure drop.

2.7 CONCENTRATION POLARIZATION

Concentration polarization is the accumulation of salts (ions) at the membrane surface. As water flows through a membrane and salts are rejected by the membrane, the retained salts can accumulate at the membrane surface where their concentration will gradually increase. The concentration build-up at the membrane will generate a diffusive flow of salts back to the bulk of the feed, but after a given period of time steady state conditions will be established. Steady-state conditions are reached when the convective salt flow to the membrane surface is balanced by the salt flux through the membrane plus the diffusive flow from the membrane to the bulk.

Under steady-state conditions, the concentration at the membrane surface (C_m) is constant. The crossflow along the membrane surface enhances back diffusion of salts to the bulk. This increase in salt concentration at the membrane surface is called concentration polarization. As a result, the concentration at the membrane surface (C_m) is higher than in the bulk (or feed water) (C_b).

This phenomenon results from: water flows through a membrane; salts (ions) are rejected; retained salts (ions) accumulate at the membrane.

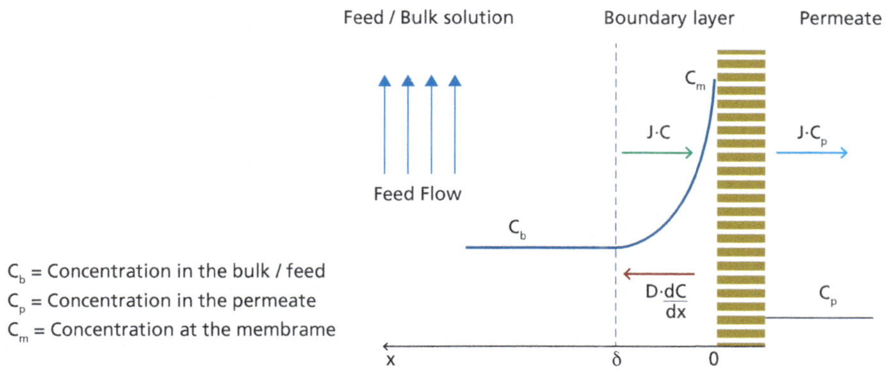

Figure 6 Concentration polarization - Concentration profile under steady state conditions (Mulder, 2003)

At steady-state the convective transport of salt to the membrane is equal to the sum of the permeate salt flow plus the diffusive back transport of salt:

$$J \cdot C_p = D \cdot \frac{dC}{dx} + J \cdot C \qquad \text{Eq. 2.49}$$

With the following boundary conditions:

When x = 0 then $C = C_m$

When x = δ then $C = C_b$

After integration:

$$\ln = \left(\frac{C_m - C_p}{C_b - C_p} \right) = \frac{J \cdot \delta}{D} \qquad \text{Eq. 2.50}$$

Or after re-arranging the previous formula we have

$$\frac{C_m - C_p}{C_b - C_p} = e^{\left(\frac{J \cdot \delta}{D}\right)} \qquad \text{Eq. 2.51}$$

The ratio of the diffusion coefficient (D) and the thickness of the boundary layer (δ) is called the mass transfer coefficient (k).

$$k = \frac{D}{\delta} \qquad \text{Eq. 2.52}$$

The intrinsic retention (R_{int}) of the membrane is the salt concentration at the membrane minus the concentration of salt in permeate over the salt concentration at the membrane.

$$R_{int} = 1 - \frac{C_p}{C_m} \qquad \text{Eq. 2.53}$$

Then, C_m / C_b becomes:

$$\frac{C_m}{C_b} = \frac{e^{\left(\frac{J}{k}\right)}}{R_{int} + (1 - R_{int}) \cdot e^{\left(\frac{J}{k}\right)}} \qquad \text{Eq. 2.54}$$

The ratio C_m / C_b is called the **concentration polarization factor**. This ratio increases (i.e., the concentration at the membrane (C_m) surfaces increases) with increasing flux (J), with increasing retention (R_{int}) and with decreasing mass transfer coefficient (k).

When the salt is completely retained by the membrane ($R_{int} = 1$ and $C_p = 0$), then

$$\frac{C_m}{C_b} = e^{\left(\frac{J}{k}\right)} \qquad \text{Eq. 2.55}$$

This is the basic equation for concentration polarization which demonstrates the two factors (the flux "J" and the mass transfer coefficient "k") and their origin (membrane part "J", hydrodynamics "k") responsible for concentration polarization.

The pure water flux (specific permeability) is determined by the membrane used and this parameter is not subject to further change once the membrane has been selected. On the other hand, the mass transfer coefficient depends strongly on the hydrodynamics of the system and can therefore be varied and optimized.

2.7.1 Control of concentration polarization

The following actions can be considered to control concentration polarization in RO membranes:

- **Decreasing the flux** (J). The higher the permeate flow (Q_p) in an element, the higher the transport of salts (ions) to the membrane surface. As a result, accumulation will be higher and concentration polarization will be higher.
  A consequence of lowering the flux is lower capacity of the plant or more elements need to be installed.
- **Increasing feed flow**. Limited to maximum allowable feed flow to avoid membrane damage. A consequence of increasing the feed flow is the higher head loss (pressure loss) across the spacer.
- **Increasing concentrate flow by reducing recovery**.

By increasing the mass transfer coefficient (*k*) we can control the concentration polarization in the RO system. *k* is mainly determined by the diffusion coefficient and the flow velocity. Because the diffusivity of solutes cannot be increased (only by changing the temperature), *k* can only be increased by **increasing the feed velocity** along the membrane or by changing the module shape and dimensions (decreasing module length or increasing the hydraulic diameter). The higher the cross flow along the membrane surface the higher the back diffusion. As a result, accumulation will be reduced and concentration polarization will be lower.

2.7.2 Effects of concentration polarization

Concentration polarization has several negative effects on the performance of reverse osmosis systems. These effects are described below.

- Higher osmotic pressure at membrane surface than in bulk feed water, resulting in lower Net Driving Pressure. Consequently, higher feed pressure is required to maintain same flux (capacity).
- Increase salt transport (Q_s) due to higher salt concentration at membrane surface. As a result, a lower salt rejection (higher salt passage). Higher C_p.
- Due to higher concentration of sparingly soluble salts (e.g., calcium carbonate, calcium sulphate) at the membrane surface, possibility of precipitation (scaling) will increase.
- Reduced water transport through the membrane (Q_w).
- Higher rate of fouling due to suspended and colloidal matter, organic polymers, due to accumulation at membrane surface.

2.7.3 Concentration polarization factor

The concentration polarization factor (β) can be calculated with the following formula:

$$\beta = \frac{C_m}{C_b} = e^{\left(\frac{J}{k}\right)} \qquad \text{Eq. 2.56}$$

In practice, the formula is simplified to:

$$\beta = K_p \cdot e^{\left(\frac{Q_p}{Q_{f\,avg}}\right)} \qquad \text{Eq. 2.57}$$

Where: K_p is a proportionality constant depending on the module geometry.
This simplification is justified by the fact that: *i*) Q_p is proportional to J, and *ii*) $Q_{f\,\text{avg}}$ is the average feed flow and is proportional to *k* and *k* is almost proportional to the cross flow velocity (*v*).

Using the arithmetic average of feed and concentrate flow as average feed flow, the concentration polarization factor can be expressed as a function of the permeate recovery rate of a membrane element R_i.

$$\beta = K_p \cdot e^{\left(\frac{2 \cdot R_i}{2 - R_i}\right)} \qquad \text{Eq. 2.58}$$

Where:

β = CPF = concentration polarization factor
K_p constant depending on type (manufacturer) membrane (usually 0.99)
Q_p permeate flow of an element (m^3/h)
Q_c concentrate flow in an element (m^3/h)
R_i recovery of a membrane element

The value of the concentration polarization factor of 1.2, which is the recommended Hydranautics limit, corresponds to 18 % permeate recovery for a 40 inches long membrane element (Hydranautics, 2001). Equations 2.57 and 2.58 and frequently applied by Koch and Hydranautics membranes manufacturers (Hydranautics, 2001).

Example 12 – Concentration polarization 1

In a reverse osmosis element, the feed flow equals: 8 m^3/h; product flow equals 1 m^3/h; the average TDS (total dissolved salts) in this element equals 43,000 mg/L.

Question: What is the average concentration polarization factor in this element? What is the average salt concentration (TDS) at the membrane surface? Assume: $K_p = 0.99$

Answers:

$\beta = 0.99 \times \exp^{(1/8)} = 1.12$

$1.12 \times 43{,}000 = 48{,}160$ mg/L

Dupont FILMTEC (2020) applies for their elements the formula:

$$\beta = e^{(0.7 \cdot R)} \qquad \text{Eq. 2.59}$$

Where:

β Concentration polarization factor
R Recovery

The recommended recovery (by Dupont) per RO element varies with the quality of the feed water e.g.:

- Seawater 10-12 %
- Filtered treated domestic wastewater 10-12 %
- Pre-treated surface water 15-18 %
- Softened well water 19-25 %

Example 13 – Concentration polarization 2

Six spiral wound RO elements (8 inch) in one vessel are fed with 10 m^3/h. The productivity of the first and the last element are assumed to be the same namely 1 m^3/h.

Question: Calculate the Concentration Polarization Factor in the first and the last element.

Answer:

Use the formula: $CPF = K_p \cdot \exp^{(Qp/Qc)}$

Where: Q_p is the permeate flow; Qc is the concentrate flow (leaving an element)

Calculate the concentrate flow (leaving the element) in the first and the last element.

The concentrate flow in the first element is: $Q_c = Q_f - Q_p$

$Q_c = 10 - 1 = 9\ m^3/h$

In the last element the concentrate flow is: $Q_c = Q_f - Q_p = 10 - 6 \times 1 = 4\ m^3/h$

Substitute in formula: $CPF = K_p \times \exp^{(Qp/Qc)}$

then we get for:

the first element: $CPF = 0.99 \times (\exp)^{1/9} = 1.10$

the last element: $CPF = 0.99 \times (\exp)^{1/4} = 1.27$

In sea water RO systems, the concentration polarization will decrease with increasing recovery. Consequently, the CPF is the last element is lower than in the first element. The reason for that is that the flux is reducing dramatically with increasing recovery.

2.8 MASS TRANSFER COEFFICIENT

The mass transfer coefficient (k), is related to the Sherwood number (Sh).

$$Sh = \frac{k \cdot d_n}{D} = a \cdot Re^b \cdot Sc^c \qquad \text{Eq. 2.60}$$

Where: *a, b* and c are constants. With

$$Re = \frac{\rho \cdot v \cdot d_n}{\eta} \qquad \text{Eq. 2.61}$$

And

$$Sc = \frac{\eta}{\rho \cdot D} \qquad \text{Eq. 2.62}$$

Where:

k Mass transfer coefficient (m/s)

Re Reynolds number, dimensionless

Sc Schmidt number, dimensionless

d_h Hydraulic diameter (m)

υ Flow velocity (m/s)

ρ Water density (kg/m^3)

η Dynamic viscosity (kg/m.s)

D Diffusion coefficient (m^2/s)

Correlations for mass transfer coefficients depend on physical characteristics of the system and the flow conditions (e.g., laminar or turbulent). The mass transfer coefficient (k) is mainly a function of the feed **flow velocity** (v), the **diffusion coefficient** of the solute (D), the density and the module shape and dimensions. Of these parameters, flow velocity and diffusion coefficient are the most important.

Table 3 Mass transfer coefficients in various flow regimes

	Laminar	Turbulent
Tube	$Sh = \frac{k \cdot d_n}{D} = 1.62 \cdot \left(\frac{Re \cdot Sc \cdot d_h}{L} \right)^{0.33}$	$Sh = 0.04 \cdot Re^{0.75} \cdot Sc^{0.33}$
Channel	$Sh = 1.85 \cdot \left(\frac{Re \cdot Sc \cdot d_h}{L} \right)^{0.33}$	$Sh = 0.04 \cdot Re^{0.75} \cdot Sc^{0.33}$

In the channel of the feed spacer of a spiral-wound RO element, Schock and Miquel (1987) found that the mass transfer coefficient could be predicted by the following equation, when calculations for the velocity in the channel and the hydraulic diameter took the presence of the spacer into account:

$$k = 0.023 \cdot \frac{D}{d_n} \cdot Re^{0.875} \cdot Sc^{0.25} \qquad \text{Eq. 2.63}$$

2.9 TEMPERATURE AND WATER QUALITY

Temperature has an effect on K_w. The higher is the water temperature the higher the permeability will be. The change in permeability is about 3 % per °C. K_w is linked with the viscosity of water.

$$TCF = 1.03^{(t-25)} \qquad \text{Eq. 2.64}$$

When dealing with the membrane permeability, the correction will be as follows:

$$K_{wt} = K_{25\,°C} \cdot 1.03^{(t-25)} \qquad \text{Eq. 2.65}$$

Where:

TCF temperature correction factor
t temperature in °C
K_{wt} membrane permeability at temperature "t"
K_{w25} membrane permeability at 25 °C

As a result of the temperature effect of viscosity and therefore on membrane permeability, the required pressure to achieve or keep a certain flux (capacity) will be lower at higher temperatures.

Example 14 – Effect of temperature on required feed pressure

A brackish water element operates at 15 °C (t_1) and a feed pressure of 20 bar (P_1). In summer season the water temperature increases to 30 °C (P_2).

Question: What will happen with the required feed pressure in summer time (P_2), if we want to keep the product flow constant? Assume that osmotic pressure can be neglected.

Answer: At higher temperature the viscosity is lower.
As a consequence the permeability will be higher. So the required pressure will be ... bar.
How much lower will be the pressure?
To calculate, we apply the temperature correction factor (TFC).

$TFC = (1.03)^{(t1 - t2)}$

Combined with: $P_2 = P_1 \times TFC$

$P_2 = 20 \times (1.03)^{(15 - 30)}$ bar $\{P_1 = 20\text{ bar}, t_1 = 15\ °C \text{ and } t_2 = 30\ °C\}$

$P_2 = 20 \times 0.63$ bar

$P_2 = 12.6$ bar, which is (20-12.6)/20 = 36.7 % lower.

With rule of thumb of "3 % per °C" we get: 3 × 15 % = 45 % lower.

Example 15 – Where does the TFC come from?

$P_2 = P_1 \times TFC$

$TFC = (1.03)^{(t1 - t2)}$

Since: $J = P \times K_w$

follows: $P_2 = J / K_{w2}$ and $P_1 = J / K_{w1}$ so, $P_2 / P_1 = K_{w1} / K_{w2}$

and from: $K_{w1} = K_{w25oC} \times (1.03)^{(t1 - 25)}$ and $K_{w2} = K_{w25oC} \cdot (1.03)^{(t2 - 25)}$

follows: $K_{w1} / K_{w2} = (1.03)^{(t1 - t2)} = TFC$ or $P_2 = P_1 \cdot TFC$

The membrane permeability and the salt passage increase with temperature. Salt permeability is connected with the diffusion of salt ions through the membrane. The **diffusion coefficient** is defined with the following formula:

$$D = \frac{k_B \cdot T}{6 \cdot \pi \cdot \eta \cdot r} \qquad \text{Eq. 2.66}$$

Where:

k_B Boltzmann constant

T absolute temperature 273 + T°C

η viscosity of water

r radius of ion

In the diffusion coefficient equation, the viscosity is a dominant factor. Viscosity will decrease with temperature.

A frequently applied formula for normalizing the salt permeability is:

$$K_{st} = K_{s25^\circ C} \cdot 1.03^{(t-25)} \qquad \text{Eq. 2.67}$$

Where:

K_{st} Salt permeability at temperature "t"
$K_{s25^\circ C}$ Salt permeability at 25 °C
t Temperature in °C

The consequence of the previous equation is that the higher the temperature the higher the salt passage will be. Similar to the effect of temperature on required pressure, the effect on salt passage can be derived:

$$C_{pt2} = \frac{C_{pt1}}{TCF} = \frac{C_{pt1}}{1.03^{(t1-t2)}} \qquad \text{Eq. 2.68}$$

Example 16 – Normalization of permeate salinity

A RO spiral wound element is treating river water at 5 °C at a flux of 25 L/m^2.h. The salinity (sodium chloride) of the permeate is 5 mg/L.

Question: What will be the salinity in the summer period when the temperature of the water increases to 25 °C. The flux is kept at the same level.

Answer:

The salt permeability will increase with approximately 3 % per °C.

Since the flux (J) and C_f are constant, C_p will increase with approximately (25-5) = 20 °C times 3 % per °C or about 60 %.

So, C_p will increase approximately with 60 %, from 5 mg/L to 8 mg/L.

Using the TFC we get the more accurate answer:

$C_{p\,t2} = C_{p\,t1} / \text{TCF} = C_{p\,t1} / 1.03^{(t1-t2)}$; {$C_{p\,t1}$ = 5 mg/L and t = 5 °C and t = 25 °C}

$C_{p\,t2}$ = 5 mg/L / TCF = $C_{p\,t1} / 1.03^{(5-25)}$

$C_{p\,t2}$ = 5 mg/L / 0.55 = 9 mg/L

2.10 FACTORS AFFECTING REVERSE OSMOSIS PERFORMANCE

The permeate flux and the RO membrane salt rejection are important operational performance parameters of a reverse osmosis system. Dupont (2020) summarizes the parameters influencing the flux and salt rejection as follows:

- With increasing effective feed pressure, the permeate salinity will decrease (increased salt rejection) while the permeate flux will increase.
- With increasing temperature (and all other parameters are kept constant), the permeate flux and the salt passage (less salt rejection) will increase.
- With increasing recovery, the permeate flux will decrease and stop if the salt concentration reaches a value where the osmotic pressure of the concentrate is as high as the applied feed pressure. The salt rejection will reduce (more salt passage) with increasing recovery.

Permeate flux and salt rejection are important performance indicators of a RO system. The flux and salt rejection of a membrane system are mainly influenced by variable parameters including: feed pressure, water temperature, RO recovery, and feedwater salt concentration. Figure 7 illustrates the impact of each of those parameters when the other three parameters are kept constant. In practice, there is normally an overlap of two or more effects. These figures are qualitative examples of RO performance. These figures are qualitative examples of RO performance and based on the solution-diffusion model.

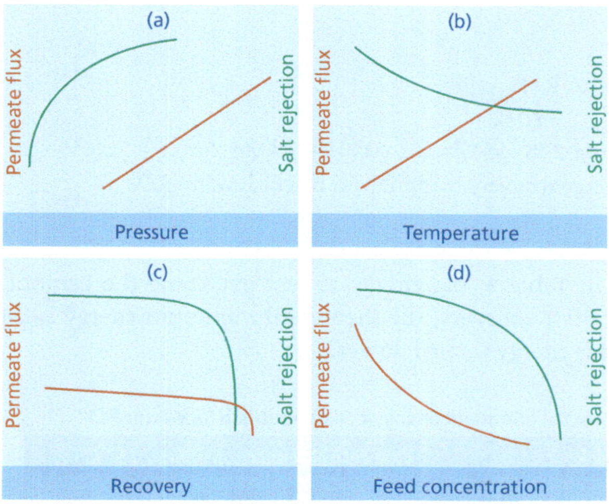

Figure 7 Impact of pressure (a), temperature (b), recovery (c), and feed concentration (d) on reverse osmosis performance (Dupont, 2020)

2.11 ENERGY CONSUMPTION

To pressurize water cost energy. The theoretical minimum energy can be calculated with the formula:

$$E = 0.0275 \cdot P \qquad \text{Eq. 2.69}$$

Taking into account the efficiency of the pump the formula changes into

$$E = \frac{0.0275 \cdot P}{N_{pump}} \qquad \text{Eq. 2.70}$$

In RO and NF the recovery is less than 100%. As a consequence, the energy consumption per m^3 water produced will be higher according the formula:

$$E = \frac{0.0275 \cdot P}{N_{pump} \cdot R} \qquad \text{Eq. 2.71}$$

Where:
E = energy consumption in kWh/m^3

P = pressure in bar
N_{pump} = efficiency pump + motor
R = recovery

Example 17 – Energy consumption in BWRO

A brackish water RO operates at a feed pressure of 15 bar. The recovery is 75 %. No energy recovery. What is the energy consumption per m^3.

Answer:

$E = (0.0275 \cdot P) / (R \cdot N_{pump})$
Assuming: $N_{pump} = 70\ \%$
We have: $E = (0.0275 \cdot 15) / (0.75 \cdot 0.7) = 0.8\ kWh/m^3$
Usually no energy recovery is applied for brackish water RO.

As a reference, in Table 4, the energy to just overcome the osmotic pressure seawater is presented. At 50 % recovery, the theoretical minimum energy is about 1 kWh/m^3. In brackish water this energy is much lower.

Table 4 Theoretical minimum energy consumption in Seawater RO

Recovery, R	Theoretical separation energy 25 °C
0%	0.71 kWh/m^3
25%	0.82 kWh/m^3
50%	0.99 kWh/m^3
75%	1.35 kWh/m^3
100%	3.1 kWh/m^3

In practice, energy recovery devices are applied to optimize the energy consumption in RO plants.

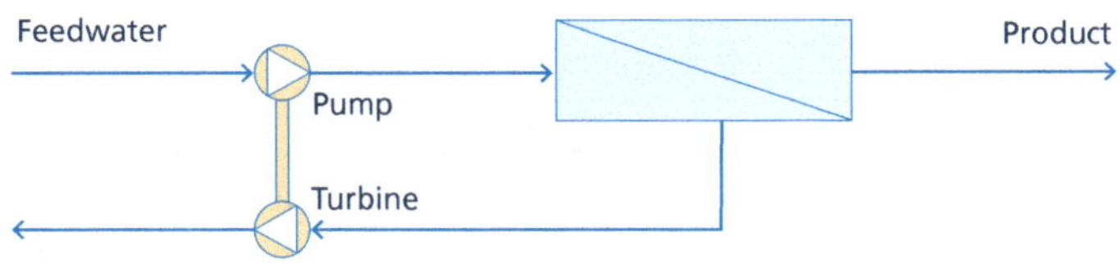

Figure 8 Schematic of an RO system with energy recovery with turbine

The energy consumption can be reduced by e.g., recovering energy from the brine with a turbine:

$$E = \frac{(0.0275 \cdot P_{feed})}{(N_{pump} \cdot R)} - \frac{(0.0275 \cdot (1-R) \cdot P_{conc} \cdot N_{turbine})}{R} \qquad \text{Eq. 2.72}$$

Where:
Pconc = Pfeed - ΔP ; (ΔP = brine pressure loss through the RO plant)
N_p = pump /generator efficiency
N_t = turbine efficiency

In practice the energy consumption equals: 0.5 – 1 kWh/m^3 for brackish (without energy recovery device); 3.0 – 4.0 kWh/m^3 for sea (with energy recovery device).

Example 18 – Energy consumption in SWRO with turbine energy recovery
A SWRO plants operates at a recovery = 50 %, the feed flow = 1500 m^3/h, the product flow = 750 m^3/h and concentrate flow = 750 m^3/h

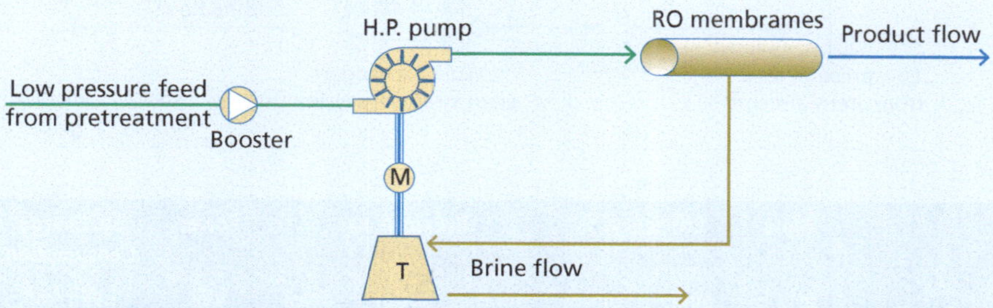

	Flow	Efficiency, %		Pressure, bar			Power	Sp. Energy Consumption
	m^3/ h	pump	motor	inlet	outlet	pres. drop	kW	kWh/m^3
Booster before HP pump	1500	83%	96%	0	11.5	11.5	595	0.794
HP pump	1500	88%	97%	11.5	74.1	62.6	2,934	
Turbine	750	87.2%		72.2	0	-72.2	-1,299	
							1,686	2.249
							2,282	**3.042**

The application of pressure exchangers is a well-established technology in SWRO desalination. An energy recovery device of the type "pressure exchanger" exchanges the pressure of the RO concentrate with a very high efficiency (> 95%) to the feed seawater as illustrated in figure 9. In this way the capacity of the feed pump can be much less (50%), having two advantages:
i) saving cost due to less high pressure pump capacity, which are very costly;
ii) saving energy, because the pressure exchanger has a higher efficiency (>95 %) than high pressure pumps (< 90 %).

Example 19 – Energy consumption in SWRO with pressure exchanger energy recovery

Same information as in the previous example.

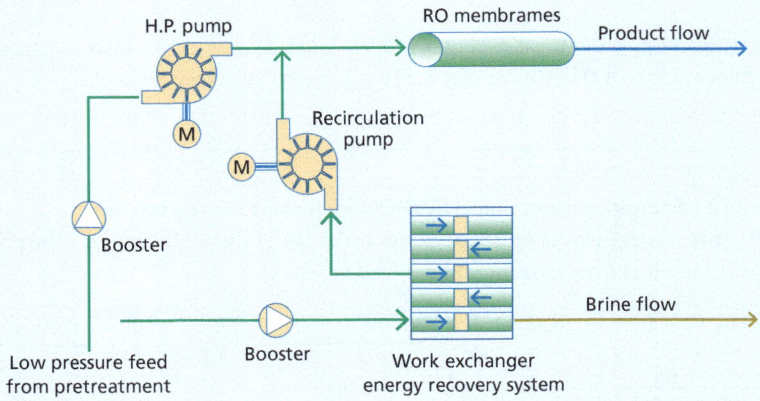

	Flow	Efficiency, %		Pressure, bar			Power	Sp. Energy Consumption
	m^3/h	pump	motor	inlet	outlet	pres. drop	kW	kWh/m^3
Booster before HP pump	750	83%	96%	0	11.5	11.5	298	0.397
HP pump	750	88%	97%	11.5	74.1	62.6	1,513	2.017
Booster before Work Exchanger System	750	83%	96%	0	1.2	1.2	31	0.041
Work Exchanger Energy Recovery System	750			72.2	71.5	0.7		
Recirculation Pump	750	80%	94%	71.5	74.1	2.6	71	0.095
Total							**1,913**	**2.55**

Pressure exchangers have several advantages, such as:

- **High efficiency** in exchanging the pressure of the brine to feed water, up to 97%. A Pelton turbine has max. 90% efficiency.
- **Replacement** of 50 - 60% of the capacity of the high-pressure feed pumps having much lower efficiency than pressure exchangers.
  Remark: High pressure feed pumps have 70-90% efficiency.
  Saving investment cost in installed high pressure pump capacity 50 - 60%.

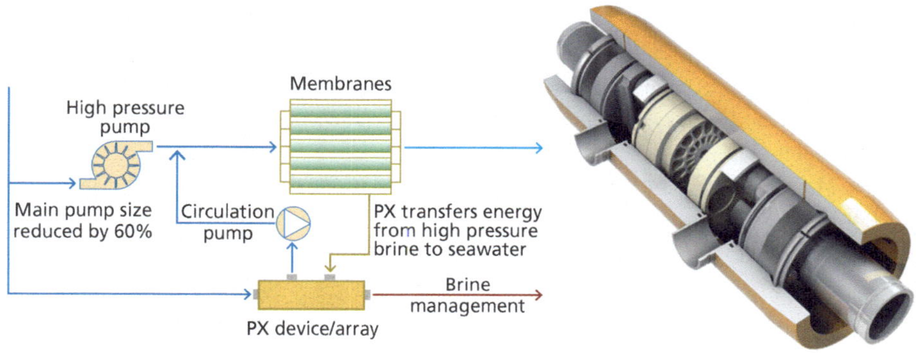

Figure 9 Schematic of a seawater RO system with a pressure exchanger (PX) and image of a pressure exchanger (Adopted from Energy Recovery, 2021)

Energy consumption represents together with the investment, the largest part of the cost of seawater RO. Seawater RO has improved tremendously over the past 20-25 years. Energy consumption has been reduced: from 8 kWh/m^3 in 1980 to 4 kWh/m^3 in 2000.
Since 2006 further reduction down to about 3 kWh/m^3 was achieved, due to lowering the feed pressure down to 60 bar and applying pressure exchangers.

2.12 SYSTEM CONFIGURATION

According a commonly applied rule of thumb for brackish water: "When conversion has to be higher than about 50 % a second stage (array) is used as well". The number of vessels in the next stage is about 50 % of the previous one. Because the ratio feed flow to permeate flow at the entrance of the next stage is the same. In the second stage about 50 % is converted in product. This brings the total conversion at about 75 %.

Modern membranes have a very high permeability. Consequently, the required pressure is much lower than in the past. Head loss due to the spacer and increasing osmotic pressure reduce the net driving pressure. As a result, the permeate flow in the last elements in a vessel are substantially lower, which allows a higher recovery before the CPF arrives at 1.21.

2.13 REFERENCES

ASTM D4516 - 00 (2000) Standard Practice for standardizing reverse osmosis performance data ASTM, West Conshohocken.

DUPONT (2020) FILMTEC™ Reverse Osmosis Membranes Technical Manual. In: Water solutions (ed), pp. 207.

Energy Recovery. (2021). PX Pressure exchanger [Online]. Available: https://energyrecovery.com/water/px-pressure-exchanger/ [Accessed 20 April 2021].

Hydranautics (2001) Terms and Equations of Reverse Osmosis.

Mulder M (2003) Basic Principles of Membrane Technology, Second edn Kluwer Academic, Dordrecht / Boston / London

Salinas Rodríguez, S. G., Schippers, J. C. & Kennedy, M. D. (2015). The process of reverse osmosis. In: Burn, S. & Gray, S. (eds.) Efficient Desalination by Reverse Osmosis: A best practice guide to RO. IWA Publishing.

Schock G, Miquel A (1987) Mass transfer and pressure loss in spiral-wound modules. Desalination 64: 339

Voet D, Voet JG, Pratt CW (2001) Fundamentals of Biochemistry (Rev. ed.), New York

Chapter 3

Fouling and pre-treatment

Jan C. Schippers, Sergio G. Salinas-Rodríguez, Maria D. Kennedy

The learning objectives of this chapter are the following:

- Define fouling and clogging in membrane systems
- Define the role of pre-treatment
- Present, discuss and propose pre-treatment processes required for RO systems depending on raw water quality.

3.1 INTRODUCTION TO FOULING

Many reverse osmosis (RO) plants run smoothly, many have suffered from membrane fouling, and many other plants either new or old still suffer from membrane fouling.
Fouling may result in a variety of problems, such as: the need for (frequent) membrane cleaning, the reduction of production capacity and/or plant availability, a higher energy consumption during treatment, a decrease in produced water quality, making RO installations less reliable, and finally a frequent replacement of the RO membranes. Figure 1 shows old RO elements and old cartridge filters from a desalination plant before their final disposal.

Figure 1 Old reverse osmosis elements and cartridge filters piling up before final disposal. (Jan C. Schippers)

The causes of fouling in RO membranes can be classified in five categories, namely:
1. Particulate fouling due to suspended and colloidal matter
2. Inorganic fouling due to iron and manganese

Editors: Sergio Salinas, Jan Schippers, Gary Amy, In Kim, Maria Kennedy
doi: 10.2166/9781780409863_0059

3. Biofouling due to growth of bacteria
4. Organic fouling due to organic compounds e.g., polymers
5. Scaling due to deposition of sparingly soluble compounds

In a RO membrane system, fouling and scaling may manifest in three ways:

i) increased differential pressure across the feed spacer in spiral wound elements due to a mechanism named "clogging", resulting in membrane damage;
ii) increased membrane resistance (decreasing normalized permeability (K_w) or mass transfer coefficient, MTC) due to deposition and/or adsorption of material on the membrane surface, resulting in higher required feed pressure to maintain capacity; and
iii) increased normalized salt passage due to concentration polarization in the foul layer, resulting in higher salinity in product water.

Clogging results in higher differential pressure (head loss) across the feed spacer resulting in mainly damage to elements due to: *i*) telescoping in spiral wound, *ii*) channelling in spiral wound, *iii*) squeezing spiral wound membrane element.

Local clogging on the RO feed spacer may occur as well, which results in uneven flow distribution, resulting in places with low or no flow at all. This yields to areas in the RO element with high conversion / recovery and with high concentration polarization. This ultimately results in in enhanced deposition of particles, local precipitation of sparingly soluble compounds, and growth and attachment of bacteria.

Figure 2 Telescoping (left), channelling (middle) and squeezing (right) of RO elements. (Jan C. Schippers)

Fouling results in an increase of membrane resistance. Due to fouling, higher RO feed pressure is required to maintain plant production capacity. As a result, the RO recovery decreases (same feed flow but less product), possibly (not always) a lower salt rejection (so higher salinity in product) due to increased concentration polarization, and increased cleaning frequency of the RO membranes which may/will result in shorter lifetime of membranes.

Concentration polarization has been discussed in chapter 2. Concentration polarization results in increased salt passage, increased membrane resistance, and reduced net driving pressure.

Concentration polarization will increase, when the cross-flow velocity close to the membrane decreases. This may occur due to uneven flow distribution and due to the foul layer. As a result of this phenomenon, accumulation of dissolved salts and organic compounds, colloidal matter, and suspended matter will occur on the surface of the membranes.

A high concentration polarization factor may ultimately lead to precipitation of sparingly soluble compounds, enhanced deposition of colloidal and suspended matter, increased salt passage due to higher concentrations at the membrane surface, and reduced net driving pressure, due to higher osmotic pressure.

To **clean the RO membranes**, and thus restore permeability, several membrane manufacturers recommend performing the cleaning in place (CIP) procedure, when: *i*) the MTC or normalized flux drops by 10%, *ii*) the normalized salt passage increases by 10%, *iii*) the normalized differential pressure (feed pressure - concentrate pressure) increases by 15%. For performing the CIP procedure, there is a wide range of chemicals that can be used, and above all, compatibility of these chemicals with the RO membrane needs to be secured.

3.2 PRE-TREATMENT

Pre-treatment steps can be implemented before the RO membranes, to preserve the performance and lifetime of RO membranes. Generally, a proper selection of pre-treatment methods for RO feed water will improve effectivity and extend the life span of the system by preventing or minimizing particulate and colloidal fouling, biological fouling and scaling as well as reduce the need for cleaning of the membranes.

The quality of surface waters shows large differences in time and also per location e.g., suspended and colloidal matter (measured by SDI) and algae. A very limited number of sources (locations) has low fouling potential and needs only cartridge filtration. Traditionally in membrane desalination systems, pre-treatment of surface water is focused on reduction of SDI. A basic cartridge filtration is always included as pre-treatment.

The majority of surface waters needs additional treatment besides cartridge filtration. A great variety of pre-treatment methods are applied (see Figure 3): e.g., artificial recharge (e.g., through shore wells / beach wells (sea water) or infiltration canals / ponds (river water)), media filtration, in-line coagulation (addition of coagulant followed by media filtration), coagulation/ sedimentation/ media filtration, coagulation/ flotation / media filtration, ultra- and microfiltration.

To meet the RO membrane manufacturers guidelines for the silt density index (SDI), a variety of conventional techniques were and still are applied. These techniques are already in use, for the production of drinking and industrial water, for many decades. We can distinguish two different water sources to illustrate these techniques: *i*) surface water: river water, lake water and seawater, and *ii*) groundwater, bank/shore filtered water (brackish or seawater). In this chapter the focus will be on surface water.

SDI is a filtration test to determine the fouling potential of RO feed water due the presence of suspended and colloidal matter. Chapter 4 describes in detail the SDI.

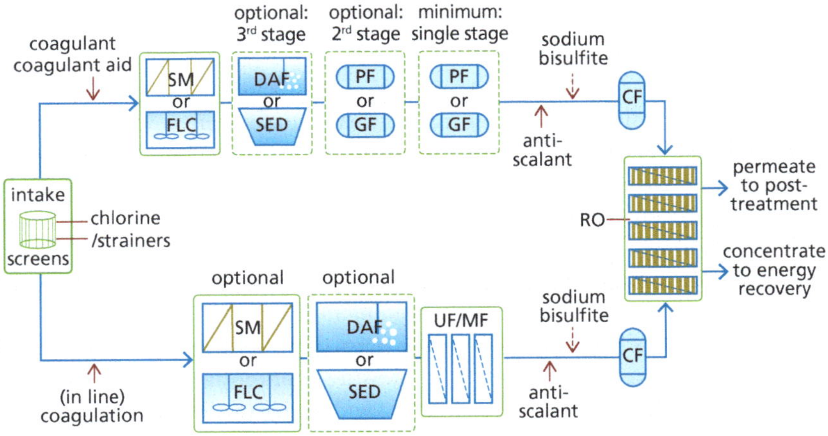

Figure 3 Simplified seawater reverse osmosis schemes. (Adapted from Lattemann, 2010). SM= sedimentation, FLC = flocculation, DAF= dissolved air flotation, PF = pressurised filter, GF = gravity filter, CF = cartridge filter, UF/MF = ultrafiltration / microfiltration

To ensure acceptable membrane cleaning frequencies, pre-treatment is a requirement. A variety of pre-treatment processes are applied, such as: conventional processes & combinations, advanced processes, and combinations of conventional with advanced processes. The focus of this chapter will be on commonly applied conventional processes as pre-treatment for reverse osmosis.

The Permasep Engineering Manual (E. I. du Pont de Nemours and Company, 1982) provides recommendations of pre-treatment for removal of particulate matter based on the SDI value of the raw water to treat as presented in Table 1. In all cases cartridge filtration (5-20 μm) just preceding the high-pressure pump is required.

Table 1 Recommended pre-treatment for removal of particulate matter. (E. I. du Pont de Nemours and Company, 1982)

SDI value	Pre-treatment
SDI < 6	Media filtration (rapid (green) sand filtration)
	Dual media filtration (anthracite/sand)
6 < SDI <50	In-line coagulation (direct filtration), which includes addition of a coagulant to water mixing, passing through media or dual media filter)
SDI > 50	Coagulation, sedimentation (or flotation), rapid sand filtration

These pre-treatment processes reduce SDI and in addition biofilm formation potential (except cartridge filtration) significantly. Chlorination combined with neutralization with sodium bisulphite was commonly applied. However, it turned out that chlorination produces large quantities of assimilable organic carbon (AOC), causing serious bio-fouling.

3.2.1 Intakes, shore wells / beach wells

Intake structures in desalination plants can be divided in indirect and direct intakes as illustrated in Figure 4. Intakes can be placed at the coast, at the bottom of the sea, or as wells. An intake is a structure with the aim to providing good water quality to the treatment plant, with low environmental impact, avoiding entrainment of materials and requiring little maintenance.

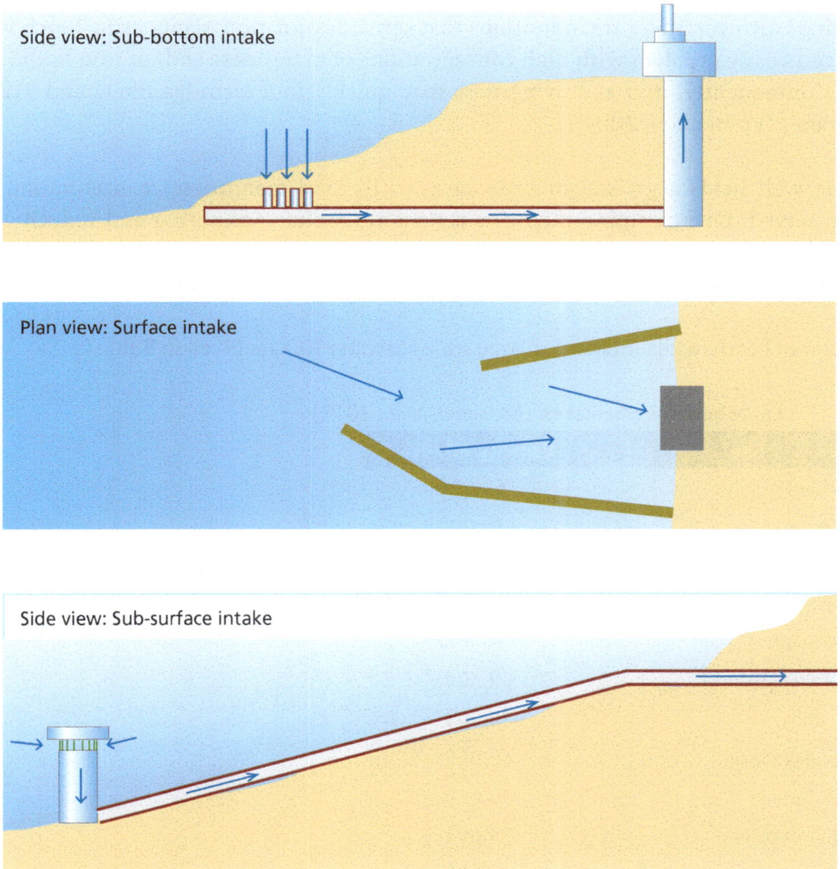

Figure 4 Intake structures for seawater desalination plants. (Adapted from Pankratz, 2006)

Beach wells produce usually water with low turbidity and low SDI. In addition, algae and transparent exo-polymeric particles (TEP) are expected to be removed effectively as well. The effect of the soil passage is responsible for the removal of these organisms and particles, because of the long residence time and small size of the pores in the soil.

Normally, beach wells are an alternative to raw water open-intakes. Also, beach wells are considered as pre-treatment prior to RO units in SWRO plants.

Beach wells are usually located on the seashore in close vicinity of the ocean. These intake facilities are relatively simple to build and the seawater they collect is pre-treated via slow filtration through the subsurface sand/seabed formations in the area of source water extraction. Consequently, raw seawater collected through beach wells has better quality in terms of solids, silt, oil and grease, natural organic contamination and aquatic micro-organisms. Sometimes, beach intakes may also yield source water of lower salinity (Voutchkov, 2004).

Typically, beach wells are assumed to eliminate the extra pre-treatment steps prior to RO units. However, there are indications that some desalination plants using beach wells may face a costly problem with high concentrations of manganese and/or iron in the feed water. Consequently, iron and manganese may quickly foul cartridge filters and SWRO membranes (Voutchkov, 2004).

In some well fields a gradual increase in iron (II) (and manganese) concentration has been observed. Overpumping might be a reason. Careful monitoring and reducing the abstraction rate might minimize the problem. The geological situation will determine the safe abstraction rate. Iron and manganese fouling is discussed in chapter 7.

Examples of beach-wells intakes for large scale seawater RO are given in Table 2.

Table 2 Large beach-wells intakes. (Missimer, *et al.*, 2013)

Site	Capacity, m^3/day
Sur (Oman)	160,000
Alicante (Spain)	130,000
Tordera (Spain)	128,000
Pembroke (Malta)	120,000
Bajo Almanzora (Spain)	120,000
Bay of Palma (Spain)	89,600
WEB (Aruba)	80,000
Lanzarote IV (Spain)	60,000
Sureste (Spain)	60,000
Blue Hills (Bahamas)	54,600
Santa Cruz de Tenerife (Spain)	50,000

A well developed and maintained beach well system can provide a constant and continuous yield, and low suspended solids in the feed RO unit. In most cases, it is possible to achieve SDI_{15} below 3 using single stage sand filtration without coagulant or even by simple cartridge or bag filters only (Wolf, *et al.*, 2005).

3.2.2 Conventional pre-treatment processes

The most commonly applied pre-treatment processes in *surface water* are: screens and strainers, chlorination, sedimentation, flotation, granular media filtration (sand filtration), coagulation enhancing sedimentation / flotation and granular media filtration, and cartridge filtration.

Commonly applied in *groundwater* treatment are the following treatment processes: aeration, granular media filtration (sand filtration), and cartridge filtration.

Other conventional processes that are applied are the following: bank / shore/beach filtration, granular activated carbon filtration, pre-coat filtration, air stripping (in the case of groundwater) to remove hydrogen sulphide. Other conventional advanced processes that are also applied as pre-treatment are micro- and ultrafiltration membranes.

The pre-treatment process to be applied depends strongly on the raw water source, on the perception of the engineer, and is also influenced by the contemporary perception of what is best technology depending on costs and environmental considerations.

There are several raw water sources, such as: *i*) surface water e.g., river, canal, brackish water, and seawater; *ii*) groundwater and beach well abstracted water, *iii*) treated domestic / industrial wastewater. Currently, media filtration (e.g., sand filtration, dual media filtration) and ultra / micro-filtration are the core of the two types of pre-treatment processes. In surface and wastewater pre-treatment frequently (intermittent) chlorination is applied.

In practice, frequently applied combinations of pre-treatment technologies are the following:

- In line coagulation / media filtration
- Coagulation / sedimentation / media filtration
- Coagulation / flotation / media filtration
- In line coagulation / ultrafiltration
- Coagulation / flotation / ultrafiltration

In almost all process schemes, cartridge filtration is applied as a final polishing step before RO.

3.2.3 Screens

Bar screens / strainers are used to protect the structure downstream against large objects which could result in *i*) obstruction of e.g., pipes, channels, pipes, and *ii*) clog filters. The thickness of the bars could be about 10 mm and the spacing of them in the range of 10 to 50 mm.

Strainers are used for smaller openings. These smaller openings are created by a wire mesh construction having openings down to 0.1 mm. Frequently for seawater the openings are in the range 2 to 3 mm.

Screens are used to protect pumps and filters against the entry of large objects e.g., fish, seaweed, jellyfish, debris, etc. The effectiveness of course screening depends on the spacing between the bars. In practice, the spacing can be considered fine (3 to 10 mm), medium (10 to 25 mm), and coarse (50 to 100 mm). Clogging of screens and filters is a potential permanent nuisance e.g., due to jellyfish.

Figure 5 illustrates the intake structure at the Gold Coast Desalination plant in Australia taking water from the Coral Sea. This plant, in operation since 2008, has a capacity of about 125000 m^3/d. The tunnel connecting the coarse screen intake to the intake shaft has a length of about 2 km.

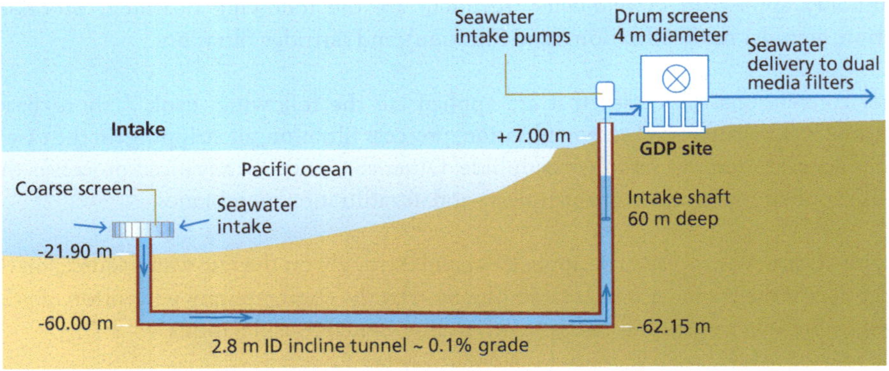

Figure 5 Intake at Gold Coast Desalination Plant (GDP), Australia. (Adapted from Baudish, 2015)

A bar screen is commonly positioned at an angle of 60º to 90º. This angular positioning increases the screening surface area and facilitates cleaning and eliminates fast head loss increase. The approach channel should be straight for at least 0.6 m ahead of the screen to produce uniform flow through the screen. The velocity in the approach channels should be at least 0.3 to 0.4 m/s to prevent accumulation of settled materials. The maximum velocity in the screens is in the range of 0.6 to 1.0 m/s.

Macro straining through perforated steel sheet or metal wire netting with openings size larger than 0.3 mm (300 µm) are used to remove suspended solids, floating or semi floating matter, fish, animal or vegetable debris, insects, twigs, algae, grass, etc. Micro straining through plastic or metal fabric with openings less than 100 µm are used to remove fine suspended matter, small fish, plankton (algae, etc.).

Automatically backwashed **disk filters** are increasingly applied e.g., as pre-treatment for ultrafiltration. Disks are available with openings from 20 to 400 µm. Ultrafiltration requires usually 100 µm.

3.2.4 Chlorination

Chlorination is commonly applied in surface water intakes to control growth of mussels, barnacles, sea anemones, hydroids. Chlorine is added to the raw water as sodium hypochlorite (NaOCl) or chlorine gas Cl_2. It hydrolyses in water to hypochlorous acid:

$$Cl_2 + H_2O \rightarrow HOCl + HCl$$

$$NaOCl + H_2O \rightarrow HOCl + NaOH$$

Where hypochlorous acid dissociates in water to hydrogen and hypochlorite ions:

$$HOCl \leftrightarrow H^+ + OCl^-$$

The sum of Cl_2, NaOCl, HOCl and OCl^- is called free residual chlorine.

Unfortunately, the most commonly applied RO membranes, based on polyamide are extremely vulnerable to chlorine and will lose their rejection properties fast. That is why chlorine is neutralized, before entering the RO with sodium meta bisulphite (SMBS). This has been applied commonly and is still applied in most plants.

Normally, **de-chlorination** is performed prior to the RO membranes to neutralize the residual chlorine in the feed water which may damage the membrane by oxidation. Sodium metabisulphite is used for de-chlorination due its high cost effectiveness. In water it reacts to sodium bisulphite:

$$Na_2S_2O_5 + H_2O \rightarrow 2NaHSO_3$$

Sodium bisulphite then reduces hypochlorous acid:

$$2NaHSO_3 + 2HOCl \rightarrow H_2SO_4 + 2HCl + Na_2SO_4$$

Typically, 3.0 mg of sodium metabisulphite is used to neutralize 1.0 mg of free chlorine, where theoretically 1.34 mg metabisulphite is necessary for 1.0 mg chlorine. Also, activated carbon is very effective to reduce residual free chlorine where water reacts with carbon and chlorine (Fritzmann, *et al.*, 2007):

$$C + 2Cl_2 + 2H_2O \rightarrow 4HCl + CO_2$$

Continous chlorination results results into severe biofouling of the RO membranes. Van der Kooij and Hijnen (1984) demonstrated that chlorination results in the formation of assimilable / biodegradable organic carbon e.g., humic acids react with chlorine to form smaller molecules which are more easily biodegradable.

Several plants lost their polyamide Thin Film Composite membranes due to failure of: sodium meta bisulphite dosing equipment, and/or chlorine monitoring equipment. A couple of plants eliminated chlorination completely to avoid *i*) the risk of damaging the membranes, *ii*) formation of disinfection by-products which are not fully removed by RO e.g., tri halo methanes, bromate, and *iii*) environmental considerations.

Besides manual cleaning, mechanical cleaning with "pigs" is successfully applied in several seawater RO intake pipes.

3.2.5 Granular media filters

The filtration process is widely used in water treatment mainly for the removal of "particulate materials". In this process, water passes through a filter medium, and particulate materials either accumulate on the surface of the medium (surface filtration) or are collected through its depth (depth filtration). A wide range of media is utilized in filtration systems.

Filtration involves two main stages: filtration stage (removal of particles by filter media), and regeneration or backwash stage (removal of deposited particles from filter media and restoration of filtration capacity).

Filtration improves the clarity of surface waters by removing algae, sediment, clay, and other organic or inorganic particulate matter. Filtration is often required in conjunction with (chemical) disinfection of surface water to ensure that water is free of pathogenic microorganisms. Groundwater is often low in microorganisms and particles but may require filtration when other treatment processes (such as oxidation or softening) generate particles that must be removed.

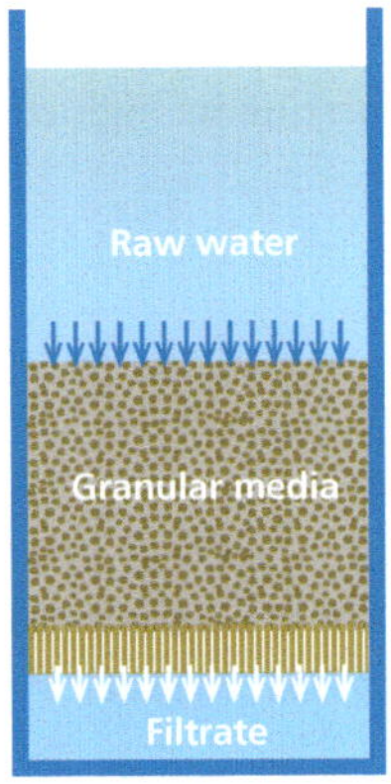

Figure 6 Simple schematic of a down flow single media filter

Filters consists of a tank or a chamber containing media (e.g., sand grains having a diameter of 0.5 – 1.0 mm) with a height of the media (filter bed) of about 1 m. The filtration rate can range between 5 to 20 m/h. The main change after filtration will be the (partly) removal of suspended and colloidal particles and lower turbidity.

Usually the direction of the flow is downwards through the filter media. The removed material accumulates in the filter bed. The filtered deposits gradually move as a front down the filter bed and finally be carried into the filtrate. Before breakthrough of particles or unacceptable head loss, the filter will be cleaned. Cleaning is achieved by backwashing with water or air, air/water followed by water. The frequency of backwashing ranges typically once per 2 – 3 days to once per 8 hours.

3.2.5.1 Filter media

Several media are applied in filtration e.g., sand, composing mainly of quartz (SiO_2), garnet e.g., almandine $Fe_3Al_2(SiO_4)_3$, anthracite (carbon). These materials are mainly natural products and used because they are rather cheap and resistant during backwashing with water and air scour. The grain size of the applied media is in the range of 0.2 – 2.0 mm.

Figure 7 Photographs of sand media, anthracite media, and garnet media (Sharma and Schippers, 2019)

Some important properties of granular filter media are: grain size distribution (effective size, uniformity coefficient), porosity, density, sphericity, hardness/attrition loss, inertness / reactivity / solubility in acid, and cost.

Natural granular materials have a nearly log-normal size distribution. The size distribution of filter media is determined by sieve analysis. In this procedure, a sample of filter media is sieved through a stack of sieves (e.g., according to ASTM or DIN standards). The weight of material retained on each sieve is measured, the cumulative % passing is calculated and then plotted as a function of sieve size.

Table 3 Typical properties of common filter media for granular bed filters. (Adapted from Tobiason, et al., 2011)

	Silica sand	Anthracite coal	Granular activated carbon	Garnet
Grain density (kg/m^3)	2650	1450 - 1730	1300 - 1500*	3600 - 4200
Loose-bed porosity	0.40 - 0.47	0.50 - 0.60	0.5	0.45 - 0.55
Sphericity	0.7 - 0.8	0.46 - 0.60	0.75	0.6

* Values for virgin carbon pores filled with water

There are two main aspects in rapid sand and other media filtration which are dominant during operation: the quality improvement or removal of unwanted constituents, and the **development of head loss** / pressure drop during filtration.

The main function in **quality improvement** is the removal of suspended and colloidal matter, usually measured as turbidity. Furthermore, the removal of dissolved substances such as: *i*) iron (II) and manganese (II) in ground water due to adsorption / catalytic oxidation, *ii*) ammonium removal due to oxidation by bacteria, and *iii*) removal of organic matter (very limited) due to biodegradation by bacteria attached at the large surface area of the grains.

There are two mechanisms involved in media filtration for particulate matter removal. Principle of media filtration is aiming at **depth filtration** instead of **surface filtration** (illustrated in Figure 8). For this purpose, physical adsorption is strived after instead of mechanical straining.

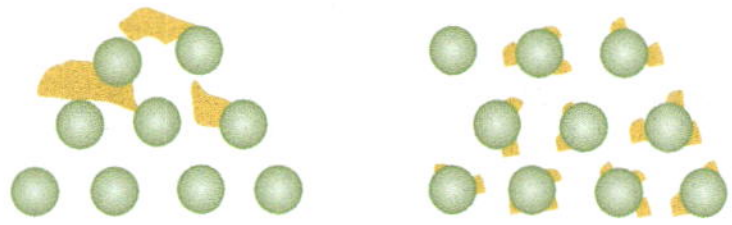

Figure 8 Schematic of mechanical straining / surface filtration (left) and physical adsorption / depth filtration (right). (Adapted from Huisman, 1986)

Why is depth filtration preferred above surface filtration? The answer to this question is related to the fact that the pore opening in granular media is 0.15 times the diameter of spheres (e.g., grains having a diameter of 1 mm, leave openings of 0.15 mm or 150 μm) as illustrated in Figure 9.

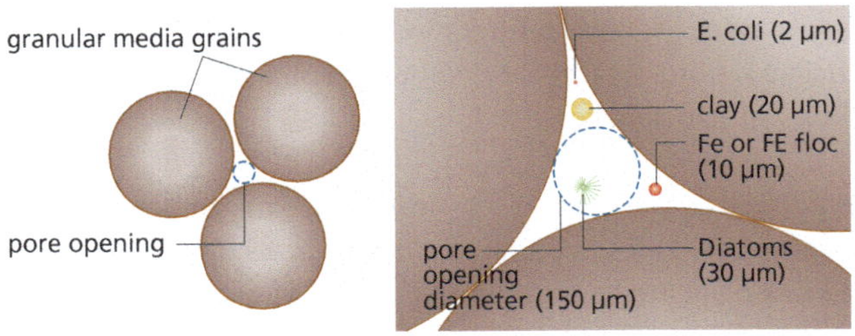

Figure 9 Surface filtration: Openings in granular media. (Adapted from Huisman, 1986)

Three mechanisms during filtration can be identified as illustrated in Figure 10, namely: interception in which particles follow the streamline of the water, gravity causing sedimentation of the particles, and diffusion due to Brownian motion. Interception and gravity are important for larger particles, while diffusion is important for smaller particles only, e.g., < 1 µm. Particles much smaller than the pores can be captured, because of the abovementioned mechanisms.

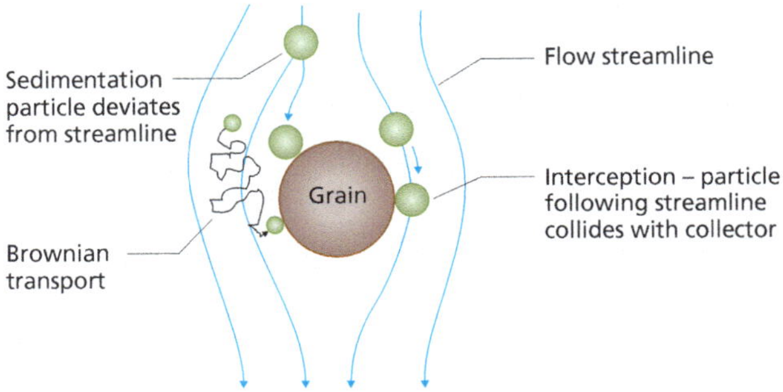

Figure 10 Schematic of transport mechanisms in media filtration (Adapted from Huisman, 1986)

Pressure drop/ head loss occurs when water flows through a granular filter bed (illustrated in Figure 11). This is due to friction between water and the surface of the grains. It is usually calculated with the Carman-Kozeny equation which is valid for laminar flow conditions (when Reynolds number is < 5; usually during filtration this value is not exceeded). The linear increase in head loss indicates that depth filtration occurs. If the line curves steeply upwards, there is an additional head loss due to a surface mat or cake on the top of the filter material.

Over time, the head loss will develop due to both mechanisms: depth filtration and due to surface filtration.

3.2.5.2 Vulnerability of media filtration

Breakthrough of turbidity in media filters is an ongoing concern. The variations in quality of the water to be treated, makes breakthrough rather unpredictable. The first filtrate water has usually a poor quality and has to be drained. Usually, turbidity meters are installed to monitoring breakthrough and SDI is measured daily. Backwashing is done as soon as the guideline for turbidity or head loss is exceeded.

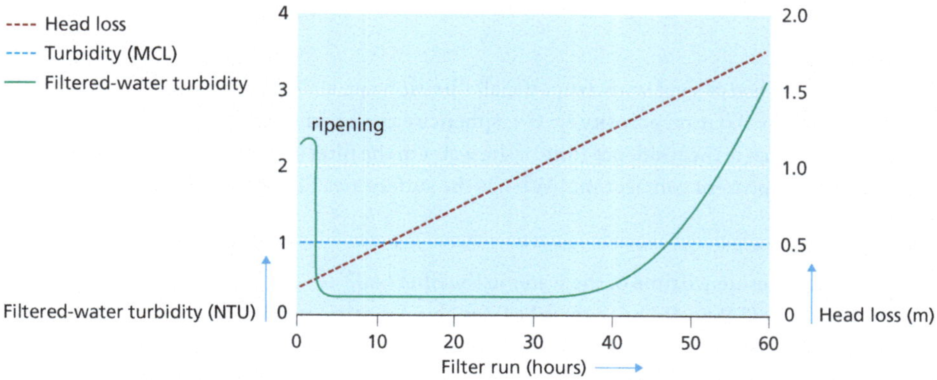

Figure 11 Breakthrough of turbidity and head loss development at constant flow rate. (Adapted from AWWA, 1995)

3.2.5.3 Filtration rate

Filtration rate is expressed as volumetric flow per unit of area of the filter bed e.g., m^3 water hour/ m^2 surface area ($m^3/m^2/h$), or more commonly used: m/h.

Example 1

A media filter has a surface area (of the filter bed) of 30 m^3. The capacity of the filter is 210 m^3/h. What is the rate of filtration?

Answer:

In practice the rate of filtration is in the range of 4 to 20 m/h depending on the quality of the water to be treated.

The **filtration rate** or velocity of flow (m/h) is defined as:

$$v = Q/A \quad \text{Eq. 3.1}$$

where Q = flow rate (m^3/h), A = area of filters (m^2).

The **interstitial velocity** is defined as:

$$v_i = v/\varepsilon \quad \text{Eq. 3.2}$$

The **empty bed contact time** (EBCT, h) is defined as:

$$EBCT = V/Q = L/v \quad \text{Eq. 3.3}$$

where: V = volume of filter (m^3), L = depth of filter bed (m)

The **residence time** t_{res} (h) is defined as:

$$t_{res} = L/v_i = \varepsilon \cdot (L/v) = \varepsilon \cdot EBCT \quad \text{Eq. 3.4}$$

Example 2

Assume a filter bed of sand with: (superficial) filtration rate v = 7 m/h, bed depth L = 1.2 m, effective size d = 0.6 mm, porosity ε = 0.4, sphericity ψ = 1 (spherical grains).

Questions: What is the residence time of the water in the filter bed? What is the empty bed contact time (apparent contact time) What is the surface area of the grains in the filter bed?

Answers:

a. What is the residence time of the water in the filter bed?

$t_{res} = (L \cdot \varepsilon) / v = (1.2\ m \cdot 0.4) / (7\ m/h) = 0.069\ h = 4.1\ min$

b. What is the empty bed contact time (apparent contact time)?

$EBCT = L/v = 1.2\ m / (7\ m/h) = 0.17\ h = 10.2\ min$

c. What is the surface area of the grains in the filter bed?

Surface area per $m^3 = 6 \cdot (1 - \varepsilon) / d = (6 \cdot 0.6) / 0.0006 = 6000\ m^2$.

3.2.5.4 Filters

Filters can be classified according to driving force: filtration under gravity (GF) and filtration under pressure (PF). According to number of mediums: single media, dual media, and multimedia. According to direction of flow: up-flow filtration, down-flow filtration. According to mode of filtration: constant rate filtration, and declining rate filtration. Figure 12 illustrates a gravity filter unit with single medium supported by a gravel layer (gravel has no filtering effect).

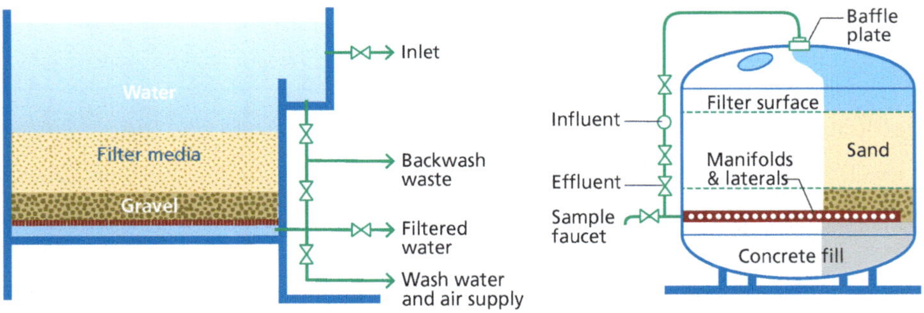

Figure 12 Schematic of a gravity filter (left) and vertical pressure filter (right). (Adapted from AWWA, 1995)

Filter underdrain systems have the function to collect the filtered water uniformly across the bottom of the filter and to distribute backwash water (and air) evenly, so that filter bed will expand without being unduly disturbed by the backwashing.

Some common systems are: pipe lateral collector, perforated tile bottom, wheeler filter bottom, porous plate bottom, and false-bottom. The pipe lateral, perforated tile and wheeler bottom systems require a gravel bed to prevent filter media from flowing into the underdrains and to distribute the backwash water evenly (illustrated in Figure 13). Newer systems allow fine media to be placed directly on the filter bottom so that a gravel layer will not be required (see Figure 14).

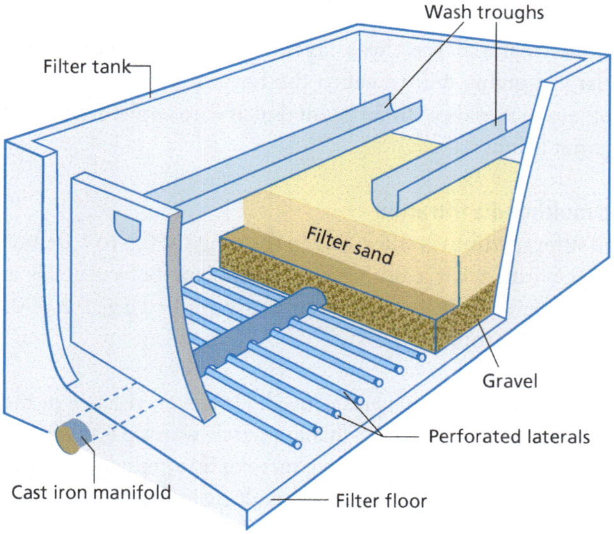

Figure 13 Schematic of perforated lateral systems of under drains in media filters. (Adapted from AWWA, 1995)

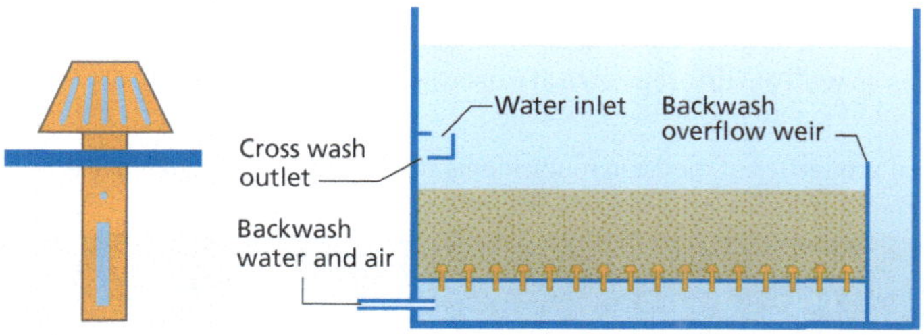

Figure 14 Schematic of a nozzle (left) and nozzle type suspended floor (false bottom) (right). Typically, 50 – 100 nozzles/m^2 are applied with a slit width 0.25 to 1 mm (normally 0.3 to 0.5 mm). (Adapted from AWWA, 1995)

The potential advantages of false filter bottoms are: no need for gravel support layers, consequently space is more effectively used. Gravel layers tend to be unstable and mix with other layers. The hydraulics conditions during backwashing are more controlled in a false bottom. "Dead" zones in gravel layers might develop accumulation of sludge and anaerobic conditions, resulting production of taste and odour compounds by bacteria.

3.2.5.5 Media and quality effluent

The smaller the grains size of the media the better the removal of particles, lower turbidity and SDI.

Using media with a small size grains result in rapid clogging of the filter bed. As a result, frequent backwashing is necessary. To overcome this problem dual media and multimedia filters are applied. In these filters, material of the largest diameter is on top, smaller material is below and the smallest is at the bottom.

Using the same filter material for these layers will not be successful because during backwashing the largest grains will travel to the bottom and the smallest to the top. To overcome this problem materials with different density are applied e.g., anthracite (lowest), sand (medium), garnet (highest).

3.2.5.6 Dual and multimedia filtration

The most common filter medium is silica sand in the range of 0.5 to 1.0 mm and an effective diameter of about 0.6 mm with a uniformity coefficient between 1.3 and 1.7 mm. In seawater pre-treatment smaller grain size is frequently applied e.g., 0.3 to 0.6 mm sand and anthracite 1.2 to 1.6 mm on top.

In many filters, only a single grade of sand is used. However, the use of media of different size has become increasingly popular. In this approach water passes first through coarser media, and progressively filtration is through course to fine grain layer. The advantage is the more efficient use of the whole filter bed depth resulting in: increasing removal of particles, longer filter runs due to decreased head loss development, or higher rates of filtration in seawater 10 m/h in first stage and up to 20 m/h in the second stage.

Just placing a sand having a larger size on top of the small sized sand grains does not work. This is because after backwashing the larger particles will remain at the bottom of the filter. Since larger grains have higher settling velocities. To overcome this problem, media of different densities as well as different size are used.

Table 4 Typical properties of media in multi-media filters. (Adapted from Sharma and Schippers, 2019)

Type	Position in bed	Media	Depth of layer (m)	Media density (g/mL)	Media effective size (mm)
Dual media	Top	Anthracite	0.2	1.5	1.5
	Bottom	Silica sand	0.6	2.6	0.6
Multi-media	Top	Anthracite	0.2	1.5	1.5
	Middle	Silica sand	0.4	2.6	0.6
	Bottom	Garnet sand	0.2	4.2	0.4

3.2.6 Inline coagulation (direct filtration)

It is performed by adding a coagulant, usually low amounts e.g., 1 to 2 mg/L to avoid rapid clogging of the filter bed, followed by rapid mixing, and immediately passing through a media filter to remove the micro flocs formed. Aluminium and iron (III) salts are usually applied as coagulant. However, some organic polymers are applied as well.

Direct filtration refers to the situation that almost directly after the addition of the coagulant, the water is filtered through the filter bed. No flocculation and so sedimentation will occur. The purpose of coagulation is to enhance the effect of sedimentation or filtration of the small particles, and to improve in this way the product water quality e.g., turbidity, SDI. Coagulation is achieved by making the particles larger by agglomeration or enmeshment of the particles by the forming flocs. For this purpose, aluminium or iron salts are added and mixed with the water.

3.2.6.1 Commonly applied coagulants

Aluminium and iron (III) sulphate or chloride salts are very well soluble at low pH levels. Dissolving these salts (in high concentrations) in water will result (from its own) into a low pH of the solution. At moderate pH values, 6 to 8, aluminium and iron are not well soluble. Adding a small amount of aluminium or iron sulphate will not change the pH of the water to be treated, which is usually in the range of 6 to 8. As a result aluminium hydroxide ($Al(OH)_3$) and ferric hydroxide ($Fe(OH)_3$) will precipitate and form flocs.

When Al^{3+} or Fe^{3+} are added to the water to be treated, they will hydrolyse (combine with water) to form hydroxides and H+.

$$Al^{3+} + 3H_2O \rightarrow Al(OH)_3 + 3H^+$$
$$Fe^{3+} + 3H_2O \rightarrow Fe(OH)_3 + 3H^+$$

These hydroxides are insoluble and will form flocs. Flocs use to form mainly in the filter bed, since the residence time in the connecting pipes is very short.

3.2.7 Flocculation – sedimentation – media filtration

The process comprises adding a coagulant, effective mixing, and formation of the flocs by gentle mixing (30 – 45 minutes), settling of the major part of the flocs up to 90% - 95%, and media filtration. Applied in case the concentration of particulates is high (and / or high coagulant dose is required) and results in very rapid clogging of in-line filters.

The **formation of the flocs** is an essential element in the process, because flocs will not grow from their own to the required size. Gentle mixing is a requirement. A great variety of flocculation systems are applied e.g., systems making use of: paddles, one or more chambers; baffled chambers; sludge blanket clarifiers.

Sedimentation is a solid-liquid separation process utilizing gravitational settling to remove suspended solids from the water (also called clarification). It is one of the cheapest and easiest way of removing suspended solids. Sedimentation tanks are also known as sedimentation basins, settling tanks, settling basins or clarifiers.

Most of the suspended particles present in the water have a specific gravity > 1 kg/m^3. In still water, these particles will therefore, tend to settle down under gravity. *Plain sedimentation* is when impurities are separated from water by the action of gravity alone. *Coagulant aided sedimentation* is when the particles are too small to be removed by gravity and coagulants are added to increase size by agglomeration of the particles.

In water treatment, sedimentation is applied for the removal of particulate material, flocculated impurities and precipitates, such as: plain settling of particulates from surface waters, and settling of coagulated precipitates in softening.

Sedimentation tanks are generally rectangular or circular basins with influent baffles and outlet weirs. In the recent years, innovations like tube settlers, plate separators, up flow clarifiers, pulsators and others are being applied, with the aim of achieving equivalent or higher removal efficiency with lower costs and/or land requirements.

The selection of a sedimentation tank depends on several aspects, such as the type of suspended matter to be removed, the overall treatment process train and role of sedimentation, the topography, ground condition of plant site; land availability and future plant expansion, the potential for hydraulic shock loading and degree of fluctuation of influent water quality, the nature and amount of sludge that will be produced, the local climatic and geological conditions, the capital and operation and maintenance costs of sedimentation tanks, and the time period available for design and building of the treatment plant (Sharma, 2019).

3.2.8 Dissolved air flotation

Dissolved air flotation (DAF) is a water treatment process that clarifies waters by the removal of suspended matter. Usually coagulants e.g., ferric or aluminium sulphate, are applied. Removal is achieved by dissolved air in water under pressure and then releasing the air in a flotation tank. The released air forms tiny bubbles (20 – 100 μm) which adhere to the suspended matter to float to the surface where it will be removed by e.g., a sludge scraper. A separate sludge treatment unit is usually required.

DAF is widely applied in industrial wastewater applications. Froth flotation is frequently applied in metal mining e.g., copper, gold. In water and wastewater treatment commonly, dissolved air flotation is applied. For this purpose, a part of the water is saturated with air at a pressure of 4 to 6 bar. Tiny air bubbles are formed when pressure is released in nozzles. Up to 10 % of the water is recycled.

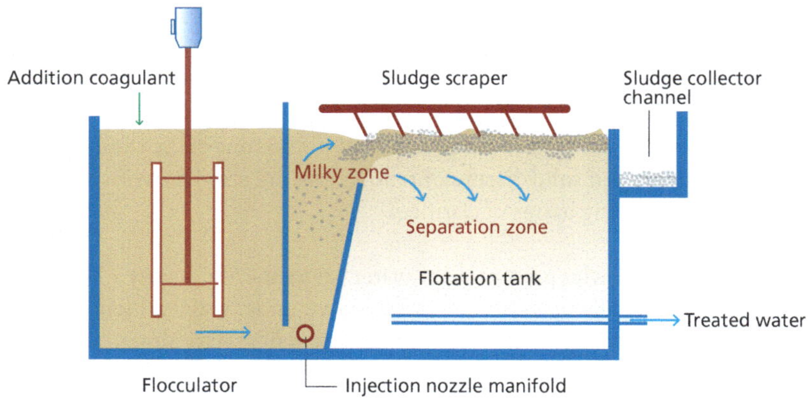

Figure 15 Principle of dissolved air flotation unit (Adapted from Alemayehu, 2010)

In the "reaction zone", tiny air bubbles are introduced and mixed with water carrying flocs. Contact between flocs and air bubbles. In the "separation zone" solid-liquid separation takes place effectively due to rising aggregates of air bubbles and flocs. Rising velocity of single air bubbles equals according Stokes (e.g., for 10 μm diameter air bubbles the rising velocity is 0.2 m/h). Separated flocs form at the top a sludge layer, which is continuously or intermittently removed using mechanical or hydraulic desludging. This is illustrated in Figure 15.

Table 5 Design parameters for conventional and high rate flotation (Alemayehu, 2010)

Parameters	Conventional	High rate
Detention time (min)	10-20	10-15
Mixing intensity (G, s^{-1})	50-100	
Contact zone loading rate (m/h)	100-200	120-300
Contact zone detention time (min)	1-2.5	1-2
Hydraulic loading	5-15	10-30
Separation zone loading rate (m/h)	6-18	20-40
Basin depth (m)	2.0-3.5	2.5-4.5
Recycle rate (%)	6-12	
Saturator gauge pressure (kPa)	400-600	
Saturator efficiency (%)	80-95 packed	

In *conventional pre-treatment*, the application of coagulation / sedimentation or flotation are always followed by media filtration to remove escaping flocs adequately and finally cartridge filtration to polish the water quality. In *advanced pre-treatment*, the application of coagulation / flotation followed by ultra / microfiltration is applied as well, and finally cartridge filtration to polish the water quality is frequently applied as well.

Example 3

Why is coagulation / sedimentation or coagulation / flotation applied in pre-treatment?

Answers

In conventional pre-treatment

- To avoid rapid clogging of media filters
- To maintain/improve product water quality (SDI / MFI, algae)
- To reduce biofouling potential
- To remove effectively oil products.

In advanced pre-treatment

- To avoid rapid fouling of ultra/microfiltration membranes;
- To reduce biofouling potential

Flotation is commonly recommended to handle algal blooms as an additional process.

The justification for applying dissolved air flotation is the assumption that this technique is able to handle high concentrations of algae (during Red Tide). This expectation is based on the fact that algae tend to float, which makes it difficult force them to settle. During normal circumstances flotation is not needed and can be bypassed. Once an algal bloom (Red Tide) the floatation has to be started. Currently, several flotation plants are contracted/under construction; however, very limited / no experience is available during algal blooms.

3.2.9 Cartridge filtration

Cartridge filtration is inherited from the period that DuPont's "Hollow Fine Fibres Permeators" dominated the market. It was used and is still used to polish the effluent from pre-treatment systems. Later on, it turned out to be very useful as well, in protecting the high-pressure pumps against media escaping from filters and ground water wells. In most designs, cartridge filtration is incorporated.

Cartridges are made of organic polymers e.g., poly propylene with an applied pore size between 1 to 25 μm.

3.2.10 Membrane pre-treatment

Microfiltration (MF) and ultrafiltration (UF) membranes are able to produce water with very low SDI values independent of the raw water quality.

MF/UF is an emerging technology in river, and sea water and treated domestic wastewater pre-treatment. MF/UF is applied on surface water and treated domestic waste water or as polishing step after conventional pre-treatment of RO feedwater. Some of the properties of MF/UF membranes of relevance in water treatment are: permeability (clean water), pore size or molecular weight cut-off (MWCO), surface porosity, hydrophobicity (or hydrophilicity), surface/pore charge, chemical tolerance (pH, chlorine). These properties will influence the amount of membrane area required during treatment, the fouling development over time, type and conditions of cleaning to be performed to restore permeability.

MF/UF membranes are pressure or vacuum driven separation processes making use of membranes having small pores. MF membranes have pores in the range of 0.1 – 0.2 μm (exceptions up to 10 μm). UF membranes have pores in the range of 0.01 – 0.05 μm. Most membranes use for water treatment have pores of approximately 0.02 μm (equivalent to about 100,000 dalton MWCO). Membranes having small pores are usually not characterized by pore size, but by MWCO.

The effect of MF and UF depends on the size of the particles in the water and the pores size (see Table 6).

Table 6 Removal of suspended, colloidal and dissolved matter by MF and UF

Matter	Size	Removal
Suspended	> 1 μm	Completely
Colloidal	0.001 – 1 μm	Partly/completely
Dissolved (Salts)	< 0.001μm	Not

UF membranes with small pores have the ability to retain (dissolved) organic polymers e.g., biopolymers. That is why historically UF membranes have been characterized by MWCO. The concept of MWCO (90 % of a target compounds rejected) is a measure of the removal characteristic of membranes in terms of molecular mass (weight) rather than size.

Small particles are morphologically difficult to define (flexible structure). So, it is useful to apply MWCO for UF membrane characterization. Moreover, it is very difficult to measure the size of small pores. Scanning Electron Microscopy (SEM) is used for this purpose.

Pores in membranes are not uniform in size. Therefore, all membranes have a distribution of pore sizes. This distribution will vary according to the membrane material and manufacturing process. Nominal pore size is equal to average pore size, while absolute pore size is equal to maximum pore size.

The surface porosity is the part of surface which is "covered" by pores. Porosity, pore size and pore size distribution can be determined (manually) by analysing processed images of scanning electron microscopy (SEM), transmission electron microscopy (TEM) or atomic force microscopy (AFM).

MF & UF membranes are made of organic polymers and inorganic materials such as ceramic, glass or metal. Membranes made of organic polymers dominate in the water treatment market. Materials commonly applied are poly-ether sulphone (PES), poly-vinylidene fluoride (PVDF), poly-acrylonitrile (PAN), and modified – with other polymers – to making them more hydrophilic and consequently more permeable for water.

Synthetic organic polymeric membranes can be divided into two classes i.e., hydrophobic & hydrophilic. Hydrophilic polymers such as cellulose and its derivatives have been used widely for the manufacture of MF and UF membranes. Other examples of hydrophilic polymeric materials are: cellulose esters, polyamide, polycarbonate, poly-sulphone, poly-ethersulphone, poly-ether-imide.

Hydrophobic membranes such as: polytetrafluoroethylene (PTFE), polyvinylidene fluoride (PVDF), Teflon, polyethylene (PE), or polypropylene (PP) are commonly used for MF and UF membranes as well.

Whether particles can pass a membrane or not, depends mainly on the size of the particles and their flexibility (some biopolymers e.g., seawater do not have a rigid structure), the size of the pores, the pore size distribution, the pores in the gel/cake layer on the membrane surface (a "new" membrane is dynamically formed on the membrane surface). Size exclusion mechanism or sieving mechanism is assumed to be dominant.

After fouling occurs, remedial actions are necessary to restore the permeability of the membranes. Usual actions are presented in Table 7. Membrane cleaning is achieved by frequent hydraulic cleaning by backwashing with permeate or by frequent hydraulic cleaning by backwashing supported with air; followed by chemically enhanced backwashing (CEB) whereby a chemical is added to the permeate. Incidentally chemical cleaning in place (CIP) is performed.

Table 7 Remedial actions after fouling in MF and UF systems

Type of cleaning	Description	Frequency
Backwashing	Water or water supported with air (air scour), backwash flux e.g., 250 $L/m^2/h$	Once or twice per hour, depending on fouling rate.
Chemical enhanced backwashing (CEB) (backwash and soak)	A low dose (about 200 mg/L) of oxidant e.g., sodium hypochlorite is automatically injected into the permeate during backwashing, to enhance the hydraulic cleaning. Firstly, a backwash with permeate (approx. 30 s) is performed to remove accumulated particles from the hollow fibers. Secondly, a short soak (e.g., 10-15 min) with a low dose of oxidant to remove adsorbed foulants from the membrane. Finally, another short backwash (with permeate) to remove the chemicals from the systems.	Daily, weekly, depending on fouling rate. Criterion: Maximum pressure exceeded e.g., 2 bar or fixed frequency e.g., once per day.
Cleaning in place (CIP)	Compared to a CEB, the chemical dose is higher when performing a chemical cleaning (ca. 400 mg/L), and the duration of chemical cleaning is longer i.e., few hours. Involves labour to make up the chemicals, fill and flush the system, etc.	Weekly, monthly, yearly, depending on fouling rate.

Backwashing is performed by automatically by reversing the flow of permeate (about every 20 – 40 min, during about 30 seconds and about 2.5 times filtration flux). Enhanced Backwash can restore the permeability further due to the applied chemicals. The ideal situation regarding backwash flux and frequency is to use a high backwash flux as frequently as possible and during a long period. However, such a practice results in a very low net flux, because a large volume of permeate is consumed in backwashing. Therefore, it is necessary to optimizing the backwash flux, frequency and time.

Example 4

A UF plant is filtering clean water (no fouling!) with a capacity of 200 m^3/h at temperature of 25 °C (t_1) at 0.2 bar with membrane surface area of 2000 m^2.

Questions: What is the flux? What is the permeability of the membranes at 25 °C? What will be the required pressure, when the temperature drops to 5 °C (t_2)?

Answers

a) Flux = Q_w / A = 200 m^3/h / 2000 m^2 = 0.1 $m^3/h/m^2$ = 100 $L/m^2/h$

b) Permeability = J / P = 100 $L/m^2/h$ / 0.2 bar = 500 $L/m^2/h/bar$

c) We need higher pressure because...

Rule of thumb: 3% per C: (25 – 5) · 3 % = 60% of 0.2 bar = 0.12 bar higher or total 0.2 + 0.12 = 0.32 bar

Formula: Pressure at 5 °C equals:

$$P_{t2} = P_{t1} \cdot TFC = P_{t1} \cdot (1.03)^{(t1-t2)}$$
$$= 0.2 \text{ bar} \cdot (1.03)^{(25-5)} = 0.2 \cdot 1.8 = 0.36 \text{ bar}$$

Normalized permeability (K_{w20} or mass transfer coefficient, MTC)) and normalized clean water flux at 20 °C and 1 bar, are commonly used to characterize the performance of MF/UF membranes and are expressed as: $L/m^2/h/bar$ at 20 °C. When water, containing suspended and colloidal matter, is filtered through a MF/UF membrane, the total resistance (membrane + pore blocking + foul layer = Rtotal) will increase due to depositing of suspended and colloidal particles on the membrane and/or in the pores. As a result, the flux will decrease when the pressure is kept constant. The required pressure will increase, when the flux (filtration rate / capacity) is kept constant, which is common practice.

The flux (or filtration rate) is a key parameter in design and operation of membrane systems. Allowable flux is governed by: the fouling potential of the feed water, pre-treatment applied; effectiveness of backwashing, effectiveness of cleaning. In practice the flux maintained, ranges from 10 to 120 $L/m^2/h$ to control fouling.

3.2.11 Comparison between conventional and membrane pre-treatment

Table 8 compares both pre-treatment processes in terms of operation and water quality. In media filters, coagulation / sedimentation or flotation are improving the feed water quality to a large extend. Moreover, the dose of coagulant is a strong tool in handling feed water quality variation. Making the extended process robust.

Ultrafiltration membranes are vulnerable to variations in feed water quality with respect to run length (backwashing). Measures taken include adapting coagulant dose in pre-treatment. Processes having ultrafiltration as a core process, are not vulnerable to variations in feed water quality with respect to SDI/MFI.

Table 8 Comparison of conventional and MF/UF pre-treatment

Parameter	Media filtration	Ultrafiltration
Filtration mechanism	Depth filtration	Surface filtration
Pore size	100 to 200 μm	0.02 μm
Flux / filtration rate	5,000 - 10,000 L/m^2/h	50 - 100 L/m^2/h
Run length	24 h	1 h
Pressure loss	0.2 - 1 bar	0.2 - 1 bar
Backwashing	30 min	1 min
Backwash flow	2.5 - 5 times filtration rate	2.5 times flux rate
Volume filtered per cycle per m^2	120,000 - 240,000 L	50 - 100 L
Detention time	2 to 4 minutes	Fraction of seconds
Biodegradation organic compound	Yes, limited	No
Removal biodegradable organic compounds	Yes, limited	Yes: polymers No: small compounds
SDI/MFI	Yes: depending process and source	Yes: independent process and source
Ripening	Need a ripening period after backwashing to required product quality	No need of ripening period after backwashing to get required product quality
Break through SDI/MFI	Potential danger	No breakthrough
Feedwater and product water quality	Vulnerable to variations in feed water quality with respect to: run length and product quality	Product quality independent from raw water quality

In media filters, three mechanisms are involved in the removal of biodegradable organic matter, and thus in biofouling control. These mechanisms are: *i*) removal of suspended biodegradable matter by filtration, *ii*) removal of suspended and colloidal matter by coagulants, and subsequent *iii*) biodegradation achieved by the biofilm of bacteria on the surface of sand grains. Pre-chlorination might disturb the biodegradation process to a large extent by inactivating the microorganisms responsible of the biodegradation process. Moreover, chlorination produces biodegradable organic compounds.

3.3 REFERENCES

Alemayehu Z (2010) Dissolved Air Flotation. Aalborg University.

AWWA (1995) Water Treatment. Principles and practivces of water supply operations, 2nd edn AWWA, Denver, CO

Baudish P (2015) Design Considerations for Tunnelled Seawater Intakes Springer International Publishing, Cham, pp. 19-38.

E. I. du Pont de Nemours and Company (1982) Permasep Engineering Manual, Bulletin 504, Wilmington, Delaware

Fritzmann C, Löwenberg J, Wintgens T, Melin T (2007) State-of-the-art of reverse osmosis desalination. Desalination 216: 1-76

Huisman I (1986) Rapid Filtration IHE Delft, Delft

Lattemann S (2010) Development of an environmental impact assessment and decision support system for seawater desalination plants CRC Press/Balkema, Delft

Missimer TM, Ghaffour N, Dehwah AHA, Rachman R, Maliva RG, Amy G (2013) Subsurface intakes for seawater reverse osmosis facilities: Capacity limitation, water quality improvement, and economics. Desalination 322: 37-51 DOI https://doi.org/10.1016/j.desal.2013.04.021

Pankratz T (2006) A review of seawater intake, pretreatment and discharge technologiesProceedings, of the International Desalination Association Seminar on Water Desalination Technologies, Tehran, Iran.

Sharma S (2019) Sedimentation IHE Delft, Delft, The Netherlands

Sharma S, Schippers J (2019) Filtration IHE Delft, Delft, The Netherlands

Tobiason J, Cleasby J, Logsdon G, O'Melia C (2011) Granular media filtration. Water quality and treatment: a handbook on drinking water

Van der Kooij D, Hijnen WAM (1984) Substrate utilization by an oxalate-consuming Spirillum species in relation to its growth in ozonated water. Appl Environ Microbiol 47: 551-559

Voutchkov N (2004) Beach wells versus open surface intakeWater & Wastewater International.

Wolf PH, Siverns S, Monti S (2005) UF membranes for RO desalination pretreatment. Desalination 182: 293-300

Chapter 4

Particulate fouling

Sergio G. Salinas-Rodriguez, Siobhan F. E. Boerlage,

Jan C. Schippers, Maria D. Kennedy

The learning objectives of this chapter are the following:

- Define particulate fouling in membrane systems
- Define and apply fouling indices for assessing particulate fouling
- Present and discuss the basic equations governing particulate fouling at constant pressure and at constant flux
- Present, discuss and apply the prediction model of particulate fouling in reverse osmosis and ultrafiltration systems
- Understand the theoretical background of fouling indices and fouling prediction as well as the assumptions involved and weaknesses of these indices

4.1 INTRODUCTION

Particulate fouling has plagued reverse osmosis (RO) systems since their first use in desalination and remains a persistent issue today for RO and other pressure driven systems such as microfiltration, ultrafiltration, and nanofiltration.

In the early sixties the Du Pont Company/Permasep Product (now DuPont Dow) successfully launched the hollow fine fibre (HFF) permeator onto the desalination market, where it dominated for several decades. A well-known weakness of this HFF permeator was its vulnerability to fouling. Initially this vulnerability was attributed to suspended and colloidal matter in the feed water i.e., particulate fouling. Therefore, Du Pont developed, the Silt Density Index (SDI), initially called the Fouling Index, as a parameter to characterize the fouling potential of RO feedwater for permeators (see Figure 1). The fouling mechanism turned out to be more complicated than just fouling of the membrane surface as was initially assumed. Gradually it became clear that the fouling was initiated by local clogging of the woven or non-woven fabric between the fibres, which is needed to ensure equal flow distribution of the feed water.

Editors: Sergio Salinas, Jan Schippers, Gary Amy, In Kim, Maria Kennedy
doi: 10.2166/9781780409863_0085

This primary fouling mechanism disturbs the flow pattern resulting in localised low flow, causing high concentration polarization and higher recovery rates in the fouled area. This then leads to higher osmotic pressure, deposition of suspended and colloidal particles and scaling of sparingly soluble salts e.g., *calcium sulphate*, reducing the permeate flow.

Figure 1 Fabric in hollow fine fibre permeator of Du Pont Permasep (Jan C. Schippers)

In the eighties it became clear that biofouling also frequently occurred resulting in the same phenomena and exacerbated fouling. The synergistic effects of these types of fouling made the fouling problem even more complicated.

In the nineties, spiral wound elements were gaining ground in the market, claiming to be less vulnerable to fouling, which was reflected in their less stringent SDI guidelines i.e., a maximum SDI_{15} of 5 was allowed in membrane guarantees with SDI_{15} of < 3 preferred. Whereas, SDI guidelines for Permasep permeators were SDI_{15} < 3 and preferably SDI_{15} of < 1.

Spiral wound elements indeed were less vulnerable to clogging than the hollow fine fibre permeators, which can be attributed to differences in design and wide spacing between the spacer and the membrane surface. The same holds for the hollow fibre element used today, which has cross wound fibres with wide spacing between these fibres.

While the SDI is a useful tool in characterizing the particulate fouling potential of RO feedwater when it comes to clogging of fabric in permeators, spacers in spiral wound elements and the new type of hollow fibre elements, it may not account for the direct fouling of the membrane surface itself which results in permeability decline. This raises the question: Is the SDI a useful tool in predicting particulate fouling of the membrane as well? This chapter examines this question and traces the development of the SDI and MFI from the early sixties to 2021.

Parameters like suspended matter, turbidity and particle counts are unreliable for assessing particulate fouling potential (Boerlage *et al.*, 2017, Schippers *et al.*, 2014, Boerlage, 2007 Boerlage, 2001). For this purpose, the SDI is commonly applied as a measure of the fouling potential due to particles in a feedwater. Aluminium hydroxide particles are also measured in the test when, e.g., Alum is used as coagulant. In general, measuring the concentration of all individual colloidal and suspended particles is very difficult; thus a "sum parameter" is applied.

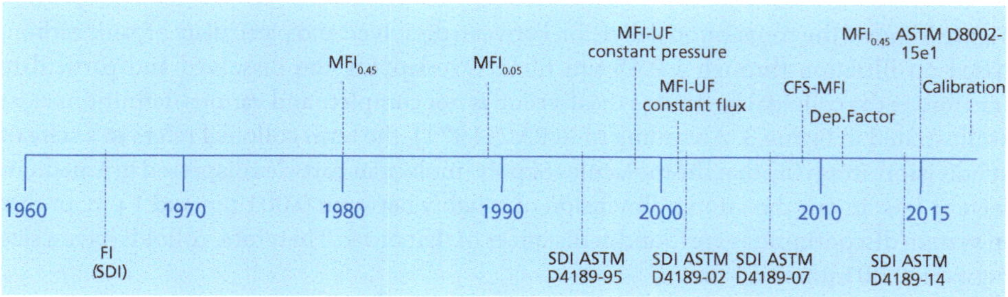

Figure 2 Historical development of fouling indices for particulate fouling assessment (based on Salinas-Rodriguez, 2011).

FI = fouling index, SDI = silt density index, MFI = modified fouling index, MFI-UF = modified fouling index ultrafiltration, MFI constant flux - deposition factor, CFS = cross flow sampler, ASTM = American society for testing and materials

The historical development of fouling indices is presented in Figure 2. SDI has a long history in water treatment and has been universally used since its inception in the 1960s, while the MFI indices are less known though gaining preference in water treatment. All these indices are explained in detail in the following sections.

4.2 PARTICLES

Particulate fouling is caused by different types of suspended (> 1 μm) and colloidal particles (< 1 μm), such as: clay minerals, organic materials, coagulants e.g., $Fe(OH)_3$, $Al(OH)_3$, algae, bacteria (not growing), extra cellular polymer substances (EPS) and/or Transparent Exo polymer Particles (TEP, see chapter 6).

There are two general types of particles in natural waters, hydrophobic (water repelling) and hydrophilic (water attracting). Hydrophobic particulates have a well-defined interface between the water and solid phases and have a low affinity for water molecules. In addition, hydrophobic particles are thermodynamically unstable and will aggregate irreversibly over time (Crittenden *et al.*, 2005).

Hydrophilic particulates such as clays, metal hydroxides, proteins, or humic acids have polar or ionized surface functional groups. Many inorganic particles in natural waters, including hydrated metal oxides (iron or aluminium oxides), silica (SiO_2), and asbestos fibres, are hydrophilic because water molecules will bind to the polar or ionized surface functional groups (Stumm and Morgan, 1996). Many organic particulates are also hydrophilic and include a wide diversity of bio-colloids (humic acids, viruses) and suspended living or dead microorganisms (bacteria, protozoa, algae). Because bio-colloids can adsorb on the surfaces of inorganic particulates, the particles in natural water often exhibit heterogeneous surface properties.

In nature, most colloids and particles are negatively charged. This knowledge has been used by membrane manufacturers to influence the surface charge of the membranes so as to repel suspended particles (Belfort *et al.*, 1994).

Figure 3 shows the conventional division between dissolved and particulate organic carbon, based on filtration through a 0.45 μm filter. Overlapping the dissolved and particulate fractions is the colloidal fraction as the division is not complete and various definitions exist as illustrated in Figure 3. According to IUPAC (1971), the term colloidal refers to a state of subdivision, implying that the molecules or poly-molecular particles dispersed in a medium have at least in one direction a dimension of roughly between 0.001 μm and 1 μm, or that in system discontinuities are found at distances of that order. Therefore, colloids have a size between 0.001 μm and 1 μm.

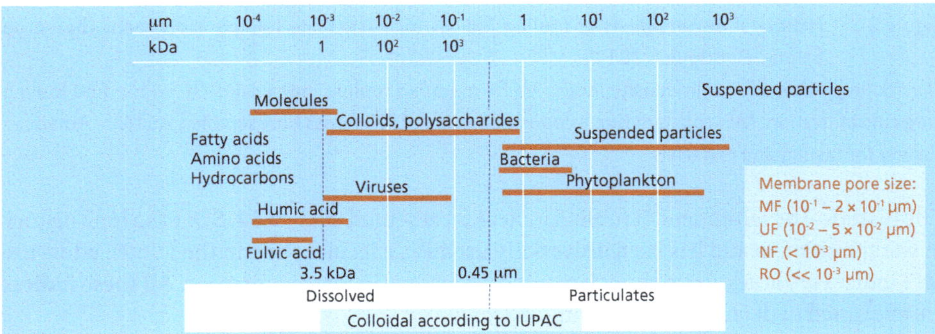

Figure 3 Continuum of particles, colloids and dissolved organic carbon in natural waters

A system containing colloidal particles is said to be stable if during the period of observation, it is slow in changing its state of dispersion (Crittenden *et al.*, 2005).

4.3 PARTICULATE FOULING EQUATION

The flow through a reverse osmosis membrane can be described by:

$$Q_w = \frac{dV}{dt} = (\Delta P - \Delta \pi) \cdot K_w \cdot A \qquad \text{Eq. 4.1}$$

where:

Q_w = permeate flow (e.g., m^3/h)
V = total filtered volume water (permeate) (L or m^3)
t = time (e.g., hour, minute, second)
ΔP = differential pressure (pressure feed - pressure permeate)
$\Delta \pi$ = difference osmotic pressure
(osmotic pressure feed – osmotic pressure permeate)
K_w = permeability constant for water ($m^3/m^2 \cdot$ s-bar)
A = surface area of the membrane(s) (m^2)

Q_w/A = permeate flow through membrane surface area ($m^3/m^2/h$)
= filtration rate ($m^3/m^2/h$), used in rapid sand filtration
= flux ($L/m^2/h$) used in membrane filtration
(ΔP - $\Delta\pi$) = net driving pressure (NDP)

In membrane technology, flux is defined as the ratio of the permeate flow and surface area of the membrane. It is expressed as:

$$J = \frac{Q_w}{A} = \frac{1}{A} \cdot \frac{dV}{dt}$$ Eq. 4.2

To simplify the equations, we assume that $\Delta\pi$ is negligible. This assumption is valid for low salinity water only. Then,

$$J = \frac{1}{A} \cdot \frac{dV}{dt} = \Delta P \cdot K_w$$ Eq. 4.3

Frequently the concept of resistance (R) is used, instead of permeability:

$$K_w = \frac{1}{\eta \cdot R_T}$$ Eq. 4.4

Where: η is the viscosity of the water and R_T is the total resistance [sum of membrane resistance (R_m), pore blocking (R_p) and cake formation (R_c)].

$$R_t = R_m + R_b + R_c$$ Eq. 4.5

Replacing Eq. 4.4 and Eq. 4.5 in Eq. 3:

$$J = \frac{1}{\eta} \cdot \frac{\Delta P}{R_m + R_b + R_c}$$ Eq. 4.6

When we assume that pore blocking does not play a dominant role in RO, then fouling is mainly due to cake formation. As a consequence:

$$J = \frac{1}{\eta} \cdot \frac{\Delta P}{R_m + R_c}$$ Eq. 4.7

Permeability of the clean filter media (R_m) is a function of filter properties such as filter thickness (Δx), surface porosity (ε), pore radius (r_p), and tortuosity (τ) and can be defined using Poiseuille's Law:

$$R_m = \frac{8 \cdot \Delta x \cdot \tau}{\varepsilon \cdot r_p^2}$$ Eq. 4.8

The cake resistance (R_c) component in (membrane) filtration can be defined following the Ruth equation (Ruth *et al.*, 1933), using the concept of "specific cake resistance" per unit weight (α) (Equation 4.9). Ruth showed that the resistance of the cake formed during constant pressure filtration is proportional to the amount of cake deposited at the filter medium provided the retention of particles and α are constant. Cake resistance is defined as:

$$R_c = I \cdot \frac{V}{A} \qquad \text{Eq. 4.9}$$

and the fouling index (I) is:

$$I = \alpha \cdot C_b \qquad \text{Eq. 4.10}$$

Where: I is a measure of the fouling characteristics of the water ($1/m^2$). The value of I is a function of the nature of the particles and is proportional to their concentration. C_b is the concentration of particles per unit volume of filtrate (e.g., mg/L) and α is the specific cake resistance per mg cake per m^2 membrane ($m^3/mg/m^2$).

The specific cake resistance is constant for incompressible cakes under constant pressure filtration and can be calculated according to the Carman-Kozeny relationship (Equation 4.11) (Carman, 1937 & 1938). Carman (1937, 1938) derived Equation 4.11 for the specific resistance of a cake composed of spherical particles of diameter d_p from the Kozeny equation including a factor for tortuosity of the voids within the cake. According to the Carmen relationship a reduction in the porosity of the cake (ε) or a decrease in particle diameter size (d_p) increases the specific resistance of the deposited cake.

$$\alpha_c = \frac{180 \cdot (1-\varepsilon)}{\rho_p \cdot d_p^2 \cdot \varepsilon^3} \qquad \text{Eq 4.11}$$

As porosity is to the power three it plays a dominant role. The more compact a cake, the higher the specific cake resistance, and therefore the higher the cake resistance and a higher pressure required to overcome this resistance.

4.3.1 Constant pressure filtration

Combining Equations 4.7, 4.9 and 4.10 and integrating at constant ΔP from t = 0 to t = t, assuming time independent permeability and uniform porosity characteristics throughout the depth of the cake (i.e., no compression of the cake), results in the well-known filtration equation:

$$\frac{t}{V} = \frac{\eta \cdot Rm}{\Delta P \cdot A} + \frac{\eta \cdot I}{2 \cdot \Delta P \cdot A^2} \cdot V \qquad \text{Eq. 4.12}$$

4.3.2 Constant flux filtration

Reverse osmosis plants typically operate at constant capacity and recovery. So, the flux is constant. When membranes foul, the pressure needs to be increased, in order to maintain a constant capacity (and flux) in the system. Rewriting Eq. 4.7:

$$J = \frac{1}{\eta} \cdot \frac{\Delta P_t}{R_m + R_c} = constant \qquad \text{Eq. 4.13}$$

Where: ΔP_t is the pressure at time "t" (which will increase). Rearranging Eq. 2 because flux is constant:

$$\frac{V}{A} = J \cdot t \qquad \text{Eq. 4.14}$$

and substituting Eq. 4.14 in Eq. 4.9:

$$R_c = I \cdot \frac{V}{A} = I \cdot J \cdot t \qquad \text{Eq. 4.15}$$

This results in:

$$J = \frac{1}{\eta} \cdot \frac{\Delta P_t}{R_m + I \cdot J \cdot t} \qquad \text{Eq. 4.16}$$

Rearranging the previous equation, we obtain:

$$\Delta P_t = \eta \cdot R_m \cdot J + \eta \cdot I \cdot J^2 \cdot t \qquad \text{Eq. 4.17}$$

Thus, the increase of pressure ΔP_t across the membrane is linearly proportional with time, with the fouling index (I) and with the flux to the power two (J^2). As a consequence, flux has a very dominant effect on the development of ΔP_t.

This is equation is valid for "dead-end" filtration. In "cross-flow" filtration only a part of the particles will deposit on the membrane surface due to the shear-force of the cross-flowing water. Therefore, "I" has to be corrected with a deposition factor "Ω". This factor is the fraction of particles which actually deposit on the membrane surface ($\Omega \leq 1$). Then, Eq. 4.17 becomes:

$$\Delta P_t = \eta \cdot R_m \cdot J + \eta \cdot \Omega \cdot I \cdot J^2 \cdot t \qquad \text{Eq. 4.18}$$

The phenomenon, that the increase of pressure is proportional to $(\text{flux})^2$ explains, why manufactures of spiral wound element recommend lower design fluxes with feedwater having a higher fouling potential.

Table 1 Recommended flux rates per type of RO feedwater. (Nitto Hydranautics, 2020)

Feedwater	Flux, L/m²/h
Surface water	14-24
Well water	24-31
RO permeate	34-51

4.3.2.1 Cross-flow and dead-end filtration

Historically, tubular membranes were used and operated in "cross-flow" mode, to control membrane fouling, as the higher the cross-flow velocity the lower the rate of fouling observed in the membrane system. At high cross-flow velocities, a major part of the suspended / colloidal particles will not deposit on the membrane surface, due to high shear forces. Unfortunately, high cross velocities result in high energy consumption, due to high head los in tubular membranes.

In UF and MF "dead-end" and "cross-flow" filtration are applied, while in RO and NF "cross-flow" filtration is applied only. In "dead-end" membrane filtration, all rejected particles present in the feed water will deposit on the membrane surface. In "cross-flow" membrane filtration, a part of the rejected particles will deposit; particles present in the concentrate will leave the module.

4.3.3 Modelling particle deposition in RO

Only a fraction of the RO feedwater is forced to pass through the membrane in cross flow filtration. This fraction of water depends on the recovery at which the RO unit operates. In dead-end filtration all the particles bigger than the membrane's pores will be retained while in the case of cross-flow, only the fraction of water passing through the membrane is affected and the associated fraction of particles may accumulated on the membrane surface (see Figure 4).

The deposition factor was first proposed by Schippers *et al.* (1980, 1981) in a model to predict flux decline in reverse osmosis systems. It was defined as the fraction of particles deposited, which are present in the water that passes through the reverse osmosis membrane.

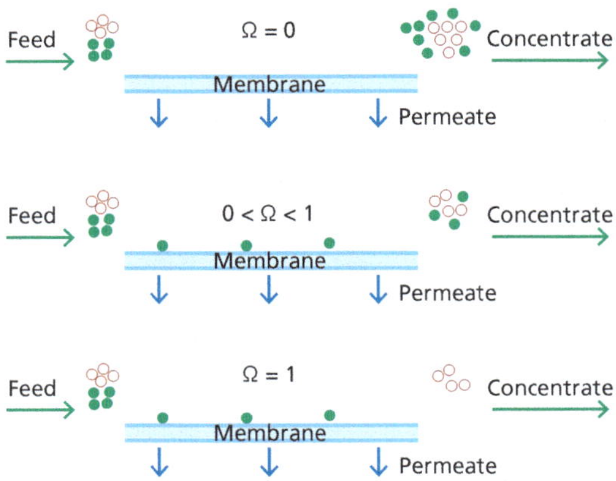

Figure 4 Particle deposition in cross flow filtration on permeable surfaces (Salinas Rodríguez, 2011)

Figure 4 shows schematically the particle deposition on a membrane surface considering 50 % recovery. Empty circles represent the fraction of particles that are not accumulated on the membrane; and full circles represent the fraction of particles that might be accumulated on the membrane.

In this sense, it is important to accurately measure the concentration of particles that are entering the RO unit, as well as the concentration of particles leaving the plant in the permeate and concentrate streams. This can be performed by doing a "mass" balance.

4.3.3.1 Mass balance equations

A schematic of a RO unit is presented in Figure 5. From this, a flow balance and mass balance can be performed as in Eq. 4.19 and Eq. 4.20, respectively.

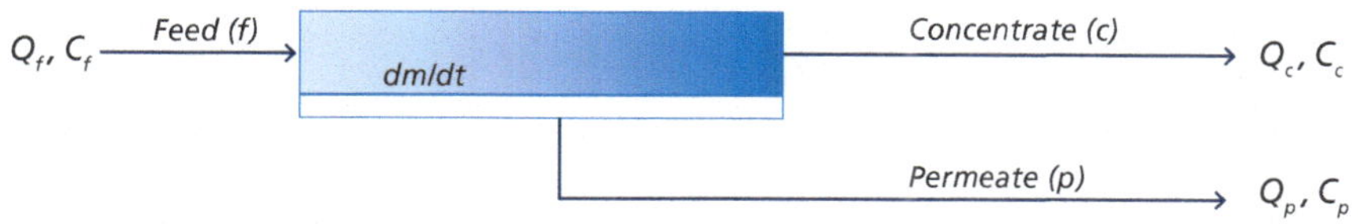

Figure 5 RO membrane mass balance schematic

$$Q_f = Q_p + Q_c \qquad \text{Eq. 4.19}$$

$$Q_f \cdot C_f = Q_p \cdot C_p + Q_c \cdot C_c \qquad \text{Eq. 4.20}$$

In these equations, it is important to notice that only the permeate water (Q_p) has passed through the membrane and therefore it is only from this volume of water that the membrane is rejecting ions, organic matter and particles. The rest of the water (concentrate) passes tangentially along the membrane without any change.

To consider the particles being accumulated/deposited on the membrane, the term dm/dt is introduced in the mass balance. Then we have:

$$Q_f \cdot C_f = Q_p \cdot C_p + Q_c \cdot C_c + \frac{dm}{dt} \qquad \text{Eq. 4.21}$$

In RO systems, only a part of the feed water passes through the membranes (Q_p). The extent of water passing through the membrane elements depends on the recovery of the system (R). From the part of water that passes through the membranes and where all the particles are rejected, only a fraction will accumulate ($\Omega \cdot C_f$) on the membrame surface and the fraction that does not accumulate on the membrane will remain in the concentrate stream. Thus, the fraction of particles that accumulates on the surface of the membrane can be expressed as $\Omega \cdot C_f \cdot Q_p$. Then,

$$Q_f \cdot C_f = Q_p \cdot C_p + Q_c \cdot C_c + \Omega \cdot Q_p \cdot C_f \qquad \text{Eq. 4.22}$$

Assuming that the particle concentration in the permeate is zero (100 % rejection), therefore $C_p = 0$. Consequently,

$$Q_f \cdot C_f = Q_c \cdot C_c + \Omega \cdot Q_p \cdot C_f \qquad \text{Eq. 4.23}$$

Rearranging the previous equation, we have:

$$\Omega \cdot Q_p \cdot C_f = Q_f \cdot C_f - \ \cdot C_c \qquad \text{Eq. 4.24}$$

Then,

$$\Omega = \frac{Q_f \cdot C_f - Q_c \cdot C_c}{Q_p \cdot C_f} \qquad \text{Eq. 4.25}$$

Rearranging Eq. 4.25,

$$\Omega = \frac{Q_f \cdot C_f}{Q_p \cdot C_f} - \frac{Q_c \cdot C_c}{Q_p \cdot C_f} \qquad \text{Eq. 4.26}$$

$$\Omega = \frac{Q_f}{Q_p} - \frac{Q_c}{Q_p} \cdot \frac{C_c}{C_f} \qquad \text{Eq. 4.27}$$

On the other hand, the system recovery (R) is defined as:

$$R = \frac{Q_p}{Q_f} \cdot 100 \qquad \text{Eq. 4.28}$$

Rearranging the previous equation, we have:

$$\frac{Q_f}{Q_p} = \frac{1}{R} \qquad \text{Eq. 4.29}$$

From Eq. 4.19, the concentrate flow is:

$$Q_c = Q_f - Q_p \qquad \text{Eq. 4.30}$$

then,

$$\frac{Q_c}{Q_p} = \frac{(Q_f - Q_p)}{Q_p} = \frac{Q_f}{Q_p} - 1 \qquad \text{Eq. 4.31}$$

Replacing Eq. 4.29 and Eq. 4.31 in Eq. 4.27,

$$\Omega = \frac{1}{R} - \left(\frac{Q_f}{Q_p} - 1\right) \cdot \left(\frac{C_c}{C_f}\right) \qquad \text{Eq. 4.32}$$

Replacing Eq. 4.29 in Eq. 4.32,

$$\Omega = \frac{1}{R} - \left(\frac{1}{R} - 1\right) \cdot \left(\frac{C_c}{C_f}\right) \qquad \text{Eq. 4.33}$$

and rearranging Eq. 4.33 we have,

$$\Omega = \frac{1}{R} - \left(1 - \frac{1}{R}\right) \cdot \left(\frac{C_c}{C_f}\right) \qquad \text{Eq. 4.34}$$

Then, we can obtain the deposition factor equation as function of recovery,

$$\Omega = \frac{1}{R} + \frac{C_c}{C_f} \cdot \left(1 - \frac{1}{R}\right) \qquad \text{Eq. 4.35}$$

Or as function of concentration factor (CF),

$$\Omega = \frac{1}{(CF-1)} \cdot \left(CF - \frac{C_c}{C_f}\right) \qquad \text{Eq. 4.36}$$

Where the concentration factor is:

$$CF = \frac{1}{(1-R)} \qquad \text{Eq. 4.37}$$

The formula above assumes that the particle rejection is 100 %.

There are possible scenarios from previous equations:

$\Omega = 0$ means $C_c = C_f \cdot CF$	No particles deposit
$\Omega = 1$ means $C_c = C_f$	All particles deposit
$\Omega > 1$ means $C_c < C_f$	All particles deposited + retention inside pressure vessel (e.g., spacer)
$\Omega < 0$ means $C_c > C_f \cdot CF$	Particles might be removed/sheared off inside the pressure vessel; earlier deposited particles released; particles formed by bacteria; particle size distribution influence results.

In this chapter, C_f and C_c correspond to MFI_{feed} and $MFI_{concentrate}$. Equations 4.35 and 4.36 are illustrated in Figure 6.

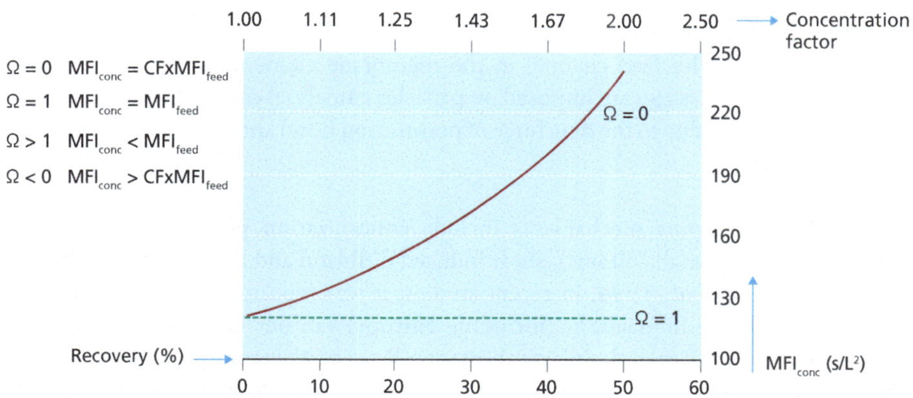

Figure 6 Deposition factor as function of RO recovery and of concentration factor (Adopted from Schippers, 1989 and Salinas-Rodriguez, 2011)

A *positive* deposition factor indicates particles are being accumulated on the membrane surface as they pass through the system while a negative factor indicates the number of particles in the concentrate exceeds the concentration in the incoming feedwater (taking into account the concentration factor) (Boerlage, 2001, Schippers, 1989).

Schippers (1989) presented results obtained in a pilot plant working with water from the IJsselmeer lake located in the north of the Netherlands to determine the deposition factor. The total recovery of the four-stage installation was 90 %. The deposition factor was obtained by measuring (at constant pressure) the MFI (0.05 μm) values of the RO feed and RO concentrate water. The deposition factor measured across each stage was generally <1 meaning that only a fraction of the particles in the feedwater attached to or deposited on the membranes. Negative values were also reported by Schippers suggesting that some particles may also have been sheared off the membrane surface during operation.

Boerlage *et al.* (2001, 2003a) presented the results of measurements with MFI-UF (constant pressure) at two locations working with fresh water from the river Rhine and from the IJssel lake. The deposition factor values for the IJssel lake plant and for the river Rhine plant were all negative. The results were attributed to changes in the composition of the cake formed on the RO membranes over time due to the forces acting on the particle in tangential flow.

Sioutopoulus *et al.* (2010) worked with colloidal organic and inorganic species to link fouling potential between UF and RO. The experimental set-up was a bench scale RO unit. The salinity levels in the water were 500, 2000, 5000 and 10000 mg/L as TDS. The range of fluxes tested was 25-40 L/m^2/h with a water recovery of 1-2 %. In this study, the deposition factor was obtained by measuring the ratio of actual fouling species deposited on the membrane over the theoretical one. The author mentioned that the theoretical mass deposition values were calculated based on the total permeate volume of each RO test. Thus, the mean deposition factor values were estimated to be 0.6, 0.9 and 1.0 for humic acids, sodium alginate and ferric oxide, respectively.

4.3.3.2 Particle deposition mechanisms

When particles enter the feed channel in the membrane element and get close to the membrane surface, two forces are imposed on particles namely: *i*) convective force towards the membrane surface (due to the drag force of permeation flow) and *ii*) the shear force (due to crossflow velocity).

The particle backtransport mechanisms include concentration polarisation (brownian diffusion, influencing small colloids), shear induced diffusion and inertial lift (influencing big particles) Belfort *et al.*, 1994. In recent studies, it was reported that particle-particle and particle-membrane interactions (including entropy, van der Waals interactions and electrostatic interactions) may also play important roles in particle transport to and/or from the membrane surface, especially in concentrated solutions of colloidal particles (Davis, 1992, Jiang, 2007).

The random movement resulting from the bombardment of particles by water molecules is defined as brownian diffusion. Shear induced diffusion occurs when individual particles undergo random displacements from the stream lines in a shear flow as they interact with and tumble over other particles (Davis and Sherwood, 1990). Belfort *et al.* (1994) mentioned that the back-diffusion of particles away from the membrane is supplemented by a lateral migration of particles due to inertial lift (also known as tubular-pinch effect).

The three backtransport mechanisms work simultaneously, and the total backtransport velocity is assumed to be the sum of them (Jiang, 2007). The contribution of the individual mechanisms depends on the particle size and crossflow velocity.

Particle's size plays a role in particle deposition on permeable surfaces; the deposition rate has been assumed to be lower for larger particles compared to smaller particles. This is due to the fact that the back transport by inertial lift is significant for larger particles (Song and Elimelech, 1995). Chellam and Wiesner (1998) reported that the cake formed in cross flow mode had a higher percentage of fine particles resulting in a higher specific cake resistance compared to the feed suspension.

Many studies have been conducted to understand factors affecting fouling of RO membranes. Results of membrane autopsies illustrate that biofouling and organic fouling may occur preferably in the first element while precipitation of salts (scaling) is expected to occur in the last elements where the concentration factor is highest. Furthermore, the fouling layer distribution may not be homogenous over the entire membrane surface.

In a RO pressure vessel, the flux distribution along the vessel is not uniform; the front elements have a higher production rate in comparison with the production rate in the rear elements. . Furthermore, the cross-flow velocity in the front and rear elements is not uniform

Furthermore, a pressure vessel may not contain identical elements. In some cases, the elements in a pressure vessel are not identical. High production membranes are placed at the front end of the vessel and high rejection membranes at the end position of the vessel. Consequently, water flow is not equally distributed through the membrane elements.

Many studies have focused on the effect of channel geometry, and shear rate on colloidal fouling in cross flow (Hoek *et al.*, 2002). All of these factors (non-uniform flux rate, cross flow velocities, geometry of spacer) make it difficult to study the deposition of particles in RO units. Therefore, measurements should be performed on site and consider retention times . Moreover, measured values are an average of the entire RO pressure vessel, which usually comprises 5-7 elements.

Furthermore, it is possible that preferential deposition of particles may occur and influence the measurements of particulate deposition through MFI. In this case the size distribution of particles in the feed water may differ from the particle size distribution in the concentrate water. Current methods to measure particles size (e.g., laser diffraction, microscopy) are limited in working with particles smaller than 0.05 μm.

4.3.4 The particulate fouling prediction model

The particulate fouling models to predict fouling developed by Schippers are based on the assumption that particulate fouling on the surface of reverse osmosis (or nanofiltration) membranes can be described by the cake filtration mechanism (Belfort and Marx, 1979, Schippers *et al.*, 1981). The relationship between the MFI measured for a feedwater and the flux decline predicted for a RO system are presented below. The relationship is based on the assumption that scaling, adsorptive blocking and biofouling do not contribute to the fouling observed on the RO membrane. Nevertheless, during the MFI test some elements contributing to biofouling might be retained by the membranes (bacteria, particulate organic matter).

RO systems operate in cross flow while the MFI(-UF) is currently a dead-end filtration test. This results in two main differences: i) in an RO system, not all of the particles are deposited on the surface of the membranes as RO units operate in cross flow, and ii) the cake formed in cross-flow RO has different characteristics than the cake formed in dead-end MF/UF, e.g., cake porosity, etc. These differences were respectively translated by Schippers and Kostense (1980) in *i*) the particle deposition factor "Ω" ($\Omega < 1$ for cross flow) as discussed earlier in section 4.3.3 and *ii*) the cake ratio factor "ψ".

Boerlage *et al.* (2003a) also made use of this model to predict particulate fouling in freshwater RO systems.

4.3.4.1 At constant pressure

The time (t_r) in which the flux of a RO membrane has decreased by a factor (e.g., $\Delta J = 15$ %) is:

$$t_r = \frac{\Delta P_r}{\eta_r \cdot J_0^2 \cdot I_r} \cdot \frac{\Delta J \cdot (2 - \Delta J)}{2 \cdot (1 - \Delta J)^2} \qquad \text{Eq. 4.38}$$

and,

$$I_r = \psi \cdot \Omega \cdot I \qquad \text{Eq. 4.39}$$

where the subscript *r* indicates that the parameter refers to filtration through a RO membrane.

4.3.4.2 At constant flux

Membrane cleaning is commonly recommended when a 15 % decrease in the normalised flux or increase in pressure drop of an installation is observed.

For a RO system operating under constant flux filtration, the time required for an increase in pressure ΔP_r to occur can be predicted by:

$$t_r = \frac{(\Delta P_r - \Delta P_{0r})}{J^2 \cdot \eta \cdot I_r} \qquad \text{Eq. 4.40}$$

The relationship between I_r and I (from the MFI measurement) was defined in Eq. 39 where the cake ratio factor (ψ) accounts for differences between the cake deposited on the MFI membrane and that deposited on the RO membrane, and the particle deposition factor (Ω) represents the ratio of the particles deposited on the RO membrane to that present in the feed water.

The prediction model equations (Eq. 4.38 and Eq. 4.41) are a function of the fouling potential of the water at RO operating conditions. The fouling index (I) plays a dominant role as its magnitude depends strongly on the pore size of the filter used, as smaller particles are retained which have greater resistance (see Eq 4.11). The smaller the filter pore size, the higher the fouling index value and thus shorter estimated running time is projected considering a percentage pressure increase.

4.4 SILT DENSITY INDEX (SDI)

The SDI was introduced by the DuPont company (Permasep Products) at the request of the U.S. Bureau of Reclamation. Initially, the test was called the Fouling Index. It was intended to characterize the fouling potential of feed water to DuPont's hollow fine fibre RO permeators (membrane elements). The target contaminants were suspended and colloidal matter. Later on, manufacturers of spiral wound elements and different hollow fibres elements recommended this test as well and formulated maximum levels for SDI to minimize suspended and colloidal fouling to enable good long-term performance. Currently, SDI < 5 has been set as a recommendation for the performance of pre-treatment systems for RO and NF, and preferably SDI< 3.

The SDI testing procedure is described in the American Society for Testing and Materials (ASTM). The latest version for SDI testing is from 2014 (code 4189-14). The method describes that the SDI test can be used as an indication of the quantity of *particulate* matter (size greater than 0.45 μm) in water and it should be used for relatively low (<1.0 NTU) turbidity waters such as well water, filtered water, or clarified effluent samples. As the nature of particulate matter in water may vary, the ASTM method indicates that the test is not an absolute measurement of the quantity of particulate matter (ASTM D4189 - 14, 2014). Furthermore, it is clearly mentioned that the test is not applicable to permeates from RO and UF systems. This recommendation is not always followed in practice as pre-treatment systems using membrane filtration are often assessed via the SDI test. In some cases, high SDI values were obtained after UF pre-treatment that could not be attributed to the "lack of integrity" of the system. In practice, high SDI indicates that fouling might occur, and low SDI does not guarantee that fouling will not occur.

SDI is measured by filtering water through a 47 mm diameter hydrophilic membrane (mixed cellulose nitrate or cellulose acetate) with 0.45 μm pores, in "dead-end" filtration mode, at constant pressure of 210 kPa (30 psi, 2 bar). A typical scheme for performing an SDI test is illustrated in Figure 7.

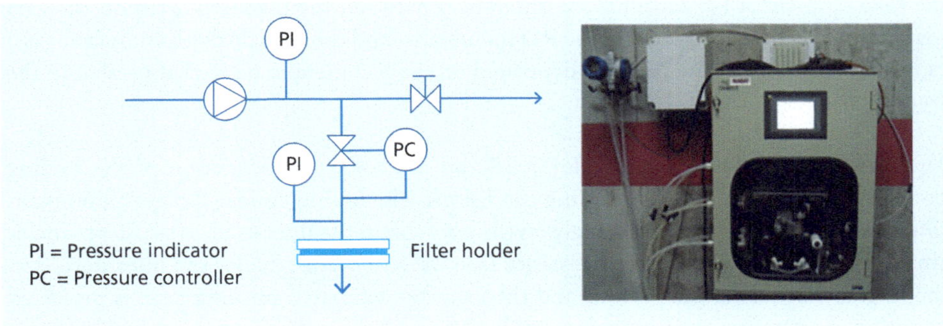

Figure 7 Scheme of an SDI apparatus (left) and picture of an automatic SDI/MFI equipment (right)

The ASTM standard provides some guidelines regarding the recommended membranes for the SDI test. The method describes that, for a range of pressures (91.4-94.7 kPa), the water flow should be around 25-50 seconds per 500 mL. Based on this information, the recommended permeability of the filters at 20 °C was calculated to be 21,911 L/m^2/h/bar to 45,405 L/m^2/h/bar.

The SD_{IT} is calculated from the following equation:

$$SDI_T = \frac{\%PF}{T} = \frac{\left(1-\frac{t_1}{t_2}\right)\cdot 100}{T} \qquad \text{Eq. 4.41}$$

Where, t_1 is filtration time of initial filtered volume (min), t_2 is the filtration time of second filtered volume (min), T is the total filtration time (min) and %PF is the percentage of plugging factor. SDI measures the decline in filtration rate expressed in percentage per minute although is usually reported without units. SDI is the percentage of permeate flow decline (filtration rate) per minute. For example: SDI = 3 means that the permeate flow reduced by 3 % per minute during the test. The filtration flow over time is illustrated in Figure 8.

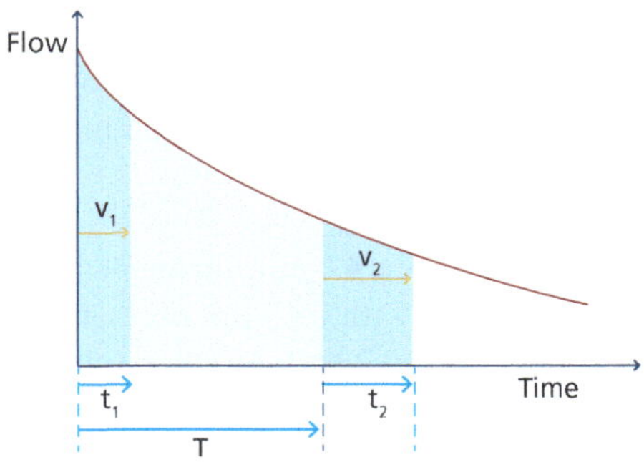

Figure 8 Schematic representation of the filtration flow over time during an SDI test

Some ASTM recommendations for performing a good SDI test are the following: *i*) flushing the equipment before use, *ii*) purging air to avoid air entrapment going at the surface of the membrane, *iii*) wetting the membrane filters before use, iv) avoid touching the membrane filters with hands. When reporting an SDI value, the following information is required to accompany the measured value: the SDI with a subscript indicating the total elapsed flow time (T) in minutes; the water temperature before and after the test; and manufacturer of the 0.45 µm membrane filter used for the test and ID.

Example 1 – Maximum value of SDI (T = 15, 10, 5, 2 min)

Has the SDI measured in about 15 minutes, a maximum value? If so, what is the maximum value?

Answer:

It has a maximum.

The maximum will occur when testing water with high particulate fouling potential. In this case, $t_2 >>> t_1$. Thus, in the SDI formula, the ratio t_1/t_2 is very low, let us say close to zero. Therefore, the SDI_{15} value will be: (1-0) / T = (1 - 0) · 100 % / 15 min = 6.67 %/min. When the fouling potential is high, then for T a shorter period has to be taken, e.g., 10, 5, and 2 minutes. Why? What will be the maximum values at different T values?

Answer:

Shorter T time periods are considered when the 0.45 µm filter gets clogged rapidly.

For T = 10 min, SDI_{10} max = 10 %/min

For T = 5 min, SDI_5 max = 20 %/min

For T = 2 min, SDI_2 max = 50 %/min

On the other hand, theoretically, when the water to be tested has no fouling potential, then $t_1 = t_2$, thus $t_1/t_2 = 1$. Replacing in the formula we have: $1-(t_1/t_2)/T = (1-1)/T = 0$.

Example 2 – Units of SDI

SDI measures the decline in filtration rate expressed in percentage per minute. The first t_1 for the first 500 mL is equivalent to the initial rate of filtration at the start and proportional with initial flux (J_0). The second t_2 for the second 500 mL is equivalent to rate of filtration after (e.g.,)15 minutes (T) and proportional with flux (J_T).

If we substitute these values in the previous formula, we obtain:

$$SDI_T = 100 \cdot \frac{\left(1 - \dfrac{J_T}{J_0}\right)}{T}$$

Or

$$SDI_T = 100 \cdot \frac{\left(\dfrac{J_0 - J_T}{J_0}\right)}{T}$$

Which is equal to percentage flux decline per minute.

SDI is not based on a fouling mechanism and can never be used to predict the rate of fouling in RO systems where cake filtration is considered the mechanism for particulate fouling. According to Boerlage (2007), the SDI is based on a mixture of filtration mechanisms; namely blocking (which is not expected for RO membranes) and cake filtration. As the test operates at 2 bar, cake compression will influence the results.

Example 3– SDI calculation of Seawater

An SDI test was performed for seawater using a filter of 25 mm diameter. The water temperature was 26 °C. Times t_1 and t_2 were collected for 125 mL (proportional to the reference filter area 47 mm) for T = 15, 10, and 5 min, as follows:

T = 15 min: $t_1 = 0.6$ min, $t_2 = 9.7$ min

T = 10 min: $t_1 = 0.6$ min, $t_2 = 4.62$ min

T = 5 min: $t_1 = 0.6$ min, $t_2 = 1.73$ min

Answers:

$SDI_{15} = (1\text{-}(0.6/9.7)) \cdot 100/15 = 6.25$ %/min. NB. Value > 75 % plugging (5 %/min)

$SDI_{10} = (1\text{-}(0.6/4.62)) \cdot 100/10 = 8.70$ %/min. NB. Value > 75 % plugging (7.5 %/min)

$SDI_5 = (1\text{-}(0.6/1.73)) \cdot 100/5 = 13.06$ %/min. NB. Value < 75 % plugging (15 %/min)

4.4.1 Weaknesses of the SDI

The limitations of the SDI test are well documented and include (Salinas Rodriguez *et al.*, 2019, Rachman *et al.*, 2013Alhadidi *et al.*, 2011c, Alhadidi *et al.*, 2011a, Nahrstedt and Camargo Schmale, 2008, Boerlage, 2007, Schippers and Verdouw, 1980):

- no correction for test water temperature;
- the result is heavily dependent on the test membrane permeability;
- not applicable for testing high fouling feed water e.g., raw water – ASTM recommends that turbidity should be < 1 NTU;
- not applicable for testing UF permeate, which is increasingly being used in desalination pre-treatment;
- no linear relation with colloidal/suspended matter;
- fouling potential of particles smaller than 0.45 µm are not measured;
- it is not based on any filtration mechanism.

It is well known that, even when the recommendations for SDI are not compromised (i.e., SDI < 3 for seawater), serious fouling may occur. This might have two principal reasons: *i*) other type(s) of fouling occurred and they are not are measured e.g., biofouling, inorganic and organic fouling, fouling due to corrosion products; *ii*) SDI has no direct predictive value in fouling RO/NF membrane systems. However, it is sometimes an indirect indicator for the fouling potential of RO/NF feed waters.

Furthermore, erratic results are reported with water supersaturated with air; different results are obtained with membranes from different manufacturers; relatively high values are reported in effluents of micro- and ultrafiltration systems. The lack of temperature correction and membrane heterogeneity may explain the non-uniform results observed in practice.

4.4.1.1 SDI versus turbidity

Turbidity is a useful parameter in monitoring suspended matter concentration in raw and RO feedwater. Unfortunately, low turbidity will not guarantee low fouling potential. Various studies have shown there is no relation between water turbidity and the SDI (Bonnelye *et al.*, 2004). Turbidity has often remained unchanged whereas, the SDI has increased indicating an increase in the particulate fouling potential of the water tested (Mosset *et al.*, 2008, Boerlage, 2007).

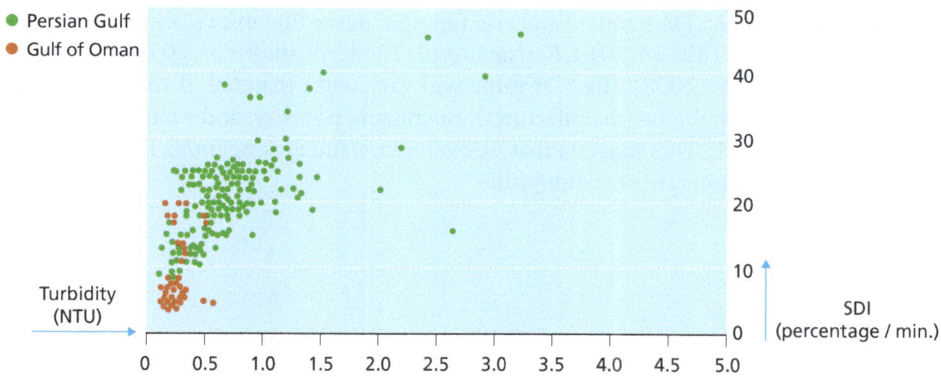

Figure 9 SDI versus turbidity for Gulf of Oman and the Persian Gulf water qualities (Bonnelye *et al.*, 2004)

As illustrated in Figure 9, even low turbidity values may result in rather high SDI values.

4.4.1.2 Non-correlation with concentration of particles

While, the non-linear relation between the measured SDI value and particle concentration means that water appears less fouling than it is, as the test filter becomes progressively plugged. This is demonstrated in Figure 10 for SDI measurements after 5-, 10-, and 15-minutes filtration for a formazine solution (Schippers, 1980). A difference between 3 mg/L and 6 mg/L gives only a marginally higher SDI. Consequently, small differences in SDI are misleading. The net result is that SDI cannot be directly compared when measured at different temperatures or for different filtration intervals and for more fouling conditions (Boerlage, 2008, Boerlage *et al.*, 2017).

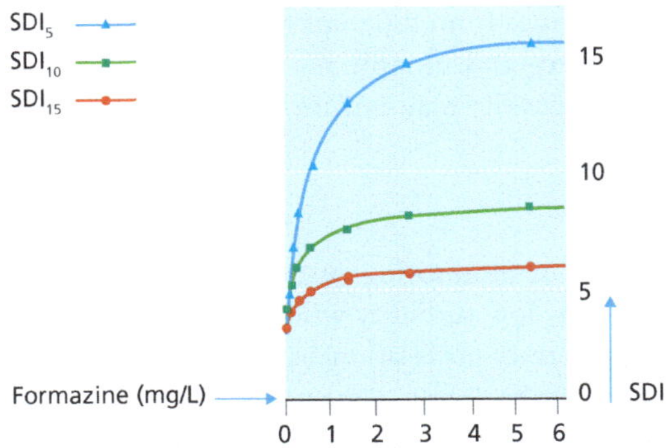

Figure 10 SDI as function of formazine concentration and filtration time (Schippers and Verdouw, 1980)

4.4.1.3 Membrane material

As mentioned in the ASTM standard and also reported in the literature (Salinas Rodriguez *et al.*, 2019, ASTM D4189 - 14, 2014, Rachman *et al.*, 2013, Alhadidi *et al.*, 2011c, Nahrstedt and Camargo Schmale, 2008), the SDI value will vary with: material of the membrane, origin of the filter membrane (manufacturer), membrane porosity, and even filters in the same production batch. This suggests that SDI values obtained using filters from different membrane manufacturers are not comparable.

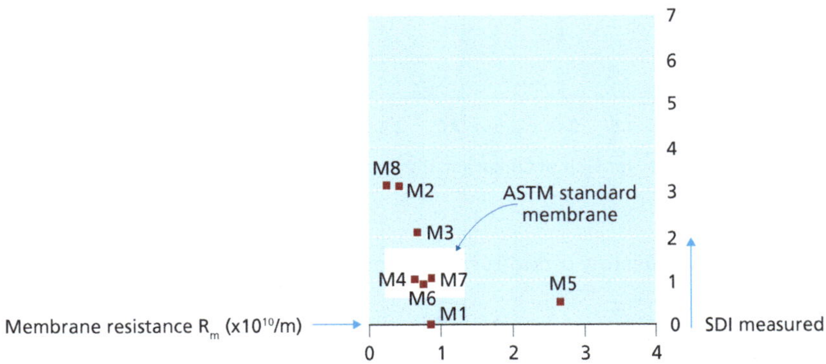

Figure 11 SDI_{15} values for eight different membranes (M1-M8) for UF permeate in a seawater UF/ RO plant. (Alhadidi *et al.*, 2012). ASTM standard membranes follow the recommendations on material and membrane resistance (reference ASTM-D4189-14). All values measured on the same day.

Mosset *et al.* compared SDI values for various hydrophilic membrane materials (nitrocellulose mixed esters, poly-vinylidene fluoride, polytetrafluoroethylene, polyacetylene). Differences of up to 300 % in SDI values were reported. Al-hadidi *et al.* (2008) reported that there is a variation in membrane properties within a same batch of manufactured membranes (acrylic copolymer, cellulose nitrate, poly-vinylideen-fluoride, poly-tetra-fluoro-ethylene). In his study, the membrane variations were in pore size and roughness up to an average of 10 % and 17 %, respectively, within a batch of membranes, while less variation was observed in bulk porosity which was less than 5 %. The variation in membranes thickness ranged from 3 to 7 % (Al-hadidi *et al.*, 2008). In a study on wastewater reuse, Escobar *et al.* found a SDI value difference of more than 100 % when using cellulose acetate and nylon membranes (Escobar *et al.*, 2009).

4.4.1.4 Water temperature

The viscosity of the water changes with temperature. Cold water has higher resistance to filtration than warm water. For this, any filtration experiment should be normalized to a reference temperature. This is not the case in SDI testing.

Alhadidi et al measured the effect of the temperature on SDI results assuming the effect to be only due to a change in the feed water viscosity. Using membrane M7, SDI was measured (see Figure 12) for different feed water temperature for 4 mg/L of AKP-15 (α-alumina) particle solution.

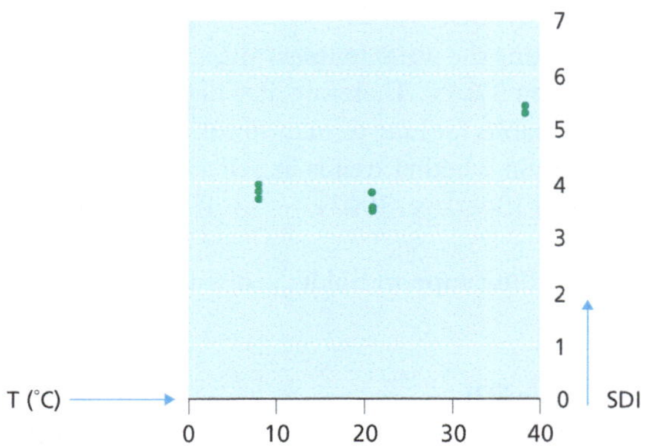

Figure 12 Measured SDI values for different temperatures of a colloidal suspension of 4 mg/L of AKP=15 (Adapted from Alhadidi *et al.*, 2011b)

The seasonal influence of temperature on SDI_{15} was examined by Boerlage (2007) using long term operational data from the Dhekelia seawater RO plant (Sallangos and Kantilaftis, 2001). The SDI_{15} was measured at the plant for the seawater feed and after pretreatment can be observed in where the seawater temperature varied seasonally from 15 to 32 ºC (Figure 13) (Sallangos and Kantilaftis, 2001). During the early years of plant operation many changes were made to plant operation including adoption of intermittent chlorination instead of continuous chlorination, variation in pretreatment chemicals and dose (Sallangos and Kantilaftis, 2001). Although, changes in the SDI_{15} shown in Figure 13 can then be attributable to both temperature and process changes, the increase in SDI_{15} in summer and decrease in winter is clearly evident (Boerlage, 2007).

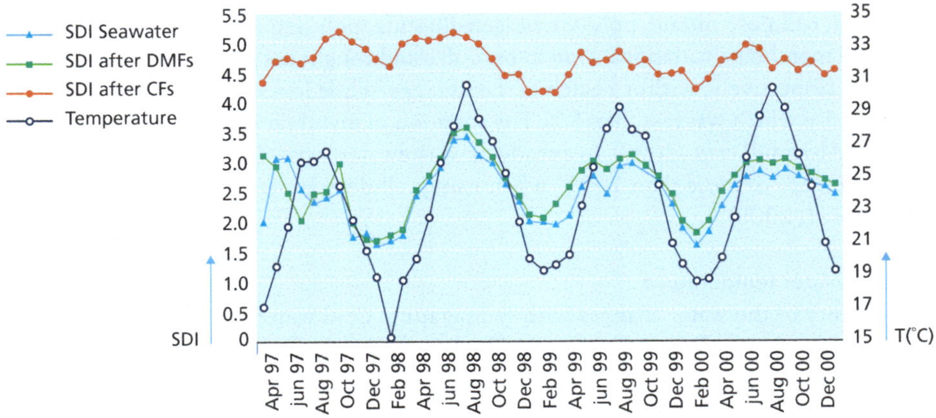

Figure 13 SDI_{15} (left hand Y axis) measured at the Dhekelia SWRO desalination plant for the seawater, after dual media filtration (DMF) and after cartridge filtration (CF) and seawater temperature (right hand Y axis) (Adapted from Sallangos and Kantilaftis, 2001)

Although, the ASTM notes the SDI will vary with temperature it supplies no correction method for the SDI, only stating the water temperature should stay constant during a test as the flow rate changes by about 3%/ºC. Therefore, it is highly recommended that feedwater temperature is included in graphs for long term monitoring of SDI over time as practiced at the Dhekelia plant, to ascertain whether trends in SDI are as a result of temperature rather than changes in water quality (Boerlage, 2007).

The effect of the membrane filter support holder is discussed together with the MFI in the following section.

4.4.2 Predictive value of the SDI

Salinas Rodriguez *et al.*, 2019, Boerlage, 2007, Schippers and Verdouw, 1980 concluded that the SDI test cannot predict the rate of fouling due to the fact that: *i*) no linear relation exists between the concentration of suspended and colloidal matter, *ii*) no correction for temperature, *iii*) the SDI is not based on any filtration mechanism, *iv*) it makes use of 0.45 μm filters while pores in RO/NF membrane are approx. 0.001 μm.

Theoretical prediction of flux decline in RO systems based on SDI results in extremely high fouling rates e.g., SDI = 3 effectively means a flux decline of 3 % per minute (Schippers *et al.*, 1981). Applying a direct correction between the SDI test flux (> 1,600 $L/m^2/h$ at the start) to a typical RO flux (which is about 20 $L/m^2/h$), predicts a flux decline of 20 % per hour. This rate of fouling is far outside the rates observed in practice.

Despite it being widely used and proven to be of great practical use, Yiantsios *et al.* (2005) criticised the SDI test as showing no clear correlation between the index value and the fouling behaviour of an RO system.

4.5 MODIFIED FOULING INDEX (MFI)

MFI has been developed to overcome the main deficiencies of SDI test. The $MFI_{0.45}$ test uses the same equipment and the same 47 mm diameter hydrophilic 0.45 μm membrane (white hydrophilic, mixed cellulose nitrate or cellulose acetate) as the SDI test. It takes into account that initially pore blocking occurs, followed by cake/gel filtration and finally cake/gel blocking and/or enhanced compression occurs. The MFI was adopted by the ASTM as a standard method in 2015 (designation D8002-15) and contrary to the SDI can be applied to measure the particulate fouling potential of ultrafiltration permeate.

The MFI makes use of the general equation describing cake filtration presented in section 4.3.1., derived in Eq. 4.12 and presented again below.

$$\frac{t}{V} = \frac{\eta \cdot Rm}{\Delta P \cdot A} + \frac{\eta \cdot I}{2 \cdot \Delta P \cdot A_m^2} \cdot V \qquad \text{Eq. 4.42}$$

Equation 4.42 gives a straight line when t/V is plotted against V and has been widely applied since suggested by Underwood in 1926 Underwood, 1926 to test for cake filtration and to obtain information on the permeability of the cake deposited. Carmen defined the gradient of the line (b) as (Carman, 1938):

$$b = \frac{\eta \cdot I}{2 \cdot \Delta P \cdot A_m^2} \qquad \text{Eq. 4.43}$$

The gradient of the line was adopted by (Schippers Schippers *et al.*, 1981) to define the Modified Fouling Index (MFI) as an index of the fouling potential of a feedwater containing particles, when fixed reference values are used for pressure (ΔP, 2 bar), water viscosity ($\eta_{20\,^\circ C}$) and (effective) membrane area A_m (13.8×10^{-4} m^2).

In the MFI (Equation 4.44), the fouling index I is the product of the specific resistance of the cake (α, see Eq 4.11)) and the concentration of particles (C_b) in the feedwater, and is assumed to be independent of pressure. An advantage of using I is that in most cases it is impossible to determine C_b and α accurately. The fouling index I is a function of the dimension and nature of the particles present in a feedwater and directly correlated to their concentration (Schippers and Verdouw, 1980).

$$MFI = \frac{\eta \cdot I}{2 \cdot \Delta P A^2} \qquad \text{Eq. 4.44}$$

Reference conditions for normalization of MFI values
P = 2 bar = 200 kPa
A = 13.8×10^{-4} m^2 (42 mm effective diameter of a 47 mm diameter filter)
At temperature 20 °C, the viscosity (η) is = 0.001 Ns/m^2.
This definition and conditions have been chosen since: MFI = 1 s/L^2 is usually in the range of approximately: SDI = 2 to 3.

Replacing the reference values in the MFI formula helps us to find the conversion factor of MFI into I, as follows:
MFI = $(0.001\ Ns/m^2) \cdot I / (2 \cdot 200000\ N/m^2 \cdot (13.8 \cdot 10^{-4}\ m^2)^2)$
MFI = $13 \times 10^{-4} \cdot I\ (s/m^6)$
MFI = $13 \times 10^{-8} \cdot I\ (s/L^2)$
or
I = $7.68 \times 10^{8} \cdot MFI\ (m^{-2})$

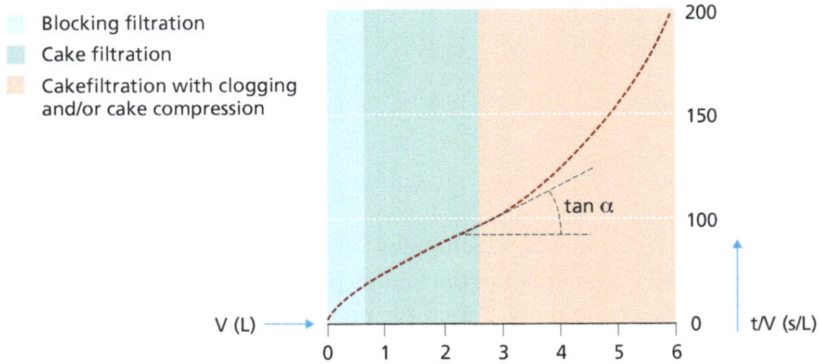

Figure 14 Filtration curve t/V versus V (Adapted from Schippers, 1989)

The I value is determined from the stage of cake/gel filtration and is defined as the minimum slope (tan α) in the curve t/V versus V. Where t = total filtration time, and V = total filtered volume (see Figure 14). The MFI can be calculated by replacing the I value in equation 4.45 and applying the reference values of pressure, membrane area, and temperature.

$$MFI = \frac{\eta \cdot 90 \cdot (1-\varepsilon) \cdot C_b}{\rho_p \cdot d_p^2 \cdot \varepsilon^3 \cdot \Delta P \cdot A_m^2} \qquad \text{Eq. 4.45}$$

Substituting the Carman-Kozeny Eq. 4.11 in Eq. 4.44 gives Eq. 4.45. This equation shows that MFI is a function of the dimension and nature of the particles forming a cake on the membrane, and directly dependent on particle concentration in water, as illustrated in Figure 15.

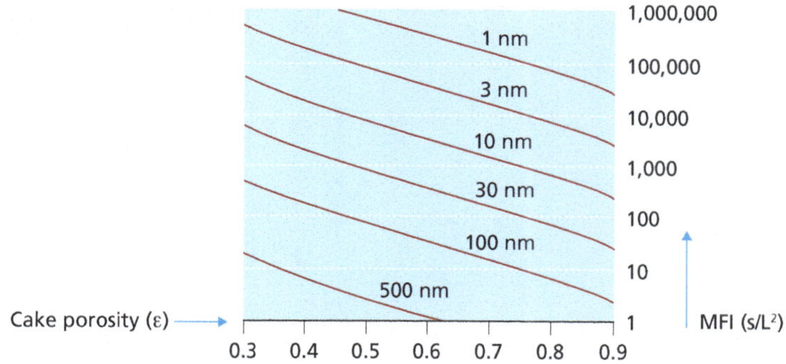

Figure 15 MFI as a function of particle size and cake porosity (C_b = 1 mg/L and ρ_p = 1430 kg/m^3) based on equation 4.45

Example 3 – Calculation of MFI value

An MFI test was performed on raw seawater at 2 bar, at 20 °C with a filter of 21 mm effective diameter. The results of the MFI test are presented in the following figure. Calculate the MFI value.

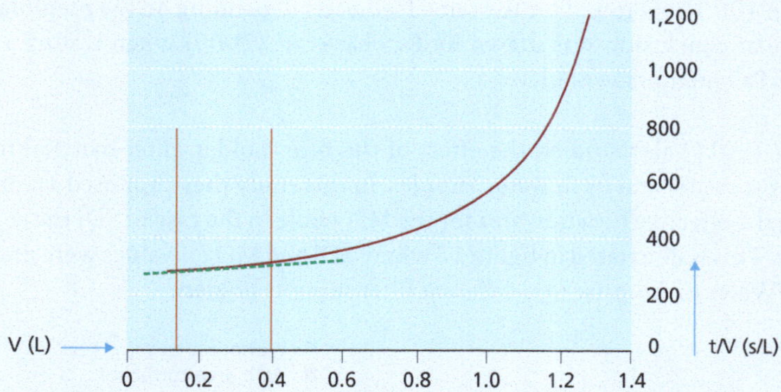

The slope of the linear region (marked by the two red lines) can be calculated.
Slope = 140 s/L^2.
Now we can calculate the $MFI_{0.45}$ value.

$$MFI = \left(\frac{\eta_{20^\circ C}}{\eta}\right)\left(\frac{\Delta P}{\Delta P_0}\right)\left(\frac{A}{A_0}\right)^2 \tan\alpha$$

By replacing the reference values and the ones used in the test, we have:
MFI = (1) · (1) · (0.0003464 / 0.00138)^2 · 140 = ~11 s/L^2

The $MFI_{0.45}$ shows a linear relation with colloidal/suspended matter concentration as illustrated in Figure 30 where the MFI values are reported for a range of formazine particles concentration.

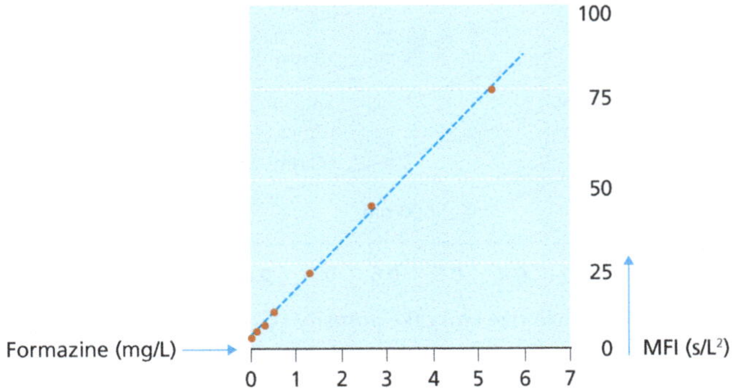

Figure 16 Relation between $MFI_{0.45}$ and formazine concentration (Adapted from Schippers and Verdouw, 1980)

4.5.1.1 Effect of membrane support holder in SDI and $MFI_{0.45}$

Nahrstedt and Camargo (2008) studied the effect of filter support on SDI and MFI values. They reported that the filter holder had a strong influence on the obtained SDI values. The type of filter holder will determine the effective membrane area during filtration. A difference of more than 100 % was found for the same feedwater depending on the membrane holder used. A similar conclusion was drawn by Escobar *et al.* (2009) when testing a Millipore holder and a Pall membrane holder.

Salinas *et al.* (2019) also studied the effect of the filter holder, filter material in SDI and MFI tests for seawater and fresh water samples. In this study they proposed a correction for considering the effective filtration area for the MFI while in the case of SDI the correction is not possible. This is illustrated in figure 17 where SDI and $MFI_{0.45}$ values were measured for Delft Canal Water making use of 4 different filter support holders.

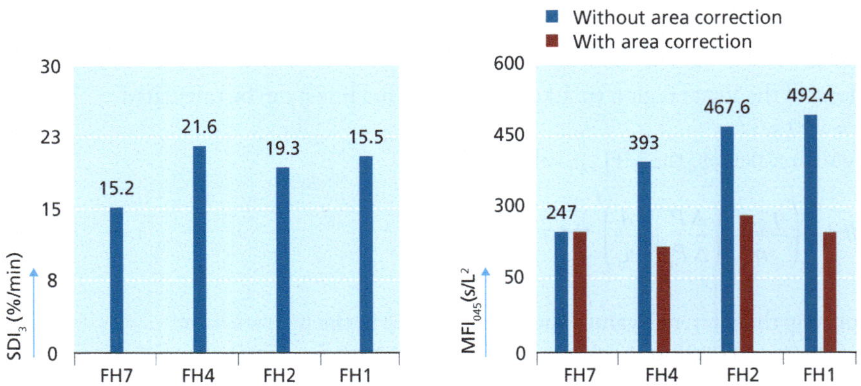

Figure 17 SDI_3 (a) and $MFI_{0.45}$ (b) values measured with various filter holders (n=10) for Delft canal water with a cellulose acetate filter. FH = filter holder (Adopted from Salinas Rodriguez *et al.*, 2019)

By correcting for the effective filter area, the $MFI_{0.45}$ results obtained with the different filter holders (Figure 17b) are closer to each other (247 s/L^2 ± 10.8 %) in comparison with the average without considering the area effect (400 s/L^2 ± 27.6 %). In the MFI formula, the area plays a significant role, hence the large variation in the MFI values for filter holders without area correction. Additionally, in the MFI the flow rate influences greatly the fouling potential of a water sample, so any effect that increases the flow rate through the membrane (like the channels in the filter support plates that reduce the effective filter area) will increase the fouling load of the membrane and consequently the measured $MFI_{0.45}$ will be higher.

4.5.2 Predicting the rate of fouling in spiral wound RO elements with $MFI_{0.45}$

Prediction of the fouling rate in RO systems, considering cake filtration as fouling mechanism is based upon equation 4.46. Rearranging this formula enables the time to be predicted in which e.g., the pressure needs to be increased by 1 bar when maintaining the flux constant.

$$t = \frac{\Delta P_t}{\eta \cdot I \cdot \Omega \cdot J^2} \qquad \text{Eq. 4.46}$$

Where: ΔP = net driving pressure (NDP) (N/m^2), η = viscosity (N · s/m^2), Rm = membrane resistance (m^{-1}), I = fouling potential derived from MFI (m^{-2}), Ω = deposition factor (-), J = flux (m^3/m^2·s), t = time (s).

From the measured MFI, the I value can be easily calculated, namely: $I = 7.61\times10^8 \times MFI$ (m^{-2}). Alternatively we can express Eq. 4.46 as function of MFI, as follows:

$$t_r = \frac{\eta_{20^\circ C} \cdot (\Delta P_r - \Delta P_{0r})}{2 \cdot MFI \cdot \Delta P_0 \cdot A_0^2 \cdot \psi \cdot \Omega \cdot J_0^2 \cdot \eta_r} \qquad \text{Eq. 4.47}$$

Equation 4.47 has been plotted in Figure 30.

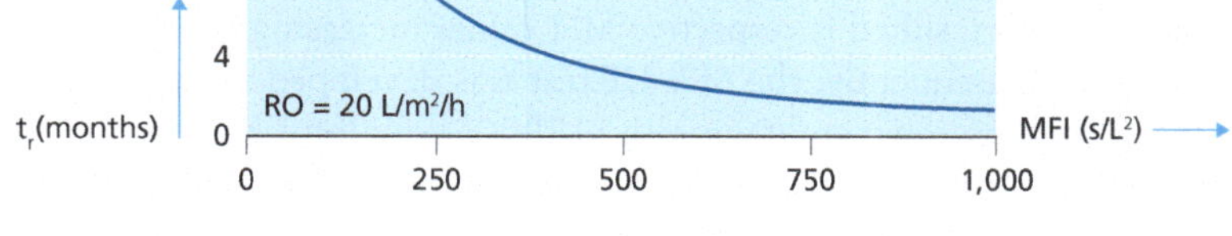

Figure 18 Projections based on MFI values. Time (in months) for an increase in Net Driving Pressure (ΔNDP_r - ΔNDP_{0r}) = 1 bar in a RO system operating at flux = 20 L/m^2/h, Ω = 1 (worst case), T = 20 °C (Adopted from Salinas Rodriguez *et al.*, 2015).

Example 4 - Predicting rate of fouling in spiral wound RO elements with $MFI_{0.45}$

Calculate the operation time to get 10 % increase in pressure (1.0 bar), due to particulate fouling? Assumptions: filtration mechanism is cake filtration; Operating pressure is 10 bar (clean membranes); average flux equals 20 $L/m^2/h$; $MFI_{0.45}$ = 1 s/L^2; deposition factor $\Omega = 1$; temperature = 20 °C.

Answer:

We can use the Eq. 4.47. Replacing values, we have:

$t_r = (\eta_{20} \cdot 1) / (2 \cdot 1 \cdot 2 \cdot (0.00138)^2 \cdot 1 \times (20/1000)^2 \cdot \eta_r)$

$t_r \sim 50{,}000$ days ~ 135 years

Following the same procedure for MFI values ranging from 10 to 10 000, we have the time for observing a 1 bar pressure increase:

MFI (s/L^2)	1	10	100	1,000	10,000
Time (days)	50,000	5,000	500	50	5

$MFI_{0.45}$ values in the range of 1 to 200 s/L^2 are not expected to cause high rates of fouling in spiral wound RO systems due to deposition of particles on the membrane surface. (An $MFI_{0.45}$ = 200 s/L^2 might result in pressure increase of 1 bar in 250 days (assuming deposition factor $\Omega = 1$).

High SDI and $MFI_{0.45}$ might indicate – for some water types – high fouling potential. However, low levels of these parameters do not guarantee low fouling potential. Particles much smaller in size than 0.45 μm, are most likely responsible for membrane fouling the surface of RO membranes, due to deposition and attachment. As a consequence, SDI and $MFI_{0.45}$ cannot predict adequately the rate of fouling of the surface of RO membrane due to particles. Both might have predictive value in clogging fibre bundles and spacers in spiral wound elements. However, the potential predictive value of this has to be verified.

4.6 MODIFIED FOULING INDEX – ULTRAFILTRATION (MFI-UF)

4.6.1 MFI-UF constant pressure

Based on the above, it was concluded that particles much smaller than 0.45 μm were responsible for the fouling rate observed in practice. This was supported by the measurement of MFI with membranes of different pore sizes varying from 0.8 μm down to 0.05 μm for RO feed water which resulted is respective MFI values increasing from 4 to 4,500 s/L^2 (Schippers 1981). Consequently, the MFI-UF test was developed using a hollow fibre poly-acrylonitrile ultrafiltration membrane with a 13000 Da molecular weight cut off (PAN 13 kDa) to capture these smaller particles (Boerlage *et al.*, 2000). The pores of the PAN 13 kDa membrane are circa 1000 times smaller than the pores of the existing $MFI_{0.45}$ as seen in Figure 14 and can therefore retain smaller particles (Boerlage, 2007).

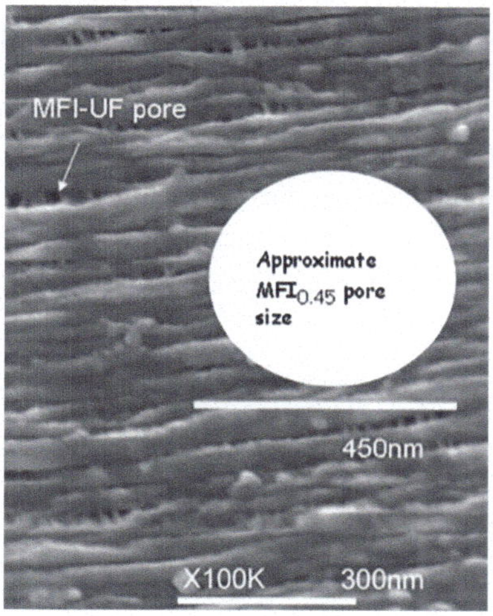

Figure 19 Scanning electron micrograph of the MFI-UF PAN 13 kDa membrane showing pore size comparison to $MFI_{-0.45}$ membrane pore (× 100,000 magnification) (Boerlage, 2007)

Brackish water measurements with the MFI-UF test using PAN 13 kDa membranes demonstrated that the cake/gel formed on the membrane surface was quite compressible (See Figure 20). However, it can be proven theoretically and experimentally that the linearity between t/V and *t* remains, under the condition that the membrane resistance is much lower than the cake resistance. The pressure dependency is illustrated in Figure 30 for: diluted canal water, tap water, treated river water with coagulation/sedimentation/rapid sand filtration, after increasing levels of pre-treatment; ozonation, biological activated carbon filtration and slow sand filtration

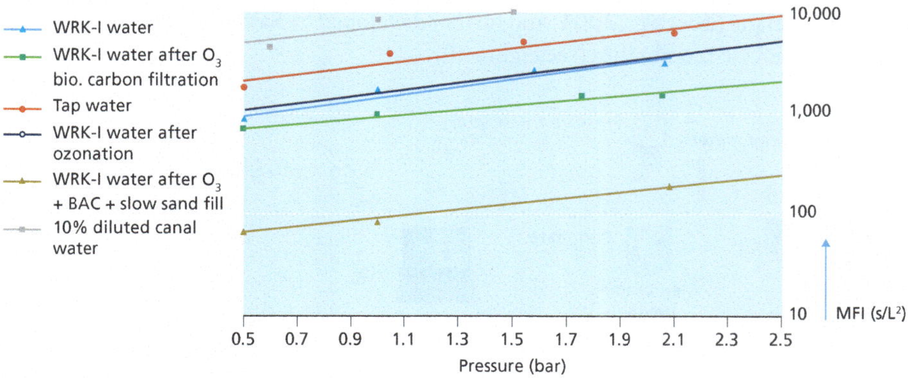

Figure 20 MFI-UF (PAN 13 kDa) as a function of pressure for different sources (Adapted from Boerlage *et al.*, 2003b) WRK-I = Conventionally pretreated (coagulation, sedimentation, and rapid sand filtration) River Rhine water (WRK-I)

The consequence of the pressure dependency of MFI-UF at constant pressure filtration is that accurately predicting the development of the pressure increase in RO membranes operating at constant flux (or constant pressure) is not possible. The reason is that the pressure loss across the cake is unknown and cannot be calculated. This pressure loss is at the start zero and will gradually increase due to the growing fouling/cake layer and in addition increasing compression (see Figure 21). As a consequence, the value of the fouling potential I (MFI) to be applied is increasing during the filtration process.

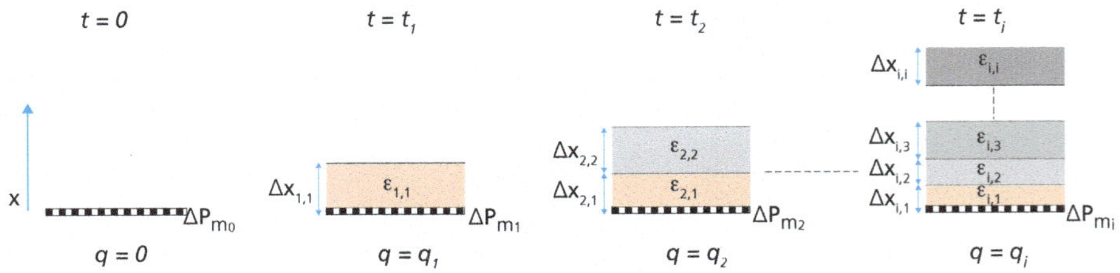

Figure 21 Schematic of the cake layer development over time, illustrating that the thickness and porosity of the sub-layers is different from each other and also changing over time (Adopted from Tung, 2008)

Due to this compressibility, accurate prediction of fouling in RO was not possible using the new MFI-UF test in constant pressure mode. Hence, the MFI-UF test was developed in constant flux mode, whereby pressure increase to maintain constant flux over time is recorded.

4.6.2 MFI-UF constant flux

MFI-UF measured at constant flux has been developed, because the index measured at constant pressure is not correct and cannot be used in prediction calculations. The MFI-UF constant flux test was proposed initially using a hollow fibre PAN 13 kDa membrane as the reference membrane (Boerlage, 2004) and further developed by Salinas (2011) using flat 25 mm diameter PES UF membranes (see Figure 30).

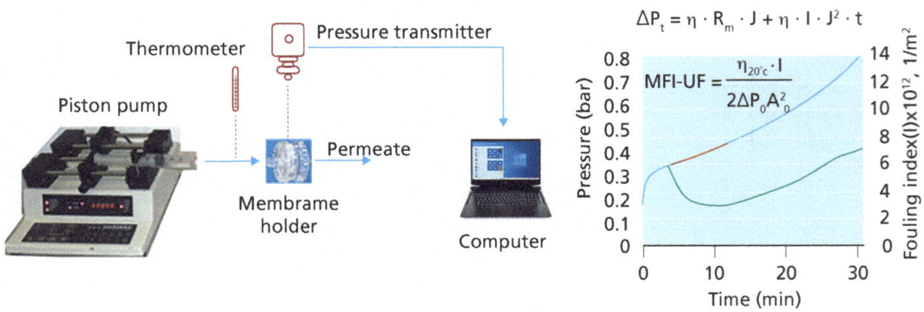

Figure 22 Filtration set-up to measure MFI-UF at constant flux (Salinas Rodríguez, 2011). Filtration flux can vary between 10 and 300 $L/m^2/h$.

4.6.2.1 Membranes

The MFI-UF test can be performed with a range of UF membranes, for instance the following membranes from Millipore, made of regenerated cellulose and polyether sulfone, have been tested: 100 kDa, 50 kDa, 30 kDa, and 10 kDa for various types of water: fresh water, sea water, pretreated water, RO feedwater, RO concentrate, RO permeate, etc.

The application of various UF membrane filters in a pilot scale desalination plant (Jacobahaven, Netherlands) is illustrated in Figure 30 where 100 kDa, 50 kDa and 10 kDa PES filters were used. The Amiad strainer showed only a small reduction in MFI-UF as expected with a relatively large aperture size of 50 μm. Whereas, the reduction in MFI-UF (and fouling) observed following UF (nominal MWCO of 150 kDa) was much larger i.e., of 94 %, 93 %, and 88 % reduction for 100 kDa, 50 kDa, and 10 kDa MFI-UF test membranes, respectively (Salinas Rodríguez *et al.*, 2015).

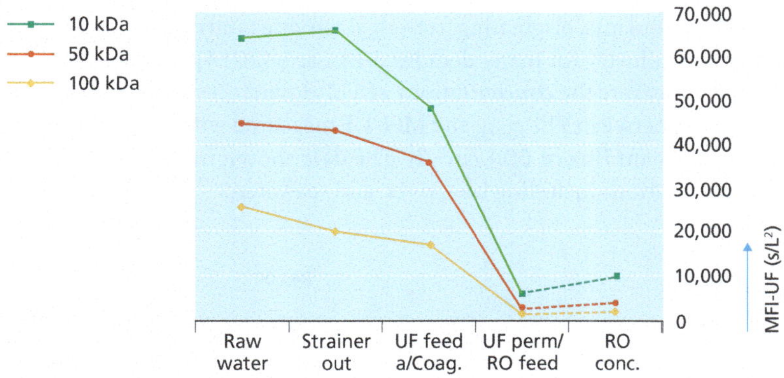

Figure 23 Effect of pre-treatment on MFI-UF at the Jacobahaven seawater pilot plant using PES test membranes of 100, 50 and 10 kDa size (Salinas Rodríguez *et al.*, 2015)

MFI-UF depends strongly on pore size; the smaller the pore size, the higher the MFI value. The MFI-UF measured with membranes having smaller pores (10 kDa) obtained higher values, than those measured with membranes with larger pores (100 kDa).

4.6.2.2 Flux rate

Theoretically the fouling potential of a water depends on the filtration flux during testing. To illustrate this effect in practice, North Sea water was tested with 5 kDa, 10 kDa, 30 kDa and 100 kDa PES membranes and the results showed higher MFI-UF values for the lower MWCO filters, and a remarkably strong dependency on flux (see Figure 24).

When assessing the fouling potential of RO feedwater, it is important to measure the MFI-UF value at the same flux at which the RO is operating, either the average flux for the whole pressure vessel or the flux for the front element.

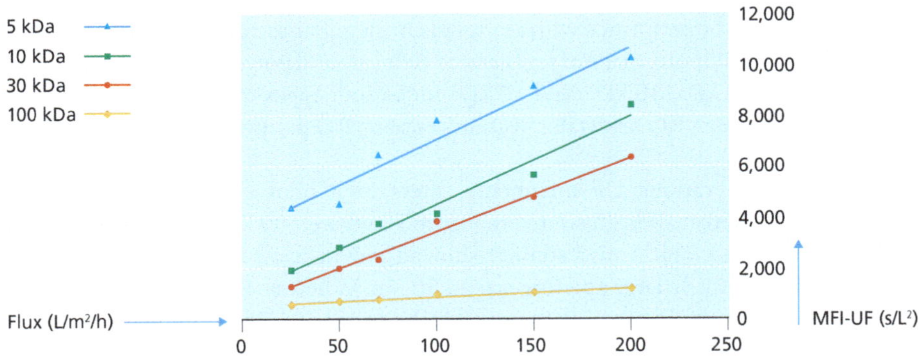

Figure 24 MFI-UF (PES, 10, 30, 10, and 5 kDa) of North Seawater measured at various flux rates

Recently transparent exopolymer particles (TEP) have been identified as potential foulants in MF, UF and RO. These foulants, originating from algal activities and other aquatic life, have been overlooked by the industry for many decades. Villacorte (2014) developed a method to semi- quantitatively measure the concentrations of TEP down to a size of 10 kDa. A good correlation was observed between $TEP_{10\,kDa}$ and MFI-UF measured with membranes having pores of 10 kDa at a constant flux of 60 $L/m^2/h$. The data shown in Figure 25 originates from 5 different plant locations including lake, river, and seawater.

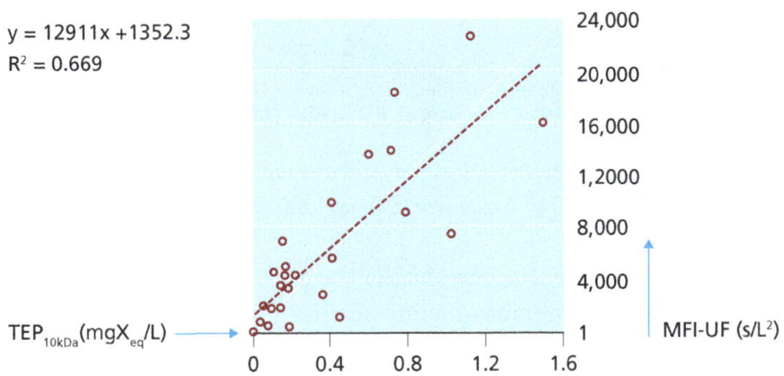

Figure 25 Relation between MFI-UF (constant flux) and TEP 10 kDa (Adapted from Villacorte, 2014)

4.6.3 Predicting pressure increase in RO systems

Generally, RO desalination plants operate at constant flux to meet production requirements. Changes in feedwater temperature are compensated for by adjusting feed pressure. Similarly, fouling resulting in an increase in membrane resistance is compensated for by increasing the feed pressure and hence net driving pressure (NDP). In this case, increase in the NDP can be predicted through equation 18 which already includes the deposition factor Ω. By re-arranging this equation, we have:

$$\frac{(NDP_r - NDP_{0r})}{t_r} = j^2 \cdot \eta \cdot \Omega \cdot I_r \qquad \text{Eq. 4.48}$$

Where: subscript "r" refers to real conditions and subscript "0" refers to initial conditions, NDP is the net driving pressure (N/m^2); Ω = deposition factor (-).
Based on equation 4.48, a theoretical "safe MFI" can be calculated assuming e.g., an allowable increase in NDP of 1 bar in 6 months. Figure 26 illustrates MFI calculated as a function of the deposition factor Ω at a flux of 10 to 20 $L/m^2/h$, which is commonly applied in seawater RO.

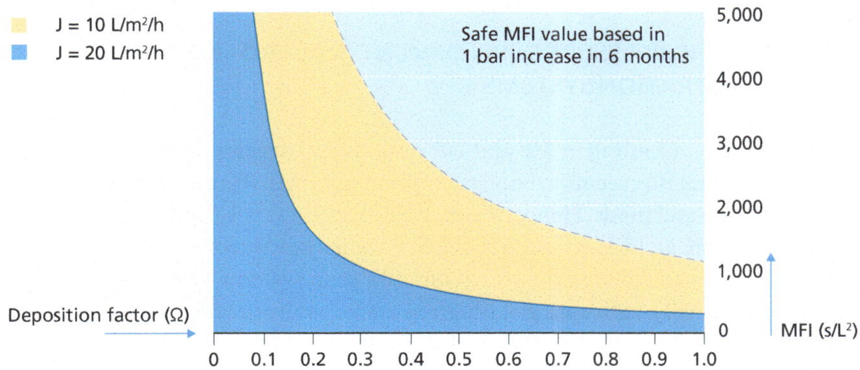

Figure 26 "Safe MFI" as a function of deposition factor and flux (Adapted from Salinas Rodríguez *et al.*, 2015)

"Safe MFI" values are heavily dependent on the deposition factor, emphasizing the need to determine deposition factors in full scale and pilot plants. An indication of the deposition factor can be obtained by measuring the MFI_{feed} in feed water and MFIconc in the concentrate and applying equation 4.35.

An alternative way of presenting the MFI constant flux prediction model is the one shown in Figure 27 which shows the predicted time necessary for a 1 bar NDP increase as a function of MFI-UF value.

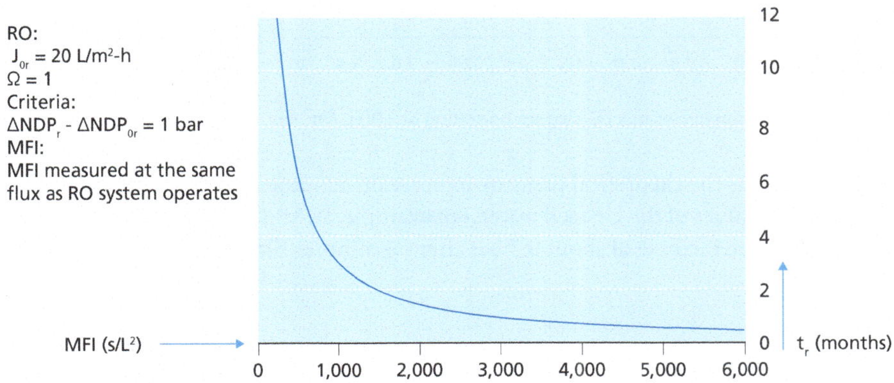

Figure 27 MFI versus time for an increase in net driving pressure of 1 bar at flux rate of 20 $L/m^2/h$ (for RO and also for MFI test).

For predicting the rate of fouling in RO and MF/UF systems (see next section), MFI -UF measured at the same and constant flux as applied in these systems has to be used.

Ideally, we should measure MFI with membranes having pores similar in size to RO membranes. Unfortunately, it is not possible to apply these membranes, because of concentration polarization in dead-end filtration, resulting in increasing osmotic pressure, which limits its scaling.

4.7 PREDICTING PRESSURE DEVELOPMENT IN MICRO- AND ULTRAFILTRATION SYSTEMS

Predicting the rate of fouling in MF and UF systems i.e., development of pressure during operation at constant flux seems to be less complicated than in RO systems, since filtration is conducted in dead-end mode. Hence, the deposition factor Ω is 1.0, completely eliminating the need to measure Ω. In addition, the MFI-UF values needs to be measured at a flux rate comparable to the actual flux in the UF. In ultrafiltration systems, the slope of the pressure development normally increases per filtration cycle, as illustrated in Figure 28 for a UF system operating at 100 $L/m^2/h$. The slope of cycle 10 is much higher than the slopes at cycle 5 and cycle 1. This increase in pressure versus time slope, is related to the effectiveness of the backwash between the cycles which does not fully restore the initial conditions. The effect of the backwash is not captured in the MFI test.

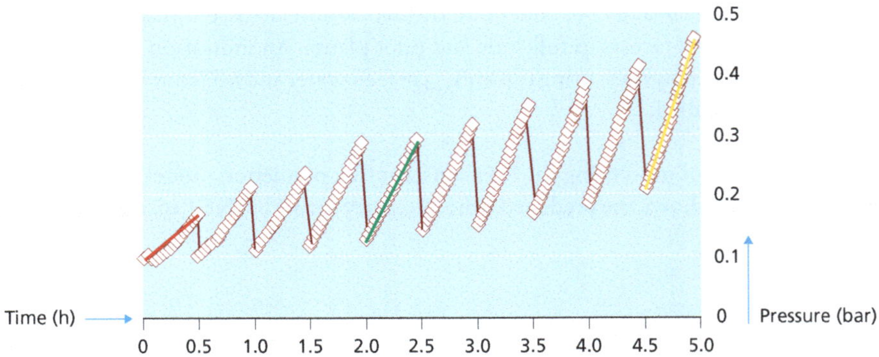

Figure 28 Filtration cycles in a UF system operating at 100 $L/m^2/h$.

Figure 29 presents the theoretical pressure increase development over the period of 1 hour for various MFI values of the UF feed water. For example, an MFI-UF value of 5000 s/L^2 will produce a pressure increase of about 0.1 bar after 60 minutes filtration.

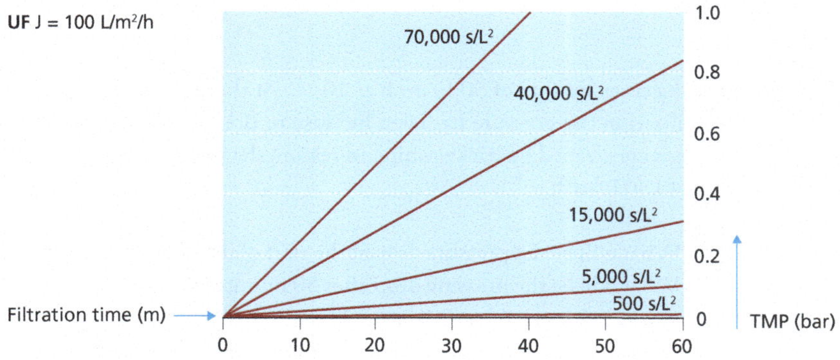

Figure 29 Pressure development (in bar) in an UF system ($J = 100\ L/m^2/h$) as a function of MFI-UF (in s/L^2) of the UF feed water and run time.

By measuring the MFI-UF values of UF feedwater at various flux rates (Figure 30 left), it is also possible to predict the pressure increase as a function of the filtration time (see Figure 30 right). This can help to optimize the operation of the ultrafiltration system.

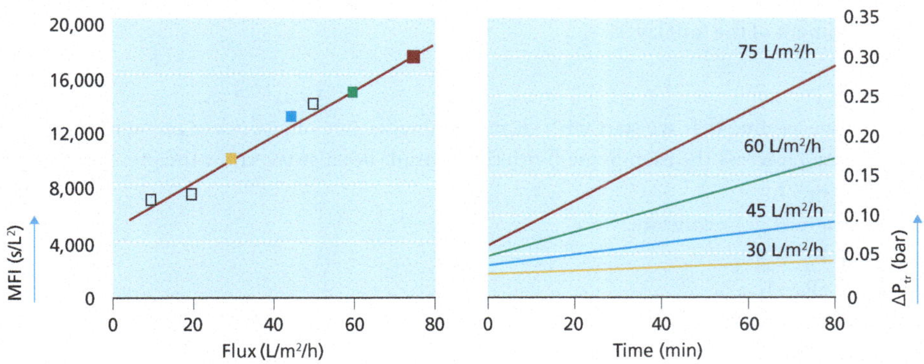

Figure 30 MFI values of UF feedwater measured at various flux rates (left) and predicted pressure increase as a function of filtration time for various filtration flux rates (right)

In UF systems, due to the flux dependency of the fouling potential (MFI-UF) the development of pressure increases over time, during one cycle, is not proportional with the flux to the power two (J^2) but almost proportional with the flux to the power three (J^3). This observation explains, at least for an important part, why fouling in UF systems tends to increase significantly with increasing flux rates.

Example 5

An UF plant runs at a constant flux of 50 L/m^2.h at 20 °C. At the start 0.1 bar is needed. After 30 minutes filtration the pressure has been increase to 0.2 bar. After this filtration cycle, the membranes are cleaned by backwashing to restore the permeability. We start a new cycle at a flux of 150 L/m^{2}h.

Question 1: What will be the pressure drop at the start of the new filtration cycle?

Question 2: What will be the pressure drop after 30 minutes filtration?
Guess: 0.2 bar, 0.6 bar, 0.9 bar, 1.2 bar, 1.6 bar?
Assume: Cake/gel filtration and no compression will occur
$\Delta P_t = P_t = \eta \cdot R_m \cdot J + \eta \cdot I \cdot J^2 \cdot t$

To answering these questions, we assume that: *i)* pressure drop across the membranes itself is not increasing during filtration and is constant, *ii)* increasing pressure drop during filtration is exclusively due to the formation of a cake/layer on the membrane surface, *iii)* cake/gel filtration without compression will occur.

To answering these questions, we look: first at the effect of higher flux on the pressure drop across the clean membranes, there after we look at the effect on the pressure drop due the development of the foul layer.

Answer 1:

When we assume that we start with clean membranes, we can calculate pressure drop at 150 L/m^2/h across the membranes with this formula because we know the pressure drop at 50 L/m^2/h:

$\Delta P_t = P_t = \eta \cdot R_m \cdot J = \text{constant}$

$P_{50} = \eta \cdot R_m \cdot J_{50}$

$P_{150} = \eta \cdot R_m \cdot J_{150}$

Where: P_{50} = pressure drop at 50 L/m^2/h and P_{150} = pressure drop at 150 L/m^2/h

J_{50} = 50 L/m^2/h and J_{150} = 150 L/m^2/h

$P_{150} / P_{50} = J_{150} / J_{50}$ = 150 L/m^2/h / 50 L/m^2/h = 3

So, the pressure-drop across the (clean) membrane = 3 · 0.1 bar = 0.3 bar

Answer 2:

We can calculate the pressure drop due to fouling at 150 L/m^{2}h with the formulas:

$P_{50Fouling} = \eta \cdot I \cdot (J_{50})^2 \cdot t$ = pressure drop due fouling at 50 L/m^{2}h

And,

$P_{150Fouling} = \eta \cdot I \cdot (J_{150})^2 \cdot t$ = pressure drop due fouling at 150 L/m^{2}h

Combining these equations results in:

$P_{150F} / P_{50F} = [\eta \cdot I \cdot (J_{150})^2 \cdot t] / [\eta \cdot I \cdot (J_{50})^2 \cdot t] = (J_{150})^2 / (J_{50})^2$

$P_{150F} / P_{50F} = [(3 \cdot J_{50})^2] / (J_{50})^2 = 9$

Consequently, the pressure loss due to the foul layer will be 9 times higher than at 50 L/m^2/h = 9 · 0.1 bar = 0.9 bar.

The total pressure loss will be 0.3 + 0.9 = 1.2 bar

Continues on page 121

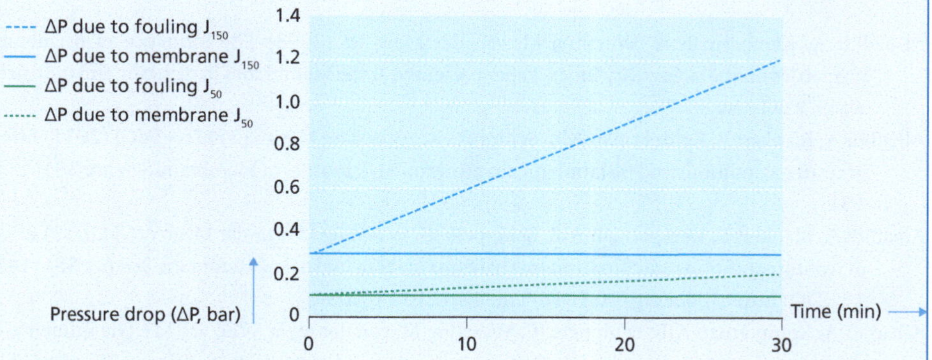

Answers without detailed calculations

Question 1. What will be the pressure at the start of a new filtration cycle?

At 150 L/m²/h pressure (loss) due to membrane resistance will be (150 / 50) = 3 times higher or 0.3 bar (because: 0.1 bar is pressure loss due (clean) membrane resistance at 50 L/m²/h)

Question 2. What will be the pressure after 30 minutes filtration?

At 150 L/m²/h pressure due to fouling after 30 minutes will be (150 / 50) times higher than at 50 L/m²/h = 9 · 0.1 bar = 0.9 bar.

Remark: 0.1 bar is the pressure loss at 50 L/m²/h, due development cake/foul layer. (0.2 bar – 0.1 bar) = 0.1 bar.

Consequently, the total pressure loss will be 0.3 + 0.9 = 1.2 bar

$\Delta P_t = \eta \cdot R_m \cdot J + \eta \cdot I \cdot J^2 \cdot t$

4.8 REFERENCES

Al-hadidi A, Kemperman A, Wessling M, van der Meer W (2008) The influence of membrane properties on the silt density index. Paper presented at the Membranes in drinking and industrial water, Toulouse

Alhadidi A, Blankert B, Kemperman AJB, Schippers JC, Wessling M, van der Meer WGJ (2011) Effect of testing conditions and filtration mechanisms on SDI. Journal of Membrane Science 381: 142-151

Alhadidi A, Blankert B, Kemperman AJB, Schippers JC, Wessling M, van der Meer WGJ (2011) Effect of testing conditions and filtration mechanisms on SDI. Journal of Membrane Science 381: 142-151 DOI https://doi.org/10.1016/j.memsci.2011.07.030

Alhadidi A, Kemperman AJB, Schippers JC, Wessling M, van der Meer WGJ (2011) The influence of membrane properties on the Silt Density Index. Journal of Membrane Science 384: 205-218 DOI https://doi.org/10.1016/j.memsci.2011.09.028

Alhadidi A, Kemperman AJB, Schurer R, Schippers JC, Wessling M, van der Meer WGJ (2012) Using SDI, SDI+ and MFI to evaluate fouling in a UF/RO desalination pilot plant. Desalination 285: 153-162 DOI https://doi.org/10.1016/j.desal.2011.09.049

Anderson DM, Boerlage SFE, Dixon MB (2017) Harmful Algal Blooms (HABs) and Desalination: A Guide to Impacts, Monitoring and Management. In: UNESCO IOCo (ed) IOC Manuals and Guides No 78 UNESCO, Paris, pp. 538.

ASTM D4189 - 14 (2014) Standard Test Method for Silt Density Index (SDI) of Water ASTM International, West Conshohocken, PA.

Belfort G, Davis RH, Zydney AL (1994) The behavior of suspensions and macromolecular solutions in crossflow microfiltration. Journal of Membrane Science 96: 1-58

Belfort G, Marx B (1979) Artificial particulate fouling of hyperfiltration membranes II. Analyses and protection from fouling. Desalination 28: 13-30

Boerlage (2008) Understanding the Silt Density Index and Modified Fouling Indices (MFI0.45 and MFI-UF) Desalination and Water Reuse Quarterly, May–June.

Boerlage SFE (2001) Scaling and Particulate Fouling in Membrane Filtration Systems Swets&Zeitlinger Publishers, Lisse

Boerlage SFE (2007) Understanding the SDI and Modified Fouling Indices (MFI0.45 and MFI-UF). Paper presented at the IDA World Congress on Desalination and Water Reuse 2007 - Desalination: Quenching a Thirst Maspalomas, Gran Canaria - Spain, 21-26 October 2007 2007

Boerlage SFE, Kennedy MD, Aniye MP, Abogrean EM, El-Hodali DEY, Tarawneh ZS, Schippers JC (2000) Modified Fouling Index ultrafiltration to compare pretreatment processes of reverse osmosis feedwater. Desalination 131: 201-214

Boerlage SFE, Kennedy MD, Aniye MP, Schippers JC (2003) Applications of the MFI-UF to measure and predict particulate fouling in RO systems. Journal of membrane science 220: 97-116

Boerlage SFE, Kennedy MD, Tarawneh Z, Abogrean E, Schippers JC (2003) The MFI-UF as a water quality test and monitor. Journal of Membrane science 211: 271-289

Bonnelye V, Sanz MA, Durand J-P, Plasse L, Gueguen F, Mazounie P (2004) Reverse osmosis on open intake seawater: pre-treatment strategy. Desalination 167: 191-200 DOI https://doi.org/10.1016/j.desal.2004.06.128

Carman PC (1937) Fluid flow through granular beds Trans Instn Chem Engrs 15: 32-48

Carman PC (1938) Fundamental principles of industrial filtration (A critical review of present knowledge). Trans Instn Chem Engrs 16: 168-188

Chellam S, Wiesner MR (1998) Evaluation of crossflow filtration models based on shear-induced diffusion and particle adhesion: Complications induced. Journal of Membrane Science 138: 83-97

Crittenden JC, Trussell RR, Hand DW, Howe KJ, Tchobanoglous G (2005) Membrane filtration. In: John Wiley & Sons I (ed) Water Treatment: Principles and Design / MWH, Second edn:955-1034.

Davis RH (1992) Modeling of fouling of crossflow microfiltration membranes. Separation and purification methods 21: 75-126

Davis RH, Sherwood JD (1990) A similarity solution for steady-state cross flow microfiltration. Chemical Engineering Science 45: 3203-3209

Escobar L, Sellerberg W, Sanchez D, Pastrana F, Wachinski A (2009) Detailed analysis of the silt density index (SDI) Results on desalination and wastewater reuse applications for reverse osmosis technology evaluation. Paper presented at the International forum on marine science and technology and economic development - Asia-Pacific Desalination conference, Qingdao, China, 8-10 July 2009 2009

Hoek EMV, Kim AS, Elimelech M (2002) Influence of Crossflow Membrane Filter Geometry and Shear Rate on Colloidal Fouling in Reverse Osmosis and Nanofiltration Separations. Environmental engineering science 19: 357-372

IUPAC (1971) Manual of symbols and terminology for physicochemical quantities and units, Appendix II definitions, Terminology and symbols in colloid and surface chemistry D. H. Everet

Jiang T (2007) Characterization and modelling of soluble microbial products in membrane bioreactors Ghent University, Ghent

Mosset A, Bonnelye V, Petry M, Sanz MA (2008) The sensitivity of SDI analysis: from RO feed water to raw water. Desalination 222: 17-23

Nahrstedt A, Camargo Schmale J (2008) New insights into SDI and MFI measurements. Water Science and Technology: Water Supply 8: 401-412

Rachman RM, Ghaffour N, Waly F, Amy GL (2013) Assessment of Silt Density Index (SDI) as fouling propensity parameter in Reverse Osmosis (RO) desalination systems. Desalination and Water Treatment 51: 1091-1103 DOI doi:10.1080/19443994.2012.699448

Ruth BF, Montillon GH, Montonna RE (1933) Studies in Filtration I. Critical Analysis of Filtration Theory. Industrial & Engineering Chemistry 25: 76-82 DOI 10.1021/ie50277a018

Salinas Rodríguez SG (2011) Particulate and organic matter fouling of SWRO systems: Characterization, modelling and applications CRC Press/Balkema, Delft

Salinas Rodríguez SG, Amy GL, Schippers JC, Kennedy MD (2015) The Modified Fouling Index Ultrafiltration Constant Flux for assessing particulate/colloidal fouling of RO systems. Desalination 365: 79-91 DOI 10.1016/j.desal.2015.02.018

Salinas Rodriguez SG, Sithole N, Dhakal N, Olive M, Schippers JC, Kennedy MD (2019) Monitoring particulate fouling of North Sea water with SDI and new ASTM MFI0.45 test. Desalination 454: 10-19 DOI https://doi.org/10.1016/j.desal.2018.12.006

Sallangos O, Kantilaftis E (2001) Operating experience of the Dhekelia seawater desalination plant. In: EDS (ed) European Conference on Desalination and the Environment: Water Shortage EDS, Lemesos, Cyprus.

Schippers JC (1989) Vervuiling van hyperfiltratiemembranen en verstopping van infiltratieputten Keuringinstituut voor waterleidingartikelen KIWA N.V., Rijswijk

Schippers JC, Folmer HC, Kostense A (1980) The effect of pre-treatment of river rhine water on fouling of spiral wound reverse osmosis membranes7th International Symposium on Fresh water from the Sea, pp. 297-306.

Schippers JC, Hanemaayer JH, Smolders CA, Kostense A (1981) Predicting flux decline or reverse osmosis membranes. Desalination 38: 339-348

Schippers JC, Salinas Rodriguez SG, Boerlage SFE, Kennedy MD (2014) Why MFI is edging SDI as a fouling indexDesalination & Water Reuse Faversham House with the cooperation of the International Desalination Association, UK, pp. 28-32.

Schippers JC, Verdouw J (1980) The modified fouling index, a method of determining the fouling characteristics of water. Desalination 32: 137-148

Sioutopoulos DC, Yiantsios SG, Karabelas AJ (2010) Relation between fouling characteristics of RO and UF membranes in experiments with colloidal organic and inorganic species. Journal of Membrane Science 350: 62-82

Song LF, Elimelech M (1995) Particle deposition onto a permeable surface in laminar flow. Journal of Colloid and Interface Science 173: 165-180

Stumm W, Morgan JJ (1996) Aquatic chemistry: chemical equilibria and rates in natural waters, 3rd edn Wiley interscience publication, New York

Tung, K.-L., Li, Y.-L., Hwang, K.-J. & Lu, W.-M. (2008). Analysis and prediction of fouling layer structure in microfiltration. Desalination, 234, 99-106.

Underwood AJV (1926) A critical review of published experiments on filtration. Trans Inst Chem Eng 4: 19

Villacorte LO (2014) Algal Blooms and Membrane Based Desalination Technology CRC Press/ Balkema, Leiden

Yiantsios SG, Sioutopoulos D, Karabelas AJ (2005) Colloidal fouling of RO membranes: an overview of key issues and efforts to develop improved prediction techniques. Desalination 183: 257-272

Chapter 5

Organic and biological fouling

In S. Kim, Abayomi Babatunde Alayande and Thanh-Tin Nguyen

The main learning objectives of this chapter are the following:

- Understand both organic and biofouling in the RO process
- Have an overview of the impact of organic and biofouling on RO performance and suggest appropriate pretreatment processes
- Know the various RO feedwater biofouling potential prediction techniques
- Be able to apply conventional and new membrane cleaning strategies
- Know the various analytical tools for membrane fouling characterization
- Have an overview of future organic and biofouling mitigation strategies

5.1 WHAT IS ORGANIC FOULING AND BIOFOULING?

Membrane fouling is the attachment, accumulation, or adsorption of unwanted material on the surface of the membrane or in its pores to impede the effective functioning/performance of the membrane (Malaeb and Ayoub, 2011). Fouling of membranes is detrimental to the system because it causes severe flux decline, increases the cost of membrane replacement, increases energy demand, increases cleaning frequency, shorten membrane lifetime, and affects the water quality of the permeate. Membrane fouling has been a major bottleneck that hinders the wide application of membrane-based technologies for water treatment, especially seawater desalination. Fouling can occur in several forms: (1) organic fouling caused by macromolecular organic compounds such as polysaccharides, protein, and humic acid; (2) inorganic fouling involving the scaling with the crystallization of sparingly soluble mineral salts; (3) colloidal fouling with the deposition of particles; (4) biofouling from bacteria attachment. In this section, organic fouling is introduced.

 Seawater Reverse Osmosis Desalination
Editors: Sergio Salinas, Jan Schippers, Gary Amy, In Kim, Maria Kennedy
doi: 10.2166/9781780409863_0125

Table 1 The standard organic matter compounds

Type	Name	Molecular weight and shape	Charged	Reference
Humic substance	Suwannee River Humic Acid (HA)	0.5-5 kDa, globular molecule (linear under high pH)	Negatively charged	(Drouiche, et al., 2009, She, *et al.*, 2016, Tang,*et al.*, 2011)
	Suwannee River Fulvic Acid (FA)	0.5-5 kDa globular molecule (linear under high pH)	Negatively charged	
	Aldrich humic acid (AHA)	> 100 kDa globular molecule (linear under high pH)	Negatively charged	
Protein	Bovine serum albumin (BSA)	~66.4 kDa	pHIEP= 4.7	(She, Wang, Fane and Tang, 2016, Venugopal and Dharmalingam, 2012)
	Bovine immunoglobulin G	155 kDa	pHIEP= 6.6	
	Bovine hemoglobin	68 kDa	pHIEP= 7.1	
	Bovine pancreas ribonuclease A	13.7 kDa	pHIEP= 7.8	
	Lysozyme	14.4 kDa	pHIEP= 11	
Polysaccharides	Alginic acid sodium salt (Alginate)	200-2000 kDa, extended random coil	Negatively charged	(Ibáñez, *et al.*, 2013, She, Wang, Fane and Tang, 2016)
	Xanthan and gellan	100-2500 kDa, linear	Negatively charged	
	Schizophyllan	400-500 kDa, rigid rod-like	Neutral	
Amino Acid	Tyrosine	181.19 g/mol	-	(Drouiche, Grib, Abdi, Lounici, Pauss and Mameri, 2009)
	Tryptophan	204.23 g/mol	-	
Others	Transparent exopolymer particles (TEP)	Transparent, sticky and amorphous substances. Exists in different forms (e.g. strings, disks, sheets, fibers)	-	(She, Wang, Fane and Tang, 2016)

Note: pH IEP is the pH at which a particular molecule carries no net electrical charge or electrically neutral in the statistical mean. "-"indicates not available

Organic fouling is the dissolved components and colloids such as humic and fulvic acids, hydrophilic and hydrophobic materials and proteins which would attach to the membrane by adsorption. Organic fouling is simply caused by bulk organic matter (OM) present in the feed water that may be adsorbed onto the membrane surface as gel layer. Moreover, the biodegradable organic matter (BOM) deposited on the membrane surface can be utilized by micro-organisms as nutrients; thereby contributing to biological growth. These organic matters, namely organic macromolecules usually include in polysaccharides, humic substances, proteins lipids, nucleic acids and amino acids, organic acids, and cell components (Jeong, et al., 2016, Wang, *et al.*, 2014). Natural organic matter (NOM), a complex mixture of organic compounds often presents in surface water and seawater (Kim and Dempsey, 2013). As the previous studies, NOM can be either hydrophobic component (e.g., humic substance) or hydrophilic component (e.g., carbohydrates, proteins, sugars and amino acids) (Matilainen, *et al.*, 2011). In detail, NOM consists of a range of different compounds which are the different molecular size and charged (Table 1).

Figure 1 presents the number of publications related to three RO organic foulants commonly studied in the past 10 years. As denoted in Figure 1, a large number of research interests involved in bovine serum albumin (BSA), alginate and humic acid as organic foulants. As mentioned beforehand (Table 1), BSA is a type of protein whilst alginate is a typical representative of the polysaccharide. The summary of the representative studies on RO membrane organic fouling behavior under different scenarios is presented in Table 2. This summary gives a conclusion that feeds water chemistry, foulant-surface interaction, and foulant-foulant interactions are three important factors affecting organic fouling.

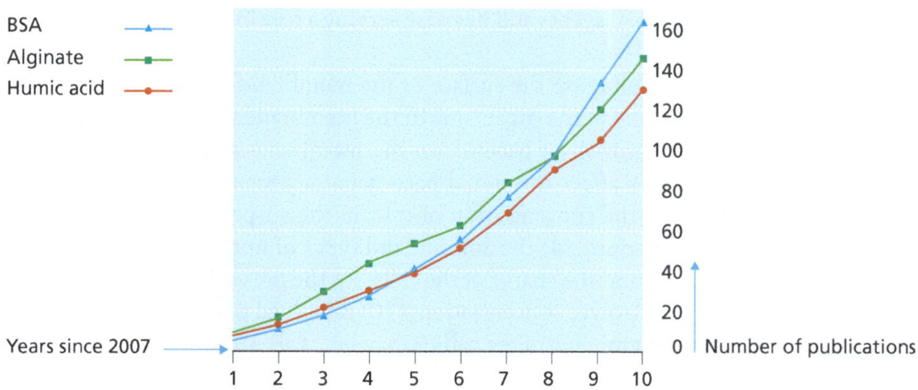

Figure 1 The number of publications related to three common RO organic foulants studied in the past 10 years (Jiang, *et al.*, 2017)

Table 2 Some studies on RO organic fouling under various scenarios

Organic matter	Category	Main findings
Alginate	Polysaccharide	Membrane fouling becomes more severe with decreasing pH, increasing ionic strength, and the addition of calcium ions (Lee, *et al.*, 2006).
Octanoic acid	Fatty acid	Either change of pH or the presence of calcium ions can induce the octanoic acid fouling behavior (Ang,*et al.*, 2006).
Humic acid	Humic substance	Fouling is induced by membrane-foulant interaction (hydrophobic interactions between organic matters and membrane) and foulant-foulant interactions (between the organic matters) (Yu, *et al.*, 2010).
Alginate, BSA, NOM, octanoic acid	Polysaccharide, protein, humic substance, fatty acid	Membrane fouling by alginate was predominant in foulant aggregate size. In detail, smaller and more compact aggregates can cause more considerable flux decline (Ang, *et al.*, 2011)
BSA	Protein	Fouling is mainly controlled by foulant-deposited-foulant interaction (Wang and Tang, 2011).

While biofouling, on the other hand, is the adhesion and reproduction of living microorganisms to the membranes and causes biofilm formation (Creber, *et al.*, 2010). Amongst the various types of fouling, biofouling is considered the biggest challenge because if a tiny fraction of microbial cells enter the system they can grow and multiply at the expense of biodegradable substances, thereby attaching to the membrane and subsequently developing to form a biofilm (Flemming, 2002). Biofouling is mainly composed of bacteria and extracellular polymeric substances (EPS). EPS are composed of polysaccharides, proteins, and nucleic acids and forms a matrix around the bacterial cells protecting them from biocides and antimicrobial agents and likewise serving a role in adhesion.

The processes of bacterial fouling on the surface of the membrane can be mainly divided into three phases: (1) transport of the organisms to the membrane surface, (2) attachment to the substratum, and (3) growth of bacteria on the membrane surface (Goosen, et al., 2005). The following factors affect the initial bacterial attachment to a surface: (1) the types of microorganism, (2) the concentration of cells in the suspension, (3) the stage in the growth cycle of the bacterium, (4) the amount and types of nutrients available for the cells, (5) the hydrophobicity and the charge of the cells, (6) the presence of a glycocalyx, (7) pH, (8) temperature, (9) electrolyte concentration, (10) dissolved organic substances, (11) characteristics of the outer membrane of the cells (Ghayeni, et al., 1998).

5.2 IMPACT OF ORGANIC FOULING AND BIOFOULING ON PLANT OPERATION

Organic fouling and biofouling have a diverse consequence on seawater desalination plant operation, most especially the membrane-based system. They have a direct effect on both the process as well as the physical components of the plant. Organic fouling and biofouling usually occur together because the organic contents provide the food necessary for biological growth. Where there is organic fouling there is most likely going to be the occurrence of biological fouling. Likewise, if the presence of organic components can be eliminated or limited in the feed water, the occurrence of biological fouling could be limited.

The major impact of both organic fouling and biofouling is membrane flux decline. As organic and biological foulants accumulate on the membrane surface, they form some films on the membrane. In most cases, these films are impermeable causing low passage of water and/or the complete blockage of water. This eventually leads to an increased system pressure to compensate for the resistance caused by the fouling layer and in turn an increased energy requirement. Microorganisms also produce some byproducts which could cause damage or degrade the membrane. Accumulation of both organic and biofouling can hinder the back diffusion of salt because of its interaction with polymeric substances produced by the microbial cells thereby enhancing concentration polarization (CP) (Herzberg, *et al.*, 2009, Matin, *et al.*, 2011). Organic and biological foulants can entrap dissolved ions, thereby increasing their concentration on and within the membrane and helping in the forward movement through the membrane (Hoek and Elimelech, 2003). Apart from the increased salt passage caused by CP, increased CP can also reduce flux by reducing the net driving pressure gradient. Also, organic and biofouling increase the frequency of cleaning and cause a huge burden on plant operation in terms of cleaning chemical cost and the need for membrane replacement.

Table 3 Summary of the studies on different pretreatment methods

Pretreatment	Objective	Capacity	Main finding
Coagulation + Granular media filtration Low-pressure membrane filtration (MF)	Organic matter Micro-organism	Pilot-scale	Using microfiltration pretreatment presented a lower silt density index (SDI) compared to coagulation + granular filtration, with the average SDI_{15} being 2.5 and 3.5, respectively. Microorganism removal in terms of bacteria and picophytoplankton was performed highly better when using low-pressure membrane filtration However, coagulation+granular media filtration exhibited a higher dissolved organic removal compared to MF (Remize, *et al.*, 2012)
Coagulation+ Sedimentation as pretreatment for ultrafiltration	Organic matter Micro-organism	Lab-scale	Ferric chloride was tested as a coagulant, and its required dose was evaluated by a Jar test analysis. A dose of 50 mg/L reduced 38% of the organic load. This chemical pretreatment can be as possible pretreatment for ultrafiltration followed by RO desalination (Friedler, *et al.*, 2008)
Granular-activated carbon (GAC)	Dissolved organic matter (DOM)	Pilot-scale	GAC pretreatment in pilot-scale columns resulted in 80–90% DOM removal. The DOM (e.g., hydrophobic and biodegradable components) which constitutes the fraction primarily causing organic fouling was mainly removed by GAC pretreatment (Gur-Reznik, *et al.*, 2008)

5.3 PRETREATMENTS

Organic matters can be removed in RO feed water using coagulation/clarification (sedimentation) or activated carbon filtration. Normally, coagulants are added to seawater (e.g., coagulation process), followed by a sedimentation process (e.g., clarifier) to remove large suspended solids, colloid and organic. However, the effluent quality from clarifier is not low enough in turbidity, which can be directly sent to an RO system; therefore, rapid sand filtration, multi-media filters or even membrane filtration (e.g., MF or UF) is used to enhance removal of suspended particles. This section introduces the studies resulting in different pretreatment methods for organic removal as shown in Table 3.

Some reports have examined the efficiency of applying UF membrane as a pretreatment step for membrane-based seawater desalination (Alhadidi, et al., 2012, Halpern, et al., 2005, Pearce, et al., 2004). Among the various membrane pretreatments for seawater desalination such as nanofiltration (NF), UF, and microfiltration (MF), UF membranes have been considered as a preferred pretreatment membranes for desalination plants because unlike the conventional pretreatments methods UF membrane pretreatment provides a lower suspended solids and biological contents; the frequency of desalination membrane cleaning is reduced, enabling longer operation time and reducing the cost of membrane cleaners; lesser fouling on desalination membranes reduces pressure drops hence, reducing energy requirement and elongate membrane life; increase water production; reduces the cost of chemical and sludge handling; reduces the plant footprint size and overall reducing capital cost (Corral, *et al.*, 2014). In a study by Busch and colleagues (Busch, *et al.*, 2009), when UF was used as pretreatment to reverse osmosis (RO) system, the UF effluent was NTU > 1, SDI < 3, and 95% recovery. When such effluent was fed into the RO system, the RO membrane was less prone to fouling making membrane cleaning frequency fewer. In a field testing program conducted at Ashdod on the Mediterranean to compare the performance of RO system using the conventional pretreatment and the UF membrane process (Glueckstern, *et al.*, 2002). The intake SDI for the Mediterranean was above 6.5 and was reduced to 2.6-3.8 and 2.1-3.0 after conventional and UF membrane pretreatments were used respectively. Recent studies have coupled coagulants with UF in order to improve the efficiency of the process (Dong, *et al.*, 2007, Li, et al., 2011). The results from these studies are remarkable, proving that the efficiency of UF membrane can be further improved by the addition of coagulants. Xia *et al.*, (2007) in a study carried out in North China reported that when hollow fiber UF membrane was coupled with powdered activated carbon (PAC) turbidity was lowered to < 0.2 NTU. The authors reported that the hybrid process of PAC/UF allowed for the removal of 41% of COD_{Mn}, 46% removal of DOC and 57% decrease in UV_{254} absorbance.

5.4 PREDICTION OF BIOFOULING POTENTIAL IN RO FEEDWATER

Biofouling in the membrane process is an unavoidable occurrence, and membrane clean-in-place (CIP) is often used to sustain water permeability. In practice, the cleaning frequency is governed by the biofouling potential of feed water and the operational conditions (flux, pressure) of the RO system (Abushaban, *et al.*, 2019). Consistent monitoring of microbial activities of the feedwater is an important aspect of desalination plant operation to respond early to any rise in microbial activity before causing fouling on the membrane. Early action in predicting biofouling potential appeared to be more important than changing operating conditions (Abushaban, Salinas-Rodriguez, Dhakal, Schippers and Kennedy, 2019). Numerous monitoring/prediction techniques are available in the SWRO process for biofouling potential measurement. Examples of such techniques are analysis based on process performance, fluorometry, ultrasonic time-domain reflectometry, biosensors/nano-sensors/microbial sensing membranes, electrical potential measurement, membrane fouling simulator, and feedwater biological parameters (Nguyen, *et al.*, 2012). Apart from analyzing the biological parameters of feedwater, most of the other methods are either not applicable in real-plant operation, cannot discriminate between biofouling potential and other types of fouling potential, or are used to monitor membrane biofouling rather than early prediction. Beyond biofouling monitoring, biofouling prediction would put measures in place to mitigate biofouling on the membrane thereby preventing operation shutdown because of CIP procedure. Biofouling potential can be predicted using biological parameters such as adenosine triphosphate (ATP) content, number of culturable microbes in feedwater quantified in colony forming units (CFU), total cell count, and assimilable organic carbon (AOM), fraction of biodegradable dissolved organic carbon (BDOC) that is more readily assimilated by microorganisms than other types of organic carbon (Abushaban *et al.*, 2019, Hobbie, *et al.*, 1977, Miller, *et al.*, 2020).

5.4.1 Colony forming units (CFU)

Cultural bacterial concentration measurement in feedwater can be used to predict biofouling potential. The number of culturable bacteria is expressed as CFU. CFU is determined by diluting feedwater as appropriate or concentrating by filtering a known quantity of feedwater through a filter (0.2 μm pore size) and cultured on a solid plate growth medium (depending on microbial type). Colonies formed on the growth medium are then enumerated by direct counting. CFU analysis is an inexpensive biofouling potential measurement that can be used for pretreatment and permeate water quality assessment. However, CFU counts do not analyze non-culturable microorganisms and laborious.

5.4.2 Total direct cell (TDC) count

TDC counting is a direct microbial enumeration technique used to quantify both culturable and non-culturable microbial cells. This could be done by directly staining the feedwater with a fluorescent dye such as acridine orange, 4',6-diamidino-2-phenylindole (DAPI) etc. or filtering it through a filter and then staining the filter. The stained water or filter can be viewed under a fluorescent microscope. Similarly, stained solution can also be measured using flow cytometry (FCM) (Farhat, *et al.*, 2018). TDC is a fast technique to enumerate the total microbial count. However, this method may not directly translate to biofouling tendency because the conditions in the feedwater may be unfavorable for the cell growth, so an increase in TDC count is not indicative of an increase in biofouling potential.

5.4.3 Adenosine triphosphate (ATP) content

ATP content measurement is a culture-independent method for biofouling potential. ATP is a component of every living microbial cell that plays a vital role in energy transfer. Increased ATP content is used to predict the biofouling potential in seawater desalination because it is fast, reliable, and accurate in seawater with low ATP level (Abushaban, *et al.*, 2019). Figure 2 illustrates the procedure of bacterial regrowth potential method using microbial adenosine triphosphate (ATP).

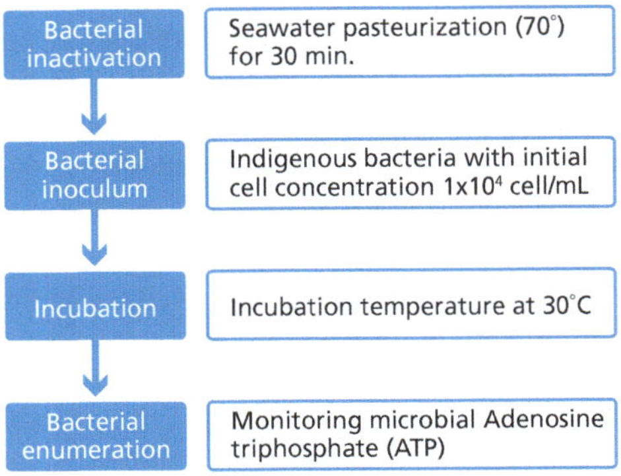

Figure 2 The procedure of measuring BGP in seawater based on microbial ATP (Abushaban *et al.*, 2019).

5.4.4 Assimilable organic carbon (AOC)

In seawater, AOC is a bioassay with two well-known bacterial strains (*Vibrio fischeri and Vibrio harveyi*) used to assess biofouling potential in water with nutrients. But due to the inability of these strains to assimilate all available nutrient for growth, the use of an indigenous microbial consortium is encouraged to diversify nutrient assimilation range compared to pure strain(s). In seawater, AOC accounts for 30% of low molecular weight compounds of DOC, which played a significant role in biofouling by enhancing microbial growth and thus boosted the production of soluble microbial products (SMPs) and extracellular polymeric substances (EPS) (Yin, *et al.*, 2020). As reported for bacterial growth in seawater, the assimilable organic carbon content ranges from about 50 to 400 μgC/L (Javier, *et al.*, 2020). However, biofouling may develop with an AOC of < 10 μgC/L (Vrouwenvelder and Van der Kooij, 2001). AOC measurement typically involves three steps: (1) inoculum preparation, (2) media preparation, and (3) AOC analysis. AOC analysis is done by analyzing total ATP content, turbidity, total CFU count, or TDC count.

5.5 MEMBRANE CLEANING

Membrane cleaning is an inevitable procedure in membrane-based desalination due to changes in operational conditions. Cleaning methods consist of physical, chemical and physiochemical. In practice, physical cleaning methods followed by chemical cleaning methods are frequently used for the maintenance operation. However, to enhance the recovery of irreversible fouling, cleaning in place (chemical solution as acid, base, chelating agents, surfactants) can be used to get more sufficient. Therefore, this section focuses on introducing the chemical cleaning method.

5.5.1 Chemical cleaning

Some studies have been done to uncover the most effective chemical cleaning agents and conditions. It was reported that the efficiency of membrane cleaning depends greatly on the cleaning chemicals, pH, temperature, cleaning time, and the cost implications. The foulants types are also important in the selection of cleaning agent (Madaeni, *et al.*, 2001). Plant operator needs to be careful in the use of chemical cleaning agents because they could produce by-products that are toxic to the ecosystem and are sometimes harmful to the membrane materials (Baker and Dudley, 1998).

Over the years, the use of acids and alkaline as membrane cleaners have to help maintain the performance of the system and the quality of water produced. They also reduce the operating cost by the reduction of energy required for water production. Compared to other cleaning agents like chlorine-based agents less membrane damage is caused by acid and alkaline cleaning. For organic fouling and biofouling, alkaline solutions such as sodium hydroxide have proven to be very effective cleaning agent because they can saponificate fat and solubilize protein (a major component in both organic and biofouling) (Al-Amoudi and Lovitt, 2007, Lee, *et al.*, 2001, Sohrabi, *et al.*, 2011). Acidic solutions such as hydrochloric acid, nitric acid, and sulfuric acid, on the other hand, are more effective in removing scaling in membrane because of their ability to dissolve precipitations (Gan, et al., 1999).

5.5.2 Acid and base coupled with chelating agents

In detail, alkaline cleaning (e.g., NaOH) is suitable for organic fouling removal. Other chemical cleaning agents include metal chelating agents, surfactants and enzymes (Chlistunoff, 2005). Another one, disinfectants (O_3), oxidants (H_2O_2, $KMnO_4$) or sequestration agents (Ethylene diamine tetraacetic acid-EDTA) are often used for biological fouling of membranes (Lin, *et al.*, 2010).

Table 4 Usual cleaning agent for organic and biological fouling (Fritzmann, *et al.*, 2007)

Types of fouling	Chemical agent
Organic	NaOH solutions, chelating agents, and surfactants
Biofilm	NaOH solutions, chelating or sequestration agents, surfactants and disinfectants

CIP: clean-in-place is a cleaning method that applies directly to on-site (Figure 3). Some advantages are obtained using CIP as follows: (1) The membrane modules are in situ when cleaning is conducted; therefore, there is no need for a second set of the membrane (2) Cleaning with CIP is faster than that with off-site (3) Less expensive compared to off-site cleaning. Effective cleaning is a function of pH, temperature, and the use of cleaning solutions. In case of using “wrong” cleaning chemical, the situation can become worse. Exemption for calcium carbonate or iron oxide (scaling) present in membrane module which must be use acid for first cleaning, most of the manufacturers always recommend using the alkaline for organic foulant cleaning. This is due to the fouling may become irreversible due to the reaction of microbial extra-cellular material to the acid condition. For maximum effectiveness, the temperature of the cleaning solutions needs to be retained above 25 °C. It means that elevating the temperature of the cleaning solution will assist in organic removal from the membrane surface. Therefore, there are limits to temperature as a function of pH, as shown in Table 5 for Dow polyamide composite membranes.

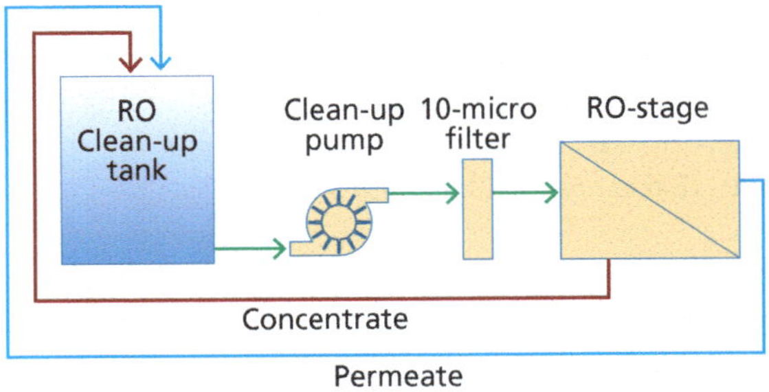

Figure 3 RO clean-up skid

Table 5 Temperature and pH limitations for DOW water solution polyamide composite membranes (Giannasi and health, 2007)

Temperature (°C)	pH
25	1-13
35	1-12
45	1-10.5

For organic and biological fouling, DOW manufacturer recommends a specific cleaning procedure as follows: (1) Make up the desired high pH cleaning solution selected Table 6 Introduction of the cleaning solution (3) Recycle the cleaning solution for 30 min (In case of changing color of the cleaning solution as cleaning, there need to dispose of the cleaning solution and replace a fresh solution) (4) Soak (5) High flow pumping (6) Flush out (7) Finally use of HCl cleaning solution at pH 2 (from step 2- 6).

Table 6 Organic and biological fouling cleaning solution (Giannasi and health, 2007)

Cleaning solution	Solution	
	Organic fouling	Biofouling
Preferred	0.1 wt% NaOH, pH 12, 30 °C maximum, followed by: 0.2% HCl, pH 2, 45 °C maximum	0.1 wt% NaOH, pH 13, 35 °C maximum
Preferred	0.1 wt% NaOH, 0.025 wt% Na-DDS, pH 12, 30 °C maximum, followed by: 0.2% HCl, pH 2, 45 °C maximum	0.1 wt% NaOH, 0.025 wt% Na-DDS, pH 13, 35 °C maximum
Alternate	0.1 wt% NaOH, 1 wt% Na_4EDTA, pH 12, 30 °C maximum, followed by: 0.2% HCl, pH 2, 45 °C maximum	0.1 wt% NaOH, 1 wt% Na_4EDTA, pH 12, 35 °C maximum

Note: NaOH is sodium hydroxide; Na-DDS is sodium salt of dodecyl sulfate (sodium lauryl sulfate); Na_4EDTA is the tetrasodium salt of ethylene diamine tetraacetic acid

Meanwhile, another manufacturer (Hydranautics) proposes the cleaning procedure for complex fouling (organic and biological fouling) the following steps: (1) Flushing with permeate water with introduction of non-oxidizing biocide (DBNPA) at the end of the flushing (2) CIP at high pH selected Table 7 Flushing with permeate until pH on the concentrate side is below pH 8.5 (4) CIP at low pH (5) Acid flushing with permeate.

Table 7 Organic and biological fouling cleaning solution (Hydranautics, 1998)

Cleaning solution	Solution	
	Organic fouling	Biofouling
Preferred	2.0 wt% STPP, 0.8 wt% Na-EDTA, pH 10	2.0 wt% STPP, 0.8 wt% Na-EDTA, pH 10
Preferred	2.0 wt% STPP, 0.025 wt% Na-DDBS, pH 10	2.0 wt% STPP, 0.025 wt% Na-DDBS, pH 10
Preferred	0.1 wt% NaOH 0.03 wt% SDS, pH 11.5	0.1 wt% NaOH 0.03 wt% SDS, pH 11.5

Note: STTP is sodium tripolyphosphate, Na-EDTA is sodium salt of ethylamine diamine tetraacetic acid, Na-DDBS is sodium salt of dodecylbenzene sulfonate, SDS is sodium dodecyl sulfate.

5.5.3 Biocides

Even though chemical cleaning is an important aspect of plant operation, and they can be relied on to remove a considerable amount of foulants on the membrane they cannot be relied on to effectively kill the microbial cells responsible for biofouling on the membrane. Microorganisms no matter how small can be deposited in other areas such as the pipes or even on the membrane and after some time with the introduction of nutrients they can recover to produce more biofilms on the membrane. Therefore the application of biocides in controlling biofouling on the membrane is a common and an important practice (Ridgway and Flemming, 1988). Nonetheless, the efficacy of biocides depends on (1) types and level of bioactivity in the system, the frequency of dosing, the contact time, pH and concentration of organics and inorganics in the system etc. (Gogate, 2007). Most of the time biocidal cleaning is usually a follow-up step after chemical cleaning procedure because biocides can reduce or eradicate viable cells leftover after chemical cleaning and prevent future reoccurrence. It is important to note that biocides are more effective in the absence of organic foulants. The presence of organic foulants would prevent the direct contact of biocides with the viable microbial cells. Biocides include all disinfecting agents capable of inactivating microorganisms. They can remove, destabilize, and disinfect biofilms on the membrane surface. Also, biocidal cleaning, if could not completely remove the foulant films, would make the leftover film permeable hence restoring the water flux. Examples of biocides include chlorine, bromine, hypochlorous acid, hypochlorite ion, chloramine, hydrogen peroxide, chlorine dioxide, and ozone (Kim, *et al.*, 2009). Biocides could be used during maintenance cleaning or clean-in-place (CIP). However, care should be taken in the use of oxidizing biocides because of their detrimental effect on the polyamide-based membrane. Also, byproducts of biocides could cause damage to desalination membranes if proper caution is not taken. Biocides should not be applied in a low dosage because microorganisms produce a large amount of EPS as a protection thus making the microorganisms resistance to biocidal attack (Baker and Dudley, 1998). This could be very detrimental to the process because the biofilm which would be eventually formed may not be easily controlled.

5.5.4 Surfactants

These are surface-active agents which may be anionic, cationic, nonionic or amphoteric. Surfactants can improve wettability and rinsability. They can solubilize the foulants by the formation of micelles around the foulant, therefore, making them easy to dislodge from the membrane surface. Sodium dodecyl sulfate (SDS) is a commonly used surfactant for membrane cleaning. Because of the amphiphilic property of SDS, the hydrophobic tail can attach to the hydrophobic organic and biological foulants, while the hydrophilic head of SDS gravitates towards water thereby helping in the dislodging of the foulants on the membrane. Likewise, surfactants could be combined with other cleaning agents. The combination of surfactants and other cleaning agents would improve the contact between foulants and cleaning agents minimize the amount of cleaning agents, water and contact time needed for effective cleaning (Trägårdh, 1989). Before the introduction of surfactants into the system, the compatibility of the surfactant, the foulant types and the membrane property is necessary to avoid the adsorption of surfactants to the membrane surface. If such happens the flux of the membrane could be brought to zero even after cleaning.

5.6 MEMBRANE FOULING CHARACTERIZATION METHODS

Examining the chemical composition, structure, and functions of these foulants and how they change in a membrane-based filtration system is worth investigating. To identify the fouling nature and establish the cause of membrane failure, membrane fouling characterization is performed. Membrane fouling characterization is usually a destructive technique and requires expensive analysis. In most cases, membrane fouling characterizations are employed as a last resort. In a general case, membrane fouling characterizations are performed for plant performance monitoring, optimization of cleaning practices, or evaluation of pilot tests.

Characterization of NOM in seawater is essential because it can provide a clearer understanding of foulant properties and foulant concentration as well as charge. This helps to optimize RO pretreatment processes. In addition, for practical operation the RO membrane system was normally removed from the canary cell for autopsy after 6-8 months of operation. Therefore, these measurements allow for further identification of potential foulants in the plant. As stated earlier, the surface of the membrane is not the only susceptible area for fouling, but also the pores of porous membranes. Among the various analytical tools used, Fourier transform infrared spectroscopy (FTIR), scanning electron microscopy (SEM), confocal laser scanning microscopy (CLSM), and atomic force microscopy (AFM) are arguably the best techniques for membrane fouling characterization.

5.6.1 Fourier Transform-Infrared (FT-IR)

FTIR and attenuated total reflection-Fourier-Transform infrared (ATR-FTIR) offer a specific structural information on the functional groups and the recognition of chemical bonds of foulants on the membrane (Kimura, *et al.*, 2005, Meng, *et al.*, 2010). The main drawback of the techniques is the need to eliminate water from the sample. Water has a wide range of infrared which would interfere with foulant detection. Therefore, it is important that the sample is dehydrated before analysis.

5.6.2 Scanning Electron Microscopy (SEM)

SEM is one of the versatile instruments for the examination and analysis of the structure of the fouling layer. Visual inspection using SEM is a qualitative analytical tool for both chemical and structural analysis of the membrane fouling. SEM also provides valuable information about the origin of the foulants. All foulants components can be visualized and analyzed using SEM. SEM is not only used for foulants analysis but can also be used to investigate the effectiveness of membrane cleaning. Environmental scanning electron microscopy (ESEM) can be used to analyze wet membrane samples without additional sample preparation; however, this technique cannot be utilized for on-line monitoring

5.6.3 Confocal scanning electron microscopy (CLSM)

CLSM provides a non-invasive 3D visualization of foulant layers on the membrane. Over the years CLSM has been used for the structural identification of biofoulants such as proteins, polysaccharides, and nucleic acids. Foulants density, thickness, and the reversibility of fouling can be examined using CLSM (Xie, *et al.*, 2015). Qualitative and quantitative estimation of foulants coverage on the membrane and types can be identified using CLSM with the aid of fluorescent dyes (Spettmann, *et al.*, 2007). CLSM provides microbial ecology on the surface of the membrane (Vanysacker, et al., 2013), the changes in the porosity of the fouling layer and membrane fouling history can be probed by CLSM (Ng and Ng, 2010). However, because CLSM involves the pretreatment of the membrane and the addition of some chemicals such as dyes, the true nature of the foulant could be compromised either by the extra washing by the chemicals or serving as a foulant on the membrane, hence, the technique cannot be a stand-alone approach for foulant characterization. Compared with SEM, CLSM has a lower magnification. Because CLSM requires fluorescence dye, there is a problem of photobleaching of fluorescence (Pawley, 1990). This is because CLSM uses a single high-energy photo to stimulate fluorescence signal when the laser cannot go through the sample, a part of the laser scatters when going through the sample. This causes the problem of photobleaching (Meng, Liao, Liang, Yang, Zhang and Song, 2010).

5.6.4 Atomic force microscopy (AFM)

AFM is a relatively new technique for the morphological analysis of foulants on the membrane and the quantification of surface forces. It is considered an extremely powerful tool for membrane fouling characterization. AFM provides a 3D imaging, however, unlike SEM, AFM requires no additional sample pretreatment. The surface coverage area, intensity, roughness, skewness and foulant height can be measured using the AFM (Zaky, *et al.*, 2013). AFM measurement can distinguish between membrane-to-foulant, foulant-to-foulant, and foulant-to-cleaning agent interaction at a molecular level (Li and Elimelech, 2004). In a report by Bowen and colleagues (Bowen, *et al.*, 1999), AFM was successfully used to study the adhesion of biological macromolecules and biological cells at the surfaces of different membranes.

5.7 PRESENT EFFORTS AND FUTURE RESEARCH DIRECTIONS

5.7.1 Membrane surface modification

Membrane roughness and hydrophobicity are among the most important factors affecting membrane fouling (Lee, *et al.*, 2008). Membranes with smooth and hydrophilic surfaces are less prone to fouling. This is because the valley structure in the rough surface membrane could serve as a harbor for bacterial attachment and growth. Similarly, the hydrophobic membrane surface can easily attract organic and biological foulants which in most cases are hydrophobic in nature. Recently, new approaches have been employed to increase the hydrophilicity of the membrane surface. There are several ways of modifying membrane surface, namely, (1) surface coating (physical modification), (2) surface grafting (chemical modification), (3) polymer blending, and (4) addition of inorganic or antimicrobial additives. The essence of surface modification is to change membrane surface properties such as surface charge, morphology, hydrophilicity, and functional groups of the membrane in such a way that it favors fouling resistance. Even though the goal of membrane surface modification is to improve antifouling properties, efforts need to be put in place so that there would not be a compromise on overall membrane performance. Among the numerous materials used in membrane surface modification include; metal oxide nanoparticles such as titanium dioxide (TiO2), silver nanoparticle, aqueous fullerene nanoparticles (Cho, *et al.*, 2005, Lyon, *et al.*, 2006, Morones, *et al.*, 2005), carbon nanotubes (CNTs) (Farahbakhsh, *et al.*, 2017, Vatanpour and Zoqi, 2017), graphene materials (He, *et al.*, 2015), and zwitterion (Ye, *et al.*, 2002), etc.

5.7.2 Biological agents

Biosurfactants are surfactants produced by microorganisms such as bacteria, yeasts, and fungi. Their hydrophilic part is composed of sugars, amino acids, or polar functional groups like a carboxylic acid group. The hydrophobic part is typically an aliphatic hydrocarbon chain of β-hydroxyl fatty acids. Biosurfactants are nowadays favored above chemical surfactants because they are biodegradable and environmentally friendly, less toxic, highly active, and stable at extreme temperature, pH, and salinity. Biosurfactants are classified according to their molecular structure as glycolipids (rhamnolipids, sophorolipids), lipopeptides (surfactin), polymeric biosurfactants (emulsan, alasan), fatty acids (3-(3-hydroxyalkanoyloxy) alkanoic acids), and phospholipids (phosphatidylethanolamine) (Abdel-Mawgoud, *et al.*, 2010). Long *et al.*, (2014) tested a biosurfactant; rhamnolipid as a cleaning agent on UF membranes. The authors reported that rhamnolipid at pH 9 displayed a superior cleaning efficiency on fouled UF membranes achieving the flux recovery ratio (FRR) of over 90% when compared with commercial cleaners and that rhamnolipid biosurfactant has no negative effect on the performance of the membrane. In another study by Kim and colleagues (Kim, *et al.*, 2015), rhamnolipid biosurfactant was used for biofouling prevention and control on reverse osmosis (RO) membrane. Other biosurfactants and biosurfactant-like compounds successfully used as membrane cleaners include surfactin (Isa, *et al.*, 2012), hydroxypropyl-beta-cyclodextrin (HP-β-CD) (Alayande, *et al.*, 2016) The use of biosurfactants as a biological cleaner on desalination membrane is a potential and a viable option to combat the problems associated with membrane fouling.

Due to the detrimental effect on the environment and the membrane caused by harsh chemical cleaning agents and biocides, research trend has been towards the discovery of non-toxic and environmentally friendly approach to controlling fouling (most especially biofouling). Progress in bacteriology has shown the importance of quorum sensing (QS) pathways in biofilm formation. QS is a process by which microorganisms communicate with each other by producing and secreting QS signal molecules that build up to a threshold level depending on the microbial population density (Dobretsov, *et al.*, 2007). Quorum quenching (QQ) is an approach to hinder or suppress such communication for biofilm not to develop. Most of the studies about QQ in the membrane-oriented process have been majorly on membrane bioreactor (MBR) and very few studies have been conducted in the membrane-based desalination process. The reason behind this could be because if QQ compounds without care are applied directly into the system, they could also serve as a foulant contributing to permeate flux decline. The use of QQ as pretreatment could only reduce biofilm formation but not necessarily the growth of microorganisms. The fate of a single bacterial cell afterward cannot be determined because a single bacterial cell can multiply to millions of bacterial cells over a specific period. Nonetheless, few reports have shown the potential of QQ approach in pressurized systems such as desalination membranes. Paul *et al.* (2009) tested the ability of Acylase I (an AHL-degrading enzyme) to inhibit biofouling on reverse osmosis membrane. A significant reduction in the A. hydrophila and Pseudomonas putida biofilm after tested with Acylase I was observed. However, at a concentration higher than the test, this enzyme does not completely prevent the formation of biofilm. Other QQ agents investigated on desalination membranes are cinnamaldehyde (CNMA) and vanillin (VA). These compounds are more effective against Gram-negative bacterial biofilms, the major biofoulant species in desalination plants (Brackman, *et al.*, 2008, Kappachery, *et al.*, 2010, Ponnusamy, *et al.*, 2013). New developments have shifted to the modification of desalination membranes with QQ agents (Wood, *et al.*, 2016).

Both organic fouling and biofouling seem to be an inherent and unavoidable problem of membrane technology. Although this chapter emphasizes on organic and biofouling, however, in real plant application fouling does not occur independently of each other. In most case, biofouling would not occur without the supply of organic matter which would serve as food source for the microorganisms. Generally, organic fouling, biofouling, inorganic, and particulate/colloidal fouling would occur together even though individual foulants have its own developmental processes. These developmental processes are interconnected, and no clear distinction could be made in the real desalination plants. Several studies have been conducted on fouling behaviors and mechanisms on desalination membranes to develop better control methods. For example, fouling prediction, membrane foulants characterization and monitoring, pretreatment, and membrane cleaning. Nevertheless, many challenges remain to be overcome in order to combat membrane fouling. One way to address this problem is to focus research on the development of novel membrane materials and synthesis approaches. It is thought that the discovery of a new membrane material and appropriate feed pretreatment could be a promising solution to the problem of organic fouling and biofouling on membrane-based desalination process. Future research should therefore seek to explore the boundless possibilities in this area of research.

5.8 REFERENCES

Abdel-Mawgoud AM, Lépine F, Déziel E (2010) Rhamnolipids: diversity of structures, microbial origins and roles. Applied microbiology and biotechnology 86: 1323-1336

Abushaban A, Salinas-Rodriguez SG, Dhakal N, Schippers JC, Kennedy MDJD (2019) Assessing pretreatment and seawater reverse osmosis performance using an ATP-based bacterial growth potential method467: 210-218

Abushaban A, Salinas-Rodriguez SG, Mangal MN, Mondal S, Goueli SA, Knezev A, Vrouwenvelder JS, Schippers JC, Kennedy MDJD (2019) ATP measurement in seawater reverse osmosis systems: Eliminating seawater matrix effects using a filtration-based method453: 1-9

Al-Amoudi A, Lovitt RW (2007) Fouling strategies and the cleaning system of NF membranes and factors affecting cleaning efficiency. Journal of Membrane Science 303: 4-28 DOI 10.1016/j.memsci.2007.06.002

Alayande AB, Kim LH, Kim IS (2016) Cleaning efficacy of hydroxypropyl-beta-cyclodextrin for biofouling reduction on reverse osmosis membranes. Biofouling 32: 359-370

Alhadidi A, Kemperman AJ, Schurer R, Schippers J, Wessling M, van der Meer WGJ (2012) Using SDI, SDI+ and MFI to evaluate fouling in a UF/RO desalination pilot plant. Desalination 285: 153-162

Ang WS, Lee S, Elimelech M (2006) Chemical and physical aspects of cleaning of organic-fouled reverse osmosis membranes. Journal of Membrane Science 272: 198-210 DOI 10.1016/j.memsci.2005.07.035

Ang WS, Tiraferri A, Chen KL, Elimelech M (2011) Fouling and cleaning of RO membranes fouled by mixtures of organic foulants simulating wastewater effluent. Journal of Membrane Science 376: 196-206 DOI 10.1016/j.memsci.2011.04.020

Baker J, Dudley L (1998) Biofouling in membrane systems—a review. Desalination 118: 81-89

Bowen WR, Hilal N, Lovitt RW, Wright CJ (1999) Characterisation of membrane surfaces: direct measurement of biological adhesion using an atomic force microscope. Journal of Membrane Science 154: 205-212

Brackman G, Defoirdt T, Miyamoto C, Bossier P, Van Calenbergh S, Nelis H, Coenye T (2008) Cinnamaldehyde and cinnamaldehyde derivatives reduce virulence in Vibrio spp. by decreasing the DNA-binding activity of the quorum sensing response regulator LuxR. BMC microbiology 8: 149

Busch M, Chu R, Kolbe U, Meng Q, Li S (2009) Ultrafiltration pretreatment to reverse osmosis for seawater desalination—three years field experience in the Wangtan Datang power plant. Desalination and water treatment 10: 1-20

Chlistunoff JJFTR (2005) Advanced chlor-alkali technology

Cho M, Chung H, Choi W, Yoon J (2005) Different inactivation behaviors of MS-2 phage and Escherichia coli in TiO2 photocatalytic disinfection. Applied and environmental microbiology 71: 270-275

Corral AF, Yenal U, Strickle R, Yan D, Holler E, Hill C, Ela WP, Arnold RG (2014) Comparison of slow sand filtration and microfiltration as pretreatments for inland desalination via reverse osmosis. Desalination 334: 1-9

Creber S, Pintelon T, Von Der Schulenburg DG, Vrouwenvelder J, Van Loosdrecht M, Johns M (2010) Magnetic resonance imaging and 3D simulation studies of biofilm accumulation and cleaning on reverse osmosis membranes. Food and Bioproducts Processing 88: 401-408

Dobretsov S, Dahms H-U, YiLi H, Wahl M, Qian P-Y (2007) The effect of quorum-sensing blockers on the formation of marine microbial communities and larval attachment. FEMS microbiology ecology 60: 177-188

Dong B-z, Yan C, Gao N-y, Fan J-c (2007) Effect of coagulation pretreatment on the fouling of ultrafiltration membrane. Journal of Environmental Sciences 19: 278-283

Drouiche N, Grib H, Abdi N, Lounici H, Pauss A, Mameri NJJohm (2009) Electrodialysis with bipolar membrane for regeneration of a spent activated carbon170: 197-202

Farahbakhsh J, Delnavaz M, Vatanpour V (2017) Investigation of raw and oxidized multiwalled carbon nanotubes in fabrication of reverse osmosis polyamide membranes for improvement in desalination and antifouling properties. Desalination 410: 1-9

Farhat N, Hammes F, Prest E, Vrouwenvelder J (2018) A uniform bacterial growth potential assay for different water types. Water Research 142: 227-235 DOI https://doi.org/10.1016/j.watres.2018.06.010

Flemming H-C (2002) Biofouling in water systems–cases, causes and countermeasures. Applied microbiology and biotechnology 59: 629-640

Friedler E, Katz I, Dosoretz CG (2008) Chlorination and coagulation as pretreatments for greywater desalination. Desalination 222: 38-49 DOI 10.1016/j.desal.2007.01.130

Fritzmann C, Löwenberg J, Wintgens T, Melin T (2007) State-of-the-art of reverse osmosis desalination. Desalination 216: 1-76 DOI 10.1016/j.desal.2006.12.009

Gan Q, Howell J, Field R, England R, Bird M, McKechinie M (1999) Synergetic cleaning procedure for a ceramic membrane fouled by beer microfiltration. Journal of membrane Science 155: 277-289

Ghayeni SS, Beatson P, Schneider R, Fane A (1998) Adhesion of waste water bacteria to reverse osmosis membranes. Journal of Membrane Science 138: 29-42

Giannasi FJIjoo, health e (2007) Ban on asbestos diaphragms in the chlorine-related chemical industry and efforts toward a worldwide ban13: 80-84

Glueckstern P, Priel M, Wilf M (2002) Field evaluation of capillary UF technology as a pretreatment for large seawater RO systems. Desalination 147: 55-62

Gogate PR (2007) Application of cavitational reactors for water disinfection: current status and path forward. Journal of environmental management 85: 801-815

Goosen M, Sablani S, Al Hinai H, Al Obeidani S, Al Belushi R, Jackson a (2005) Fouling of reverse osmosis and ultrafiltration membranes: a critical review. Separation science and technology 39: 2261-2297

Gur-Reznik S, Katz I, Dosoretz CG (2008) Removal of dissolved organic matter by granular-activated carbon adsorption as a pretreatment to reverse osmosis of membrane bioreactor effluents. Water Res 42: 1595-1605 DOI 10.1016/j.watres.2007.10.004

Halpern DF, McArdle J, Antrim B (2005) UF pretreatment for SWRO: pilot studies. Desalination 182: 323-332

He L, Dumée LF, Feng C, Velleman L, Reis R, She F, Gao W, Kong L (2015) Promoted water transport across graphene oxide–poly (amide) thin film composite membranes and their antibacterial activity. Desalination 365: 126-135

Herzberg M, Kang S, Elimelech M (2009) Role of extracellular polymeric substances (EPS) in biofouling of reverse osmosis membranes. Environmental science & technology 43: 4393-4398

Hobbie JE, Daley RJ, Jasper SJA, microbiology e (1977) Use of nuclepore filters for counting bacteria by fluorescence microscopy33: 1225-1228

Hoek EM, Elimelech M (2003) Cake-enhanced concentration polarization: a new fouling mechanism for salt-rejecting membranes. Environmental science & technology 37: 5581-5588

Foulants and Cleaning Procedures for Composite Polyamide. Hydranautics (1998), 1998
Ibáñez R, Pérez-González A, Gómez P, Urtiaga AM, Ortiz I (2013) Acid and base recovery from softened reverse osmosis (RO) brines. Experimental assessment using model concentrates. Desalination 309: 165-170 DOI https://doi.org/10.1016/j.desal.2012.10.006
Isa MHM, Frazier RA, Jauregi P (2012) Cleaning potential of surfactin on fouled ultrafiltration (UF) membranes. Sains Malaysiana 41: 1117-1124
Javier L, Farhat NM, Desmond P, Linares RV, Bucs S, Kruithof JC, Vrouwenvelder JSJWR (2020) Biofouling control by phosphorus limitation strongly depends on the assimilable organic carbon concentration183: 116051
Jeong S, Naidu G, Vollprecht R, Leiknes T, Vigneswaran S (2016) In-depth analyses of organic matters in a full-scale seawater desalination plant and an autopsy of reverse osmosis membrane. Separation and Purification Technology 162: 171-179 DOI 10.1016/j.seppur.2016.02.029
Jiang S, Li Y, Ladewig BP (2017) A review of reverse osmosis membrane fouling and control strategies. Sci Total Environ 595: 567-583 DOI 10.1016/j.scitotenv.2017.03.235
Kappachery S, Paul D, Yoon J, Kweon JH (2010) Vanillin, a potential agent to prevent biofouling of reverse osmosis membrane. Biofouling 26: 667-672
Kim D, Jung S, Sohn J, Kim H, Lee S (2009) Biocide application for controlling biofouling of SWRO membranes—an overview. Desalination 238: 43-52
Kim H-C, Dempsey BA (2013) Membrane fouling due to alginate, SMP, EfOM, humic acid, and NOM. Journal of Membrane Science 428: 190-197 DOI 10.1016/j.memsci.2012.11.004
Kim LH, Jung Y, Kim S-J, Kim C-M, Yu H-W, Park H-D, Kim IS (2015) Use of rhamnolipid biosurfactant for membrane biofouling prevention and cleaning. Biofouling 31: 211-220
Kimura K, Yamato N, Yamamura H, Watanabe Y (2005) Membrane fouling in pilot-scale membrane bioreactors (MBRs) treating municipal wastewater. Environmental science & technology 39: 6293-6299
Lee EK, Chen V, Fane A (2008) Natural organic matter (NOM) fouling in low pressure membrane filtration—effect of membranes and operation modes. Desalination 218: 257-270
Lee H, Amy G, Cho J, Yoon Y, Moon S-H, Kim IS (2001) Cleaning strategies for flux recovery of an ultrafiltration membrane fouled by natural organic matter. Water research 35: 3301-3308
Lee S, Ang WS, Elimelech M (2006) Fouling of reverse osmosis membranes by hydrophilic organic matter: implications for water reuse. Desalination 187: 313-321 DOI 10.1016/j.desal.2005.04.090
Li Q, Elimelech M (2004) Organic fouling and chemical cleaning of nanofiltration membranes: measurements and mechanisms. Environmental science & technology 38: 4683-4693
Li Y, Wang J, Zhang W, Zhang X, Chen C (2011) Effects of coagulation on submerged ultrafiltration membrane fouling caused by particles and natural organic matter (NOM). Chinese science bulletin 56: 584-590
Lin JC-T, Lee D-J, Huang C (2010) Membrane Fouling Mitigation: Membrane Cleaning. Separation Science and Technology 45: 858-872 DOI 10.1080/01496391003666940
Long X, Meng Q, Zhang G (2014) Application of biosurfactant rhamnolipid for cleaning of UF membranes. Journal of Membrane Science 457: 113-119
Lyon DY, Adams LK, Falkner JC, Alvarez PJ (2006) Antibacterial activity of fullerene water suspensions: effects of preparation method and particle size. Environmental Science & Technology 40: 4360-4366
Madaeni SS, Mohamamdi T, Moghadam MK (2001) Chemical cleaning of reverse osmosis membranes. Desalination 134: 77-82

Malaeb L, Ayoub GM (2011) Reverse osmosis technology for water treatment: state of the art review. Desalination 267: 1-8

Matilainen A, Gjessing ET, Lahtinen T, Hed L, Bhatnagar A, Sillanpaa M (2011) An overview of the methods used in the characterisation of natural organic matter (NOM) in relation to drinking water treatment. Chemosphere 83: 1431-1442 DOI 10.1016/j.chemosphere.2011.01.018

Matin A, Khan Z, Zaidi S, Boyce M (2011) Biofouling in reverse osmosis membranes for seawater desalination: phenomena and prevention. Desalination 281: 1-16

Meng F, Liao B, Liang S, Yang F, Zhang H, Song L (2010) Morphological visualization, componential characterization and microbiological identification of membrane fouling in membrane bioreactors (MBRs). Journal of membrane science 361: 1-14

Miller SE, Rodriguez RA, Nelson KLJESWR, Technology (2020) Removal and growth of microorganisms across treatment and simulated distribution at a pilot-scale direct potable reuse facility6: 1370-1387

Morones JR, Elechiguerra JL, Camacho A, Holt K, Kouri JB, Ramírez JT, Yacaman MJ (2005) The bactericidal effect of silver nanoparticles. Nanotechnology 16: 2346

Ng TCA, Ng HY (2010) Characterisation of initial fouling in aerobic submerged membrane bioreactors in relation to physico-chemical characteristics under different flux conditions. Water research 44: 2336-2348

Nguyen T, Roddick FA, Fan LJM (2012) Biofouling of water treatment membranes: a review of the underlying causes, monitoring techniques and control measures2: 804-840

Paul D, Kim YS, Ponnusamy K, Kweon JH (2009) Application of quorum quenching to inhibit biofilm formation. Environmental Engineering Science 26: 1319-1324

Pawley J (1990) Handbook of biological confocal microscopy. Plenum Press: New York 232: 150-152

Pearce G, Talo S, Chida K, Basha A, Gulamhusein A (2004) Pretreatment options for large scale SWRO plants: case studies of OF trials at Kindasa, Saudi Arabia, and conventional pretreatment in Spain. Desalination 167: 175-189

Ponnusamy K, Kappachery S, Thekeettle M, Song J, Kweon J (2013) Anti-biofouling property of vanillin on Aeromonas hydrophila initial biofilm on various membrane surfaces. World Journal of Microbiology and Biotechnology 29: 1695-1703

Remize PJ, Laroche JF, Leparc J, Schrotter JC (2012) A pilot-scale comparison of granular media filtration and low-pressure membrane filtration for seawater pretreatment. Desalination and Water Treatment 5: 6-11 DOI 10.5004/dwt.2009.557

Ridgway H, Flemming H (1988) Microbial adhesion and biofouling of reverse osmosis membranes. Reverse Osmosis Technology: Applications for High Purity Water Production, ed by BS Pakekh and M Dekker: 429-481

She Q, Wang R, Fane AG, Tang CY (2016) Membrane fouling in osmotically driven membrane processes: A review. Journal of Membrane Science 499: 201-233 DOI 10.1016/j.memsci.2015.10.040

Sohrabi M, Madaeni S, Khosravi M, Ghaedi A (2011) Chemical cleaning of reverse osmosis and nanofiltration membranes fouled by licorice aqueous solutions. Desalination 267: 93-100

Spettmann D, Eppmann S, Flemming H-C, Wingender J (2007) Simultaneous visualisation of biofouling, organic and inorganic particle fouling on separation membranes. Water science and technology 55: 207-210

Tang CY, Chong TH, Fane AG (2011) Colloidal interactions and fouling of NF and RO membranes: a review. Adv Colloid Interface Sci 164: 126-143 DOI 10.1016/j.cis.2010.10.007

Trägårdh G (1989) Membrane cleaning. Desalination 71: 325-335

Vanysacker L, Declerck P, Vankelecom I (2013) Development of a high throughput cross-flow filtration system for detailed investigation of fouling processes. Journal of membrane science 442: 168-176

Vatanpour V, Zoqi N (2017) Surface modification of commercial seawater reverse osmosis membranes by grafting of hydrophilic monomer blended with carboxylated multiwalled carbon nanotubes. Applied Surface Science 396: 1478-1489

Venugopal K, Dharmalingam S (2012) Desalination efficiency of a novel bipolar membrane based on functionalized polysulfone. Desalination 296: 37-45 DOI https://doi.org/10.1016/j.desal.2012.04.006

Vrouwenvelder J, Van der Kooij DJD (2001) Diagnosis, prediction and prevention of biofouling of NF and RO membranes139: 65-71

Wang M, Wang K-k, Jia Y-x, Ren Q-cJJoms (2014) The reclamation ofbrine generated from desalination process by bipolar membrane electrodialysis452: 54-61

Wang Y-N, Tang CY (2011) Protein fouling of nanofiltration, reverse osmosis, and ultrafiltration membranes—The role of hydrodynamic conditions, solution chemistry, and membrane properties. Journal of Membrane Science 376: 275-282 DOI 10.1016/j.memsci.2011.04.036

Wood TL, Guha R, Tang L, Geitner M, Kumar M, Wood TK (2016) Living biofouling-resistant membranes as a model for the beneficial use of engineered biofilms. Proceedings of the National Academy of Sciences 113: E2802-E2811

Xia S-j, Liu Y-n, Xing L, Yao J-j (2007) Drinking water production by ultrafiltration of Songhuajiang River with PAC adsorption. Journal of Environmental Sciences 19: 536-539

Xie M, Lee J, Nghiem LD, Elimelech M (2015) Role of pressure in organic fouling in forward osmosis and reverse osmosis. Journal of Membrane Science 493: 748-754

Ye SH, Watanabe J, Iwasaki Y, Ishihara K (2002) Novel cellulose acetate membrane blended with phospholipid polymer for hemocompatible filtration system. Journal of Membrane Science 210: 411-421

Yin W, Ho JS, Cornelissen ER, Chong THJWr (2020) Impact of isolated dissolved organic fractions from seawater on biofouling in reverse osmosis (RO) desalination process168: 115198

Yu Y, Lee S, Hong S (2010) Effect of solution chemistry on organic fouling of reverse osmosis membranes in seawater desalination. Journal of Membrane Science 351: 205-213 DOI 10.1016/j.memsci.2010.01.051

Zaky AM, Escobar IC, Gruden CL (2013) Application of atomic force microscopy for characterizing membrane biofouling in the micrometer and nanometer scales. Environmental Progress & Sustainable Energy 32: 449-457

Chapter 6

Algal blooms and RO desalination

Loreen O. Villacorte, Siobhan F. E. Boerlage and Mike Dixon

The main learning objectives of this chapter are the following:

- Have an understanding of the dynamics of algal blooms from a desalination perspective
- Have an overview of algal bloom associated issues in various processes in an RO plant
- Know the various methods for monitoring/characterizing algal blooms and their membrane fouling potential.
- Have an overview of the current and emerging technologies to mitigate impact on RO operation
- Understand the importance of proper monitoring, operational and pre-treatment strategies to protect RO plants from algal blooms

6.1 INTRODUCTION

The seasonal proliferation of algae in water sources can cause major operational challenges in reverse osmosis (RO) desalination plants. These phenomena, commonly known as algal blooms, can potentially affect practically any RO plant with open fresh or seawater sources. Their adverse effect to desalination became widely recognised over the last decade due to the recurrence of severe algal blooms in the Gulf region, which forced several seawater RO plants to reduce or shut down operation, consequently hampering the water supply in the region for up to several months (Figure 1; Berktay, 2011; Richlen *et al.*, 2010; Maniyar, 2018).

Although natural environmental processes are the main triggers of algal blooms, human activities have been increasingly linked directly or indirectly to the increase in their frequency and severity. As the global population is expected to continue its rapid growth trend, exacerbated by the effect of climate change on available freshwater resources, the demand for installing new seawater desalination plants will be higher than ever. Pollution in coastal areas due to increased human activities can trigger widespread and severe algal blooms. Hence, algal blooms and RO desalination are in a collision course for the foreseeable future and their impact should be addressed at the water source, in the intake/pretreatment processes and in the operation of the RO plant.

Editors: Sergio Salinas, Jan Schippers, Gary Amy, In Kim, Maria Kennedy
doi: 10.2166/9781780409863_0145

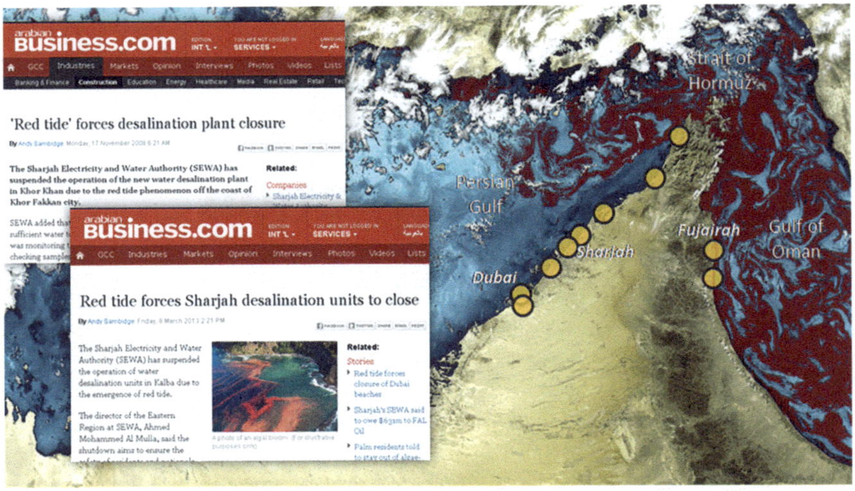

Figure 1 A "red tide" algal bloom off the coast of Oman, UAE and Iran as illustrated in this enhanced image based from the satellite image obtained by the European Space Agency MERIS FR satellite on 22 November 2008 (Planetek Hellas/ESA). Yellow points indicate locations of large RO plants in the area. Inset screenshots of online news regarding RO desalination plant shutdowns in the area in 2008 and 2013 (www.arabianbusiness.com).

Over the last decade, there has been a rapid increase in interest regarding the impact of algal blooms on RO plants which has been addressed in multi-disciplinary workshops and conferences as well as numerous scientific articles (e.g., Caron *et al.*, 2010; Villacorte *et al.*, 2015a). Recognizing the growing need from the desalination industry, a conference was organized by the Middle East Desalination Research Center (MEDRC) and the UNESCO Intergovernmental Oceanographic Commission (IOC) in 2012 bringing together academic researchers and professionals from both fields to exchange knowledge, present experience in the field and discuss strategies to address operational challenges (Anderson and McCarthy, 2012). Based on the recommendation of the conference, a guidance manual for RO plant operators on harmful algal blooms and desalination was commissioned and was released in 2017 with the contribution of 63 experts from both the academe and the industry (Anderson *et al.*, 2017).

This chapter consolidated the latest theoretical and practical knowledge on the impact of (harmful) algal blooms in RO desalination operation. Specifically, it introduces (1) the basic dynamics of marine algal blooms, (2) the mechanisms of fouling in various processes in RO plants, (3) methods for monitoring its occurrence and quantifying their potential impact to RO operation, and (4) operational and pre-treatment strategies for mitigating associated challenges.

6.2 ALGAL BLOOMS

The term "algae" refers to a diverse group of mainly photosynthetic and free floating or swimming organisms comprising thousands of species in the oceans, lakes and other open water bodies. Each species has a set of environmental conditions that favour their growth and proliferation. A continuous peak succession of dominant species can be observed over time within a specific area. These peak events are termed as "algal blooms". Such blooms can be characterised as dilute suspensions of algae cells to dense accumulations that can make the water appear coloured, often red or brown; hence, the term "red tide" is commonly used (Figure 1).

Algal blooms are critical to many aspects of aquatic ecology and mankind utilizing the aquatic resources. However, some species of algae can form blooms that are considered harmful to humans and the environment. Such blooms are classified as "harmful algal blooms" or HABs. Around 60-80 species among over 300 bloom-forming algal species have been reported to cause HABs (Smayda, 1997). Table 1 shows typical characteristics of some of these HAB species to provide examples of cell sizes, cell concentrations, and impacts. The term "HAB" is broadly applied for blooms which produce toxic compounds and those that cause harm in some other ways such as dissolved oxygen depletion (hypoxia). As the rapidly growing human population demands increased exploitation of the coastal zone (for shelter, food, recreation, and commerce), new impacts are expected to emerge going forward, and with that will come the designation of new harmful species (Anderson, 2014). An emerging example is their recurring impact to RO plants.

6.2.1 Factors triggering algal blooms

The growth and accumulation of an algal species in a mixed community of marine organisms is an exceedingly complex process involving an array of chemical, physical, and biological interactions (Anderson *et al.*, 2012). The distribution and concentration of algae in the ocean are influenced by natural physico-chemical variations such as sunlight exposure, temperature, current, salinity, nutrient load, etc. (Sellner *et al.*, 2003). Nutrients such as nitrogen (N), phosphorus (P) and silicon (Si) and some trace metals are among the most important of these factors (see Figure 2). These nutrients may originate from natural or anthropogenic sources.

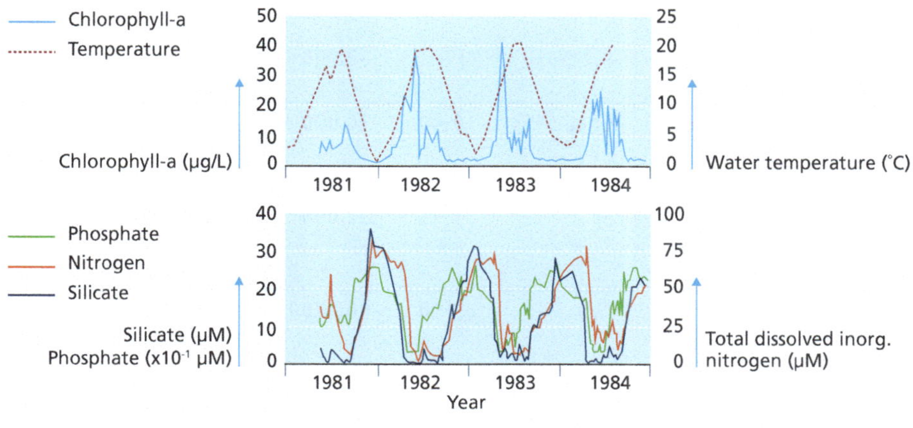

Figure 2 Typical profile of algae concentration (as chlorophyll-a), temperature and nutrient concentrations (silicate, phosphate, nitrogen) Oosterschelde bay, the Netherlands. Adapted from Wetsteyn *et al.* (1990) and van der Hoeven (1984).

Table 1 Characteristics of selected HAB species of microscopic algae and cyanobacteria in marine environment (adapted from Villacorte *et al.*, 2015a).

Group	Bloom-forming species	Cell shape(+)	Cell size μm(+)	Algal concentration #		Potential impact	References
				cells/mL	μg Chl-a/L		
Dinoflagellates	*Alexandrium spp.*	RE	25-32	7,000	-	toxins	Selina *et al.* (2006)
	Cochlodinium polykrikoides	RE	20–40	48,000	-	toxins, high biomass	Kim (2010)
	Karenia brevis	RE	20-40	37,000	-	toxins, high biomass	Tester *et al.* (2004)
	Noctiluca scintillans	Sp	200-2000	1,900	-	high biomass	Fonda-Umani *et al.* (2004)
	Prorocentrum spp.	FE	30-60	70,000	~200	high biomass	Tas and Okus (2011)
Diatoms	*Chaetoceros spp.*	OC	8-25	30,100	14	high biomass	Booth *et al.* (2002)
	Pseudo-nitzschia spp.	0.8*PP	3-100	19,000	-	toxins, high biomass	Anderson *et al.* (2010)
	Skeletonema costatum	Cy	2-25	88,000	-	high biomass	Shikata *et al.* (2008)
	Thalassiosira spp.	Cy	10-50	28,000	~100	high biomass	Maier *et al.* (2012)
Haptophytes	*Emiliania huxleyi*	Sp	2-6	115,000	-	high biomass	Berge (1962)
	Phaeocystis spp.	0.9*Sp	4-9	52,000	-	high biomass	Janse *et al.* (1996)
Raphidophytes	*Chattonella spp.*	Co+0.5*Sp	10-40	10,000	-	toxins, high biomass	Orlova et al. (2002)
	Heterosigma akashiwo	Sp	15-25	32,000	-	toxins, high biomass	Shikata *et al.* (2008)
Cyanobacteria	*Nodularia spp.*	Cy	6-100	605,200	-	toxins, high biomass	McGregor *et al.* (2012)

Legend: (+) Equivalent geometric dimensions and size range of algal cells based on Olenina *et al.* (2006); (#) High concentrations reported in reference literature. RE=rotational ellipsoid; Sp=sphere; FE=flattened ellipsoid; OC=oval cylinder; PP=parallelepiped; Cy=cylinder; Co=cone.

Specific hydrographic and meteorological conditions permit the accumulation of a certain species of algae. Hence, a progression of natural processes in a specific coastal area such as winds, tides, currents, fronts, and other processes often dictates the occurrence, scale and succession of algal blooms. Storm events tend to increase river discharges of nutrients to the sea while strong winds can induce mixing and transport of nutrients from the lower water column to the surface where they are utilised by algae (Smith *et al.*, 1990; Trainer *et al.*, 1998).

Algal blooms in temperate regions are predominantly a spring, summer, or fall phenomena. Some occur during periods when heating or fresh water runoff creates a stratified surface layer overlying colder, nutrient rich waters. This situation favours motile algae (e.g., dinoflagellates), as they are able to regulate their position and access nutrients below the pycnocline. Some of the motile algae species can reside in surface waters during the day to get sunlight and then swim to the pycnocline and below to take up nutrients at night. This strategy can explain how dense accumulations of cells can appear in surface waters that are devoid of nutrients and which would seem to be incapable of supporting such prolific growth. This vertical migration is a factor of concern to RO plants, where intakes might see episodic pulses of algal cells during the daily migrations near intakes (Villacorte *et al.*, 2015a, Boerlage *et al.*, 2017a).

Human activities can also affect the dynamics of algal blooms by increasing nutrient load in coastal seawater through direct/indirect discharge of un/poorly-treated wastewater and run-off of untreated livestock wastes and residual fertilisers from agricultural areas. Increased incidence of HABs has been shown to have substantial correlation with growing human populations, and with higher fertilizer use and livestock production (Anderson *et al.*, 2002; Sellner *et al.*, 2003). Many regions in the world that have implemented stricter environmental regulations to limit anthropogenic nutrient discharges to rivers have observed localised reduction in algal blooms, as in the case of the Seto Inland Sea in Japan (Okaichi, 1989).

6.2.2 Type of blooms

Algal blooms can take a variety of forms and harmful impacts to the environment. It can be broadly categorized into micro-algal and macro-algal blooms. Micro-algal blooms can be further sub-divided into toxic and non-toxic blooms.

6.2.2.1 Toxic micro-algal blooms

A "toxic bloom" is typically caused by toxin producing, microscopic algae which can cause illness and death in humans, fish, seabirds, marine mammals, and other oceanic life (Anderson *et al.* 2012). The harmful impact can occur when toxic algae are filtered from the water as food by shellfish which then accumulate the biotoxins to levels which can be lethal to humans or other consumers. A major public health concern is the potential for algal toxins to be retained in treated drinking water coming from RO plants. The common algal-derived toxins and associated poisoning syndromes are listed in Table 2.

6.2.2.2 Non-toxic micro-algal blooms

Non-toxic blooms generally relate to the high biomass that some blooms achieve. When a dense algal biomass begins to decay as the bloom terminates, oxygen is consumed, leading to widespread mortalities of plants and animals in the affected area. These types of blooms are sometimes linked to excessive pollution inputs but can also occur in relatively pristine waters. High-biomass blooms can disrupt RO desalination operations, as discussed in detail in Section 6.3.

6.2.2.3 Macro-algal blooms

Macro-algal blooms typically occur in nutrient-enriched estuaries and nearshore areas that are shallow enough for light to penetrate to the sea floor. These blooms can last longer than micro-algal blooms. Such blooms can cause fouling to open water intake structures of RO plants. One prominent example is the "green tides" in northeast China where floating masses of seaweed pose challenges to power plants, desalination plants, and recreational resources in the area (Smetacek and Zingone, 2013; Mathiesen, 2013).

6.2.3 Algal-derived organic matter

The natural organic matter (NOM) present in aquatic environment is a mixture of diverse organic compounds originating from both autochthonous (local input) and allochthonous (external input) sources (Leenheer and Croué, 2003). Algae are a major source of autochthonous NOM in the Earth's oceans, accounting for about half the organic matter input (Field *et al.*, 1998). These algae-derived substances are collectively known as algal (or algogenic) organic matter (AOM).

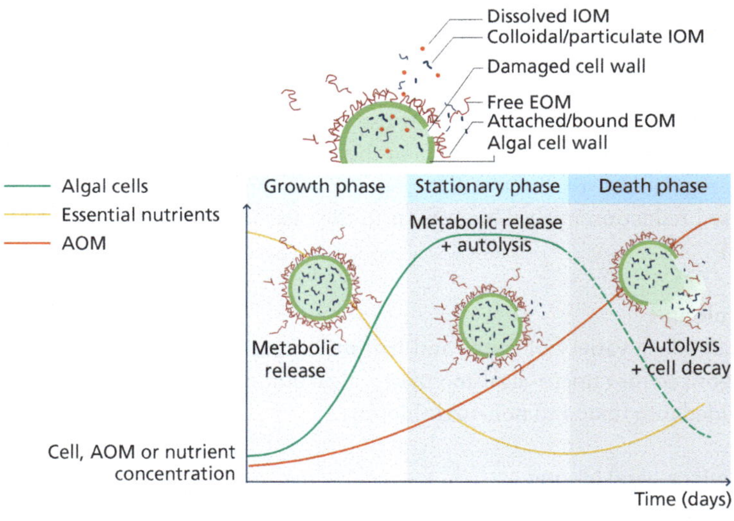

Figure 3 Graphical illustrations of how AOMs can be released into seawater by algae at different phases of a bloom and in response to the availability of essential nutrients. Adopted from Hess *et al.* (2017).

Algal blooms produce various forms and differing concentrations of AOM which includes polysaccharides, proteins, lipids, nucleic acids and other dissolved organic components (Fogg, 1983; Bhaskar and Bhosle, 2005; Decho, 1990; Myklestad, 1995). AOM can be divided to two types, namely: (1) extracellular organic matter (EOM) - organic substances released during the metabolic activity of algae and (2) intracellular organic matter (IOM)

- substances released through autolysis and/or during the process of cell decay. Figure 3 shows a general representation of various phases of an algal bloom and how EOMs and IOMs are released.

6.2.3.1 Extracellular organic matter

Algal cells excrete EOM typically in response to nutrient stress and other unfavourable conditions (e.g., light, pH and temperature) or invasion by bacteria or viruses (Fogg, 1983; Leppard, 1993; Myklestad, 1999). EOM substances can be either discrete or remained attached (bound) to the algal cell as coatings. Discrete EOMs often contain mainly polysaccharides and tend to be more hydrophilic while bound EOM contain more proteins and tend to be more hydrophobic (Qu *et al.*, 2012a).

Polysaccharides can comprise more than 80% of EOM production (Myklestad, 1995). Excessive production of EOM can cause mucilage events characterized by the appearance of a sporadic but massive accumulation of gelatinous material at and below the water surface. Major mucilage events are common in the North Sea, Adriatic Sea and other parts of the Mediterranean region but proliferation of smaller mucilage aggregates such as "marine snow" has been reported in other areas (Mingazzini and Thake, 1995; Lancelot, 1995; Rinaldi *et al.*, 1995; Gotsis-Skretas, 1995). Marine mucilages could occur in a form of marine snow (>0.5 mm diameter), strings (2-15 cm long), tapes and clouds of up to several kilometres long.

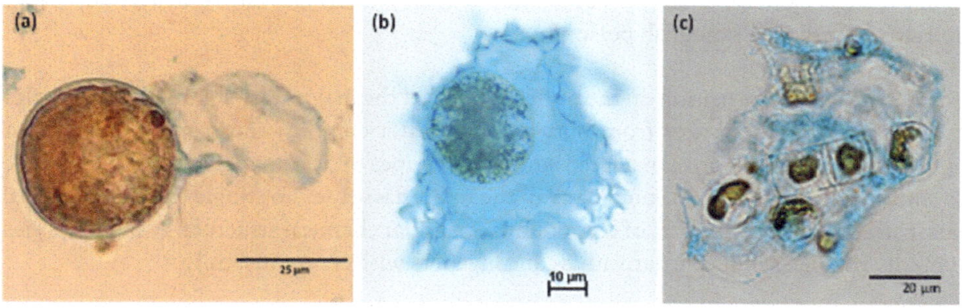

Figure 4 Optical microscope images of Alcian Blue stained (a) Alexandrium tamarense, (b) Lepidodinium chlorophorum and (c) Chaetoceros affinis. In these images, TEPs (stained blue) were released by marine algae through shedding of cell mucus (b and c) and membrane coatings (a and b). Sources: (a) and (c) Villacorte *et al.* (2015b) ; (b) Claquin *et al.* (2008).

A sub-group of EOMs comprising hydrophilic, anionic muco-polysaccharides and glycoproteins are collectively known as transparent exopolymer particles (TEP). Although the existence of these transparent material was suggested in the 1980's (Winters *et al.*, 1983), the term "TEP" has only been used since the 1990s (Alldredge *et al.*, 1993). Since then, it has been the subject of hundreds of studies, revealing their essential role in natural coagulation and sedimentation, and in many aspects of particle dynamics in aquatic systems (see review by Passow, 2002). Characteristically deformable and sticky, they have the tendency to aggregate into large flocs and to adhere to other materials (Mopper *et al.*, 1995; Figure 4). Their stickiness was reported to be between 2 to 4 orders of magnitude higher than most suspended particles in the sea (Passow, 2002). Hence, they tend to accumulate on solid-liquid interfaces and facilitate adsorption of suspended particles, including bacteria.

6.2.3.2 Intracellular organic matter

IOMs comprise mainly low molecular weight organics released from the interior of compromised, dying or deteriorating cells, which may include toxins, and taste and odour compounds.

6.2.3.2.1 Algal toxins

Marine and cyanobacterial toxins occur in various coastal environments worldwide. Risks classically described for these compounds mostly relate to acute toxicity as toxins were identified due to (*i*) poisoning events following the consumption of fish or shellfish or (*ii*) harmful effects through direct contact or exposure to aerosols. Even low densities of toxic algae may be sufficient to cause illness or death in humans, while some species can selectively kill fish by inhibiting their respiration (ichtyotoxic toxins) (Deeds *et al.*, 2004). Five of the most potent and well characterised groups of marine toxins which could appear at desalination plant intakes include; saxitoxin, domoic acid, okadaic acid, brevetoxin and azaspiracid (a comprehensive list is described in Hess *et al.* (2017)). The aforementioned toxins have been classified based on the poisoning syndromes the toxins illicit from studies of shellfish poisoning; paralytic (PSP), amnesic (ASP), diarrhetic (DSP), neurotoxic (NSP) and azaspiracid (AZP) shellfish poisoning.

Table 2 presents information on the molecular size of the major HAB toxins with examples of causative species. Toxins from both pelagic (water column) and benthic (seafloor or epiphytic) micro-algae are considered since intake pipes of desalination plants can be close to surface or close to the seafloor. Many of the toxin classes are not single chemical entities, but instead represent families of compounds of similar chemical structure. A single species typically produces multiple derivatives or congeners within a toxin family.

6.2.3.3 Taste and odour compounds

Geosmin (GSM) and *methylisoborneol* (MIB) are both non toxic volatiles produced by freshwater and marine cyanobacteria and algal species (Suurnakki *et al.*, 2015). Geosmin is a bicyclic alcohol with a distinct earthy flavour and aroma produced by a type of actinobacteria and is responsible for the earthy taste of beets and a contributor to the strong scent that occurs in the air when rain falls after a dry spell of weather or when soil is disturbed. In chemical terms, it is a lipophilic compound and a derivative of decalin. Cyanobacteria are also major producers of geosmin and MIB, another compound potentially adding to poor smelling drinking water (Polak and Provasi 1992; Suurnäkki *et al.*, 2015). MIB and GSM are rarely produced by marine HABs but will cause difficulty in treatment barriers as most treatment processes will not remove either compound. Both MIB and geosmin have smaller molecular weights than the HAB toxins presented in Table 2 (168 and 182 Da respectively).

Table 2 Characteristics of typical marine algal toxins (Villacorte *et al.*, 2015a).

Poisoning syndrome	Toxins	Molecular formula*	Molecular weight (Da)	Causative species
Paralytic shellfish poisoning (PSP)	Saxitoxins	$C_{10}H_{17}N_7O_4$	299.29	*Alexandrium spp.* *Pyrodinium bahamense* *Gymnodinium catenatum Anabaena spp.* *Aphanizomenon spp.* *Cylindrospermopsis spp.* *Lyngbya spp.* *Planktothrix spp.*
Neurotoxic shellfish poisoning (NSP)	Brevetoxins	$C_{50}H_{70}O_{14}$	895.08	*Karenia brevis* *Chattonella veruculosa* and possibly: *K. brevisculcatum,* *K. selliformis,* *K. papilionacea,* *K. mikimotoi*
Diarrhetic shellfish poisoning (DSP)	Dinophysis toxins - okadaic acid - pectenotoxins	 $C_{44}H_{68}O_{13}$ $C_{47}H_{70}O_{14}$	 805.00 859.05	*Dinophysis spp.,* *Prorocentrum lima*
Amnesic shellfish poisoning (ASP)	Domoic acid	$C_{15}H_{21}NO_6$	311.33	Pseudo-*nitzschia spp.,* *Nitzschia navis-varingica,* *Chondria armata*
Azaspiracid shellfish poisoning (AZP)	Azaspiracids	$C_{47}H_{71}NO_{12}$	842.07	*Azadinium spinosum*

* Formula given for parent toxin compound only

6.3 RO CHALLENGES DURING ALGAL BLOOMS

The operational challenges that a RO desalination plant may experience during algal blooms are as follows:

- Contamination of product water with algal toxins;
- Clogging/fouling of pretreatment system; and
- Fouling of RO membranes

6.3.1 Algal toxins

The potential breakthrough of algal-derived toxins to drinking water from RO plants is a public health concern especially in countries relying mainly on desalinated seawater as some of these toxins are highly potent (Caron *et al.*, 2010; Anderson and McCarthy, 2012; Boerlage and Nada, 2014). If the seawater becomes malodorous or fish deaths are evident during a bloom, some desalination plants may assume an algal bloom is toxic and adopt the precautionary measure of shutting down to address community perceptions related to marine algal toxins until the species is confirmed to be non-toxic or the bloom dissipates (Boerlage and Nada, 2014).

Although a WHO drinking water guideline for freshwater algal toxins (e.g., microcystins) is already established, so far there are no World Health Organisation (WHO) guidelines for marine algal toxins. Fortunately, most algal blooms do not produce toxins. Moreover, a significant portion of marine toxins are intracellular, or cell bound and therefore toxins may be removed during pretreatment processes such as DAF and UF where cells remain intact (Dixon *et al.*, 2011). Should the cells lyse or be ruptured during pretreatment e.g., through chlorination to release their toxins extracellularly (See section 6.5.2), RO remains the main barrier in toxin removal, especially extracellular toxin (Boerlage and Nada, 2014, Dixon *et al.*, 2010).

6.3.1.1 Fate of algal toxins through RO

In principle, RO is an excellent barrier for algal toxins and the removal mechanism is the same as for removal of organic micropollutants in contaminated surface water (Bellona *et al.* 2004; Verliefde *et al.*, 2007, 2009; Schoonenberg Kegel *et al.*, 2010). Based on the physico-chemical properties of the five major classes of marine toxins (see Table 2), these toxins should be efficiently removed by RO desalination processes. Although available studies on algal toxin removal by RO are so far limited to laboratory bench-scale tests and pilot studies and only for certain toxins. They all suggest that a >99% removal efficiency can be achieved with a RO membrane (Laycock *et al.*, 2012; Seubert *et al.*, 2012; Boerlage and Nada, 2014; Dixon, 2014). However, the adequacy of these rejection levels is still not conclusive, as extensive data from operational plants during toxic blooms are still not available.

As with the removal of organics, toxin removal will be governed by the properties of the RO membrane and the properties of the specific metabolite itself. Bellona *et al.* (2004) reported that in estimating the rejection of a solute by high pressure membranes (RO), properties such as molecular weight cut-off (MWCO), desalting degree, porosity, membrane morphology, and hydrophobicity of the membrane, and the molecular weight, molecular size, charge, and hydrophobicity of the solute as well as the feedwater chemistry must all be considered.

A couple of studies suggest that the pore size of RO membranes range from 0.6 to 0.7 nm while the molecular weight cut off is between 100 to 300 Da (Dixon *et al.* 2012, Sasaki *et al.*, 2013). Taking into account the aforementioned factors governing solute rejection, if a toxin is larger in molecular weight than approximately 200-300 Da (as a guide), then it will be effectively removed by RO. On the other hand, smaller molecules (50-200 Da) are more difficult to remove by RO such as taste and odour causing compounds MIB and geosmin. The charge of the molecule becomes more important for the 50-200 Da molecular weight range. If the molecule is negatively charged, then the molecule will be repelled from the negatively charged RO surface. If the molecule is positively charged, then it will be attracted to the surface of the membrane and might be sorbed into the polyamide and pass into the permeate (Bellona *et al.* 2004; Verliefde *et al.*, 2007, 2009; Schoonenberg Kegel *et al.*, 2010).

Fortunately, the most common HAB toxins are above 200 Da (see Table 2). Brevetoxin (895 Da) and okadaic acid (805 Da) are far larger than the MWCO of a RO membrane and will therefore be easily removed by size exclusion. The more challenging marine toxins to be removed are the hydrophilic low molecular weight domoic acid (311 Da) and saxitoxin (299 Da) as they are the closest of any HAB toxin to the MWCO of a RO membrane.

Consequently, they have been the focus of studies of algal toxin removal by RO. One such study was undertaken at a seawater RO pilot study in Monterey Bay, California. Due to a lack of natural toxins present, kainic acid was selected as a toxin surrogate to spike into the treatment system as it has a similar chemical structure to domoic acid, but is non-toxic (Desormeaux *et al.*, 2009). Dissolved kainic acid was spiked at concentrations 100 - 1,000 times greater than observed during blooms of domoic acid-producing algae. Removal of the toxin surrogate was greater than 99.5% for two different RO pilot systems.

In a later study, Seubert *et al.* (2012) undertook bench-scale RO experiments to explore the potential of extracellular algal toxins contaminating RO product water using domoic acid, saxitoxin and brevetoxin as test toxins. None were detectable in the desalinated product water in the laboratory studies. Intracellular and extracellular concentrations of domoic acid and saxitoxin in the intake and RO treated water from a pilot RO desalination plant in El Segundo, California were also monitored in the same study by Seubert *et al.* (2012) from 2005 to 2009. During the five-year monitoring period, domoic acid and saxitoxin were detected sporadically in the intake water but never in the RO treated water. Similar results were found by Laycock *et al.* (2012) using a small laboratory-scale RO device to study HAB toxin removal including the smaller saxitoxin and domoic acid toxins in synthetic seawater. Toxin removals of 99.4 and 99.0 %, respectively were found in the study.

6.3.2 Pre-treatment challenges

Most RO desalination plants are installed with a pre-treatment system comprising at least one filtration process preceding the RO (see Chapter 3). Granular media filters (GMF) has been used for many years as the main pre-treatment method. However, low pressure membrane filtration processes such as microfiltration (MF) and ultrafiltration (UF) are increasingly applied as alternatives. Auxiliary processes such as chlorination-dechlorination, coagulation, dissolved air flotation, micro-screens and/or cartridge filters are usually integrated to stabilise operation of the main pre-treatment process and/or enhance RO feedwater quality. During algal blooms, the high biomass load in seawater can drastically affect both the operation and effluent quality of the main pre-treatment system. GMF are particularly vulnerable to severe blooms but MF/UF can suffer operational challenges as well.

6.3.2.1 Clogging of granular media filters

The primary function of GMF in RO pre-treatment is to reduce high loads of particulate and colloidal matter. GMF relies on depth filtration to remove particles/colloids from the RO feedwater. However, when high concentrations of organic matter or particulate loads are encountered during algal blooms, (inline) coagulation is required to ensure that RO feed water of acceptable quality is produced. The coagulation process aggregates algal biomass in seawater into large flocs, which is capable of rapidly blocking the interstitial voids of the granular media and may therefore shift filtration mechanism from depth filtration to surface straining (cake filtration). As filtration rates are relatively high (5-10 m/h) in GMF, cake filtration during algal blooms can result in exponential head loss through the filters. This can drastically reduce water production flow, shorten filter runs and increase downtime due to frequent backwashing (Pankratz, 2008; Schippers, 2012). Moreover, a high biomass load in inlet water can exhaust rapidly the adsorption capacity of GMF, resulting in periodic breakthrough of biomass to the RO feedwater. Blooms caused by diatoms, green algae, flagellates, and cyanobacteria can clog GMF (Edzwald, 2010). Clogging of GMF at high algal biomass concentration of approximately 27,000 cells/mL were reported to be at least partially responsible for the shutdown of RO plants in the Gulf region (Richlen *et al.*, 2010).

6.3.2.2 Fouling of MF/UF

Surface filtration is the main mechanism of separating particles from water in MF/UF membranes. Retention of colloids inside the membrane pores can also occur particularly in the beginning of the filtration run. During algal blooms, algal cells and particulate (and part of colloidal) AOM are retained on the membrane surface while part of the colloidal AOM pass through or retained inside the pores of the membrane. Over time, a cake or gel layer accumulate on the surface which forms an additional but more dense retention layer, eventually retaining more colloidal AOM on the surface. A membrane cake comprising algal biomass characteristically has high hydraulic resistance and compressible at high pressure drop. This cake layer might only be partially removed during backwashing due to the gluey nature of AOMs (e.g., TEPs), resulting in progressively lower permeability (or higher feed pressure if operated at constant flux) in the succeeding filtration cycles.

In practice, fouling can be minimised by dosing a coagulant inline in the water before filtration (see Section 6.5.6). However, Voutchkov (2010) reported that in a submerged vacuum-driven UF system, a driving vacuum higher than 0.4 bar can cause disruption of soft-walled algal cells resulting in the release of easily biodegradable dissolved intracellular substances which might be detrimental to the operation of downstream RO. So far, such issues have not been reported nor verified in pressure-driven UF systems.

When algae and AOM accumulate on the MF/UF membrane pores and surface, the impact on membrane permeability and backwashability can be explained based on known fouling mechanisms, as illustrated in Figure 5 and explained further in the succeeding sections.

Membrane cake and pore constriction (Figure 5a). Colloidal AOMs can enter into the narrow pores of MF/UF membranes, some of which will adsorb to the pore wall and eventually cause partial blockage of permeate flow (Herman and Bredee, 1936). This can cause rapid increase in trans-membrane pressure (TMP) during the initial stage of filtration. Algal cells and large AOM can form a cake/gel layer on the surface of the membrane. Colloidal AOMs and other colloids will then fill up the large interstitial voids of the cake, narrowing the voids in the process. This may result in substantial increase in cake resistance due to the gradual reduction in cake porosity. During backwashing, the sticky AOM accumulated inside the membrane pores may not be completely removed, resulting in only partial recovery of the initial membrane permeability.

Substantial loss in effective filtration area (Figure 5b). Colloidal AOM can accumulate inside membrane pores while algal cells and large AOM can accumulate at the entrance of the pores. In both cases, some pores may be completely blocked by the material and the active filtration area (membrane surface porosity) is substantially reduced, resulting in higher localized flux for the remaining active pores (Herman and Bredee 1936). An increase in flux can cause proportional increase in membrane resistance to filtration. Additionally, non-backwashable fouling can occur if the foulants blocking the pores are not effectively removed by hydraulic backwashing.

Incomplete cake/gel removal during hydraulic backwashing (Figure 5c). Since algal cells (typically range from 2 μm to 2 mm) and a substantial fraction of AOM are much larger than the pores of commercial MF/UF membranes (<1 μm), cake/gel formation can be mainly responsible for the increase in TMP. The accumulated AOM (like TEPs) are typically sticky and tend to adhere strongly to the membrane surface. During backwashing, a layer of the cake may remain on the surface of the membrane, which will then cause additional filtration resistance in the subsequent filtration cycle.

Compression of accumulated cake/gel (Figure 5d). Filter cake/gel comprising AOM and algal cells (soft-bodied) can be compressed due to localized increase of flux. Such localized increase in flux may be a consequence of pores narrowing and/or complete blocking as described above and hence occurs in combination with these fouling mechanisms.

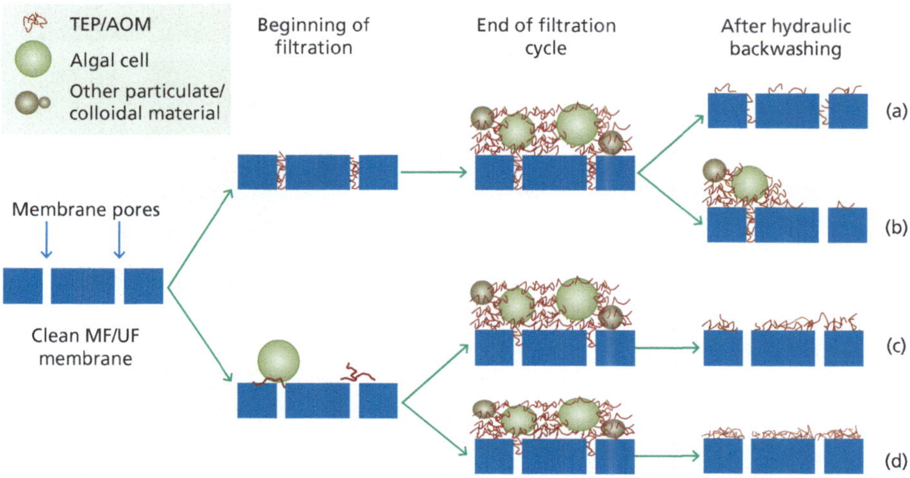

Figure 5 Possible fouling mechanisms involved due to uniform deposition of AOM and small algal cells in MF/UF. Each process is explained in detail in the text. Adopted from Villacorte (2014).

6.3.3 RO fouling

The performance of the RO during algal blooms largely depends on the effluent quality of the pretreatment system. A significant concentration of algal biomass in the feedwater can cause irreversible fouling issues in RO membranes. If algal cells and AOM are not effectively removed by the pretreatment process, they can accumulate in the RO module, clogging spacer channels and/or forming a heterogeneous and compressible cake layer on the surface of the membrane. These can lead to significant reduction in RO permeability, increased feed channel pressure drop, and increased salt passage.

Algal blooms can accelerate biofilm accumulation in RO (Villacorte *et al.*, 2017a). Some AOM components, specifically TEPs, tends to adhere and accumulate on the surface of the membranes and spacers. Such accumulation can serve as a "conditioning layer" – a platform for effective attachment and initial colonization by microbes - where bacteria can effectively utilize biodegradable nutrients from the feedwater (Berman and Holenberg, 2005; Winters and Isquith, 1979). Furthermore, TEP can be partially degraded and may later serve as a substrate for bacterial growth (Passow and Alldredge, 1994; Alldredge *et al.*, 1993). As illustrated in Figure 5, TEP (and their pre-cursors) and protobiofilms (suspended TEP with extensive microbial outgrowth and colonization) in surface water can initiate, enhance and possibly accelerate biofilm accumulation in RO membranes (Berman and Holenberg, 2005; Bar-Zeev *et al.* (2012a).

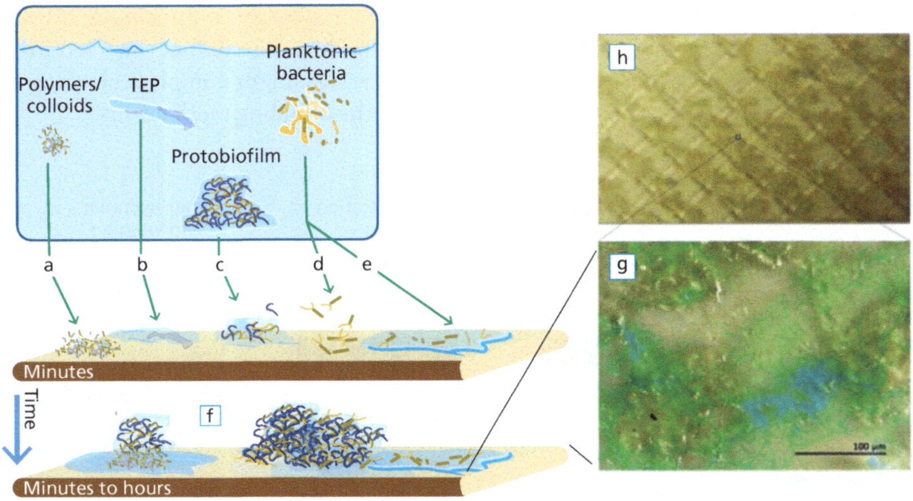

Figure 6 Schematic illustration of the possible contribution of (a) colloidal biopolymers, (b) TEP, and (c) protobiofilm in the initiation of aquatic biofilms. A number of planktonic bacteria (first colonizers) can attach (d) reversibly on clean surfaces or (e) irreversibly on TEP-conditioned surfaces. When nutrients are not limited in the water, (f) contiguous coverage of mature biofilm can develop within a short period of time (minutes to hours). Biofilm accumulation can cover a significant surface area of a (g) spiral wound membrane. Operational issues will occur when substantial accumulation (h) obstructs permeate and feed channel flow. Photos and descriptions adapted from (a-f) Bar-Zeev *et al.* (2012a), (g) Villacorte *et al.* (2009) and (h) Villacorte (2014).

Since bacteria require nutrients for energy generation and cellular synthesis, essential nutrients such as biodegradable or assimilable organic carbon (BDOC or AOC), phosphates and nitrates are likely the main factors dictating the formation and growth rate of biofilm in RO modules. During the peak of an algal bloom, some of these essential nutrients may be limited (e.g., phosphate, nitrates) due to uptake by algae. However, when the bloom reaches the death phase (see Figure 3), algal cells start to disintegrate while releasing some of these nutrients. Hence, biofouling initiated or enhanced by AOM will likely occur in RO some time (depending on available nutrients) after the termination of an algal bloom. So far, the role of AOM on membrane biofouling has only been illustrated in lab-scale experiments but some membrane autopsy studies shown abundance of TEP among the biofilm found in fouled brackish and seawater RO elements (Figure 6; Villacorte *et al.*, 2009; Villacorte, 2014).

6.4 ALGAL BLOOM MONITORING IN RO PLANTS

Regardless of their location, RO plants can be affected by algal blooms resulting in a substantial increase in the organic and solids load in the source water to be treated. Characterization of the raw water and monitoring to detect poor water quality events including HABs is critical throughout the lifetime of a desalination plant from design through to operation. It is therefore important to establish a monitoring programme to measure the potential impact of algal blooms and to assess the effectiveness of a proposed or existing pretreatment system in preventing operational issues in the RO plant. This requires methods to measure the concentration of algae, AOM, membrane fouling potential and other associated water quality changes in the raw water and treatment process streams. Proper monitoring should allow operators to respond to a bloom in a timely manner to optimize plant operation and avoid disruption to water supply (Boerlage *et al.*, 2017b). The following sections introduce the relevant water quality parameters for algal bloom monitoring in RO plants and the various methods to measure them.

6.4.1 Conventional parameters

Typical online parameters continuously measured at a RO plant intake may include temperature, conductivity, dissolved oxygen (DO), pH and turbidity. None of the aforementioned parameters are specific to algal blooms. Changes in these core physiochemical parameters can be caused by other factors such as pollution events and/or marine hydrodynamics, thus the interpretation of these water quality variables can be complex. Nonetheless, these measurements may indicate conditions that promote a bloom, such as temperature and salinity or indirect impacts from algal blooms such as low DO following degradation of a dense bloom. In conjunction with other conventional water quality tests such as SDI, TSS, the standard online water quality parameters can be useful in indicating a deterioration in feedwater quality due to HAB events and can provide timely and valuable information that action is required (Boerlage *et al.*, 2017b). Case studies in the HAB guidance manual for seawater RO plant operators have illustrated this (Boerlage *et al.*, 2017c). For instance, an increase in TSS was observed at the Abengoa pilot plant during a bloom and low DO following decomposition of a dense bloom at the La Chimba plant. Elevated SDI at the intake corresponded to algal bloom events at Fujairah 2, Barka 1, Sohar and Gas Atacama plants during bloom events (Boerlage *et al.*, 2017c). However, care needs to be taken in interpreting SDI results due to its inherent limitations (Section 6.4.4; Boerlage, 2008; Al-Hadidi, 2011).

6.4.2 Algae concentration

The magnitude of algal blooms can be measured at the intake of an RO plant either in terms of cell count or chlorophyll-a concentration. Cell count gives an indication of the relative abundance of individual species, while chlorophyll-a is a bulk parameter that includes many different, co-occurring algal species. Furthermore, spatial abundance of algae can be monitored using satellite remote sensing tools.

6.4.2.1 Cell count

Cell abundance is usually measured by microscopic enumeration using microscope and a counting chamber. A standard compound (upright) microscope is used with chamber methods (e.g., haemocytometer, Palmer-Maloney and Sedgewick-Rafter) while an inverted microscope is used when using the Utermöhl method (Karlson *et al.*, 2010). The Utermöhl method has an advantage over other methods in that algal cells can be both identified and enumerated (Edler and Elbrächter, 2010). More advanced techniques such as fluorescence microscopy, flow cytometry and molecular techniques are also available. Flow cytometry can detect autofluorescence and scattering properties of algae which can be used to distinguish different types of algae. A more advanced type of flow cytometer is fitted with a camera that generates images of each particle/organism. An automated image analysis software makes it possible to count and identify algae in-situ (Olson & Sosik, 2007).

To compensate for the size differences of bloom forming species, cell concentration can be expressed in terms of volume fraction (total cell volume per volume of water sample) instead of cell number per volume of water. Operationally, this is difficult to calculate, as it requires conversion factors on the cell volume of each species that might be encountered.

6.4.2.2 Cholorophyll-a

Chlorophyll-a is a common proxy for algal biomass. Monitoring its presence provides a useful estimate of algae concentration and its spatial and temporal variability in the water. There are various techniques to measure chlorophyll-a, including spectrophotometry, high performance liquid chromatography (HPLC), and fluorometry. The fluorometric method is widely used for quantitative analysis. Such method is time consuming and usually require an experienced analyst to generate accurate and reproducible results. To overcome these challenges, online optical sensors for chlorophyll-a determinations can be used in continuous monitoring applications but with lower accuracy. The sensor generally works by irradiating the sample with light typically at 470 nm wavelength while chlorophyll from algal cells emits light at higher wavelength (e.g., 650-700 nm). Chlorophyll-a concentration is then estimated based on the emitted light intensity after passing through a filter.

The relative amount of chlorophyll-a to algae biomass can vary substantially between species (Karlson *et al*, 2017). It can also vary between the same species growing at different seasons or water depths because they adapt to the changing levels of photosynthetic active radiation (PAR) over seasons and depths. In practice, it has been reported that for low concentrations of algae (ca. 5,000 cells/ml) no increase in chlorophyll-a above background was observed (Boerlage *et al*, 2017b). Nevertheless, chlorophyll-a is still considered to be relatively more reliable than volume fraction or biomass estimated from cell counts because the latter tends to overlook the picoplankton species which are smaller than 2 μm and undiscernible through a light microscope (Leblanc *et al.*, 2018).

6.4.2.3 Remote sensing to monitor algal bloom transport and landfall

Horizontal transport of biomass over a large area in the water body is an important feature of many algal blooms. For instance, major toxic blooms can suddenly appear at a site due to the transport of established blooms from other areas by ocean currents. Hence, advance warning of imminent outbreaks can be potentially useful in RO desalination. Large-scale algal blooms off-shore can be monitored using satellite optical sensors coupled with numerical models to forecast the transport and landfall of such blooms (Stumpf *et al.*, 2009; Wynne *et al.*, 2011). Typically, ocean colour data from satellite remote sensing and algorithms for chlorophyll-a are used to map and simulate the spread of the bloom over specific area of the sea (see Figure 7). Since this technique relies on satellite data, cloud cover and water turbidity may significantly reduce its accuracy. Moreover, estimates of algae concentration will be limited to those forming near the water surface and tends to overlook vertical migrations of algae. Although this emerging application is still subject to intensive research and verification, it has good potential in developing an early warning system for RO plants. As such, remote sensing is still far beyond the technical or financial resources of many RO plants, but regional approaches to this type of technology are being explored. Efforts are underway to combine the available remote sensing technologies into forecast systems that would be of value to multiple desalination plants within large regions (Karlson *et al.*, 2017).

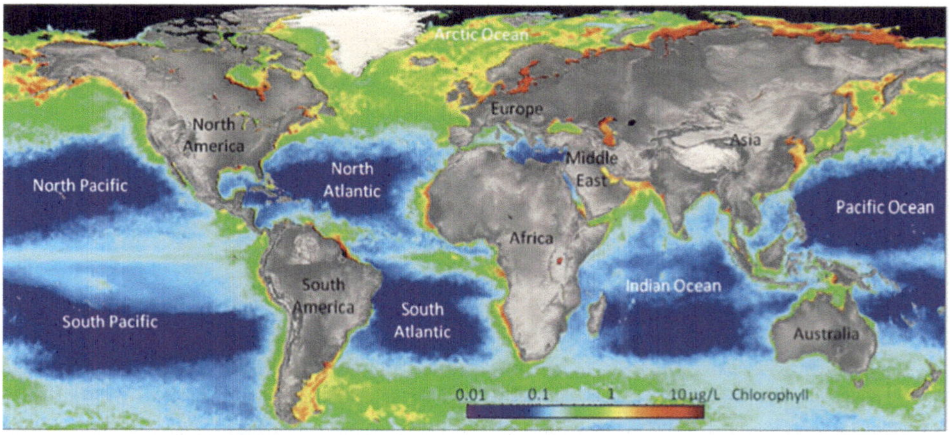

Figure 7 A composite map of average annual distribution of chlorophyll a in large surface water bodies in 2009. Figure modified from map generated by Gledhill and Buck (2012) with data from MODIS-Aqua.

6.4.3 Algal organic matter characterisation

Algal blooms are often responsible for the highest annual pulses of NOM in seawater. Significant spikes in organic carbon concentration has been recorded in coastal seawater during algal blooms (e.g., Petry *et al.*, 2007). Nevertheless, AOM produced during algal blooms may vary substantially in terms of concentration, composition and characteristics, depending on the causative species and environmental conditions. Total organic carbon (TOC) and dissolved organic carbon (DOC) are common measures of the concentration of NOM at desalination plant intakes and are often used to assess the efficiency of pretreatment processes in removing organics. Ultraviolet absorption at 254 nm (UV254) and the related specific ultraviolet absorbance (SUVA) are used to a lesser extent. More sophisticated techniques are used to characterize the composition and concentration of AOM such as liquid chromatography - organic carbon detection (LC-OCD), fluorescence excitation emission matrix (FEEM) and TEP as shown in Figure 8. In some cases, more specific AOM components such as algal toxins and taste and odor compounds are also monitored, albeit rarely, not because of its fouling potential but due to its potential impact to the product water quality of the RO plant.

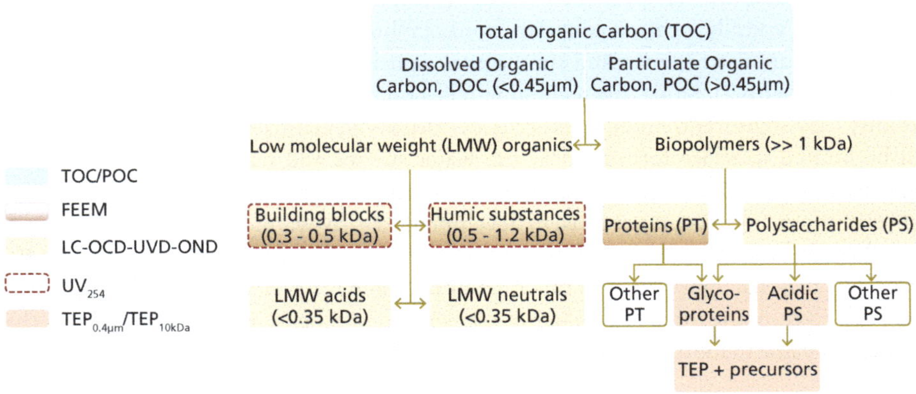

Figure 8 Major components of aquatic organic matter and corresponding analytical techniques for their identification and quantification. Legend: LC is liquid chromatography with inline detectors for organic carbon (OCD), UV absorbance at 254 nm (UVD,) and organic nitrogen (OND). FEEM is fluorescence excitation-emission matrices. TEP refers to transparent exopolymer particles measured with a 0.4 µm or 10 kDa membrane. Adopted from Boerlage *et al.* (2017b).

6.4.3.1 Liquid chromatography - organic carbon detection (LC-OCD)

LC-OCD is a semi-quantitative technique which can be used to fractionate AOM into six major sub-fractions which could be assigned to specific classes of compounds based on their retention time through a chromatogram column (Huber *et al.*, 2011). The high molecular weight fraction is classified as biopolymers, which can be further divided into polysaccharides and protein components when LC-OCD is coupled with an organic nitrogen detector (OND). The low molecular weight fractions (< 1 kDa) are sub-classified as humic-like substances, building blocks, low molecular weight acids, and neutrals. Considering that the sticky high molecular weight faction of AOM are likely to deposit/accumulate in the RO system, measuring the biopolymer fraction of organic matter in the water is a promising indicator of organic and biological fouling potential of AOM (Villacorte, 2014; Villacorte *et al.*, 2017a). Although not proven via LC-OCD analysis, one can expect biodegradable/assimilable low molecular weight acids to contribute to the biological fouling as well.

6.4.3.2 FEEM

Qualitative assessment of the presence of protein-like and humic-like organics based on fluorescence excitation-emission matrices (FEEM) has been employed for characterisation of AOM (Henderson *et al.*, 2008b; Villacorte *et al.*, 2015b). The water sample is excited by a light source to a specific wavelength at which AOM fluorophores absorb light and subsequently emit the light at longer wavelength. This technique is performed using a spectrofluorometer across a spectrum of light wavelengths. The acquired data is then plotted in a 3D fluorescence contour for analysis. This method can identify proteins and humic-like AOMs but not polysaccharides.

6.4.3.3 TEP concentration

Various methods have been introduced to quantify TEP and their precursors. The first ever TEP method is a direct method based on alcian blue staining and optical microscopic enumeration (Alldredge *et al.*, 1993). This method provides useful information on the size-frequency distribution of TEP in seawater, but not feasible for quantifying TEPs < 2 μm and their precursors. The succeeding methods based on semi-quantitative spectrophotometric techniques were able to address these issues. The method by Passow and Alldredge (1995) also known as TEP0.4μm is the most widely used. With additional sample preparation techniques (e.g., bubble adsorption, laminar shear) TEP precursors can be measured using such methods (Zhou *et al.*, 1998; Passow 2000). Two alternative methods were introduced by Arruda-Fatibello *et al.* (2004) and Thornton *et al.* (2007), which are capable of measuring both TEP and their precursors without laborious sample pretreatment. However, the former is only applicable in freshwater samples while the latter requires a dialysis step for saline samples. Further modification of the method, known as TEP10kDa, was later introduced to address various practical limitations of these methods through introduction of a concentration step by filtration through 10 kDa membrane (Villacorte *et al.*, 2015c; Villacorte *et al.*, 2017c). In principle, this method enables size fractionation of TEPs in seawater using a series of membranes with different pore sizes during the extraction step.

More recent studies have shown that TEP can be measured online using an auto-imaging technique (Thuy *et al.*, 2017) or a cross-flow filtration unit with integrated spectrophotometer (Sim *et al.*, 2018). Although still not demonstrated in RO plants, online measurement techniques like these would be the next logical step towards routine TEP monitoring during algal blooms.

6.4.3.4 HAB toxins

A wide range of analytical methods are available for detecting and measuring algal derived toxins, but they can vary greatly in terms of level of detection and sophistication. An extensive overview of these techniques and their limits of detection is presented elsewhere (Hess *et al.*, 2017). Liquid chromatography coupled to tandem mass spectrometry (LC-MS/MS) has been extensively used to measure commonly occurring algal toxins such as domoic acid, saxitoxin, microcystins and azaspiracids. It has also been adapted for detection and quantitation of multiple groups of toxins in a single analysis. For direct analysis of seawater, the sensitivity of such technique by itself might be insufficient and pre-concentration or other sample pretreatment techniques is typically required to achieve desirable detection level (Zendong *et al.*, 2015). Other methods used within various research laboratories for screening and analysis of algal toxins includes ELISA methods for microcystins, neuroblastoma cytotoxicity assay, saxiphilin and single-run HPLC methods for saxitoxins. The detection limit may vary depending on the available standards and the instrumentation used.

While the above methods are useful to determine the concentration of toxins in seawater and the final desalinated drinking water, they do not lend themselves to routine analysis. RO desalination plants are typically not equipped for such analysis and samples need to be sent to external laboratories for analysis resulting in time delays. Moreover, the risk for toxins in a HAB to be abstracted at a desalination plant intake often goes unrecognised as some toxic HAB never reach the densities to colour the water. In addition, toxic blooms are normally only short-lived intermittent phenomena in a particular location, dispersing within days (Boerlage and Nada, 2014). Therefore, in line with risk management approaches to water quality risks such as toxins, continuous monitoring of the integrity of RO membranes as a critical barrier to intra- and extracellular toxins using conductivity (as a surrogate for salt rejection) is an imperative to ensure toxin removal (see section 6.3.1.2; Boerlage and Nada, 2014).

6.4.3.5 Taste and odour compounds

The most common method currently used for quantitative analysis of organic taste and odour compounds in water is gas chromatography combined with mass spectrometry (GC/MS). As these compounds need to be detected at very low concentrations (ng/L levels), a pre-concentration method is often required. The most important methods used for the pre-concentration step are closed-loop stripping analysis (CLSA) and solid phase microextraction (SPME). CLSA is widely used for the analysis of non-polar volatile organic compounds such as geosmin and MIB, at the ng/L to µg/L level (Krasner *et al.*, 1983). The compounds are stripped from the water by a recirculating stream of air and then adsorbed from the gas phase onto a few milligrams of activated carbon. They are then extracted from the carbon with a few µL of carbon disulphide for direct analysis. This method can be applied to both raw and treated waters. The limit of detection (LOD) for CLSA is typically 1-2 ng/L. SPME has gained more popularity in recent years because it is simpler and more cost-effective method than CLSA (Huang *et al.*, 2004). The LOD for this method were reported as 1-2 ng/L and 4 ng/L for geosmin and MIB, respectively.

6.4.4 Particulate fouling potential

During and following an algal bloom, the organic and solids/particulate load in seawater can rapidly increase and hence the associated risk of fouling can increase. There are two established methods to measure the particulate fouling potential of RO feedwater, namely: Silt Density Index (SDI) and Modified Fouling Index (MFI). See Chapter 4 for the detailed information on the tests and their derivation. Despite its well documented limitations, the SDI (ASTM D8002-15e1, 2015) has been widely used in practice for the last 50 years due to its simplicity and remains the basis of many membrane guarantees and other plant performance contracts (Boerlage, 2008; Boerlage *et al.*, 2017b). The MFI was subsequently developed (Schippers, and Verdouw, 1980) to overcome the deficiencies of the SDI and can employ membranes of smaller pore sizes to capture smaller and more fouling particles such as TEP precursors (Boerlage *et al.*, 2002, Villacorte, 2014, Salinas-Rodriguez, 2015). MFI tests, using MWCO ranging from 150 kDa decreasing in size to 10 kDa, have recently been investigated for application in RO plants prone to algal blooms and are denoted as MFI-UF with the MWCO of the test membrane shown as subscript (e.g., MFI-UF_{10kDa}).

Both the SDI and MFI indices are non-specific for algal-related particulate material as inorganic and non-algal organic colloids and bacteria may also contribute to the result. Nonetheless, they can indicate the presence of a bloom at a plant intake and can to some extent be used to assess process efficiency as illustrated in case studies of plants experiencing a bloom, with more results available for the SDI than MFI (Boerlage *et al.*, 2017c). For instance, elevated SDI at the seawater intake corresponded to algal bloom events at Fujairah 2, Barka 1, Sohar and Gas Atacama plants during bloom events and alerted operators of a deterioration in seawater quality. However, severe algal bloom events can significantly increase the fouling potential of seawater at open intakes to the point that the SDI may not be measurable due to rapid plugging of the SDI test membrane. SDI results are reported according to ASTM where the time interval for the test (5,10 or 15 minutes) are indicated respectively as SDI_5, SDI_{10} and SDI_{15}. Operators may consider measuring the SDI at intervals less than the ASTM minimum of 5 minutes e.g., 3 minute or even 1-minute intervals as was done during various bloom events outlined in the HAB Manual. Care should be taken in interpreting such results as values are unreliable as the SDI is not linear with particle concentration, especially with highly fouling feedwater.

Increases in SDI downstream of the intake may indicate the failure of pretreatment steps and that operator action is required, but high SDI15 and SDI3 values such as those reported during the 2008 Gulf bloom may have underestimated the fouling potential of the feed or various desalination process streams. When assessing process performance, it should be remembered that the SDI cannot be directly compared for different filtration intervals e.g., SDI_5 for raw water and SDI_{15} after pretreatment or when measured at different temperatures as SDI test applies no temperature correction for differing feedwater temperature (Boerlage, 2008).

In comparison to the SDI, the MFI is not limited to low fouling feedwater. It can therefore be used to measure the fouling potential of seawater with a high biomass as observed during a bloom. More importantly, smaller MWCO MFI-UF test membranes on the order of 10 kDa can capture some of the TEP precursors (ranging in size from a few nm up to 0.4 μm) as well as TEP (>0.4 μm) present in a bloom which cause fouling on both UF and RO membranes (Boerlage *et al.*, 2017b, Villacorte *et al.*, 2015d).

Preliminary applications of the MFI-UF with a range of MWCO test membranes have proven promising in assessing the efficiency of pretreatment processes for particles of various sizes during algal blooms or determining coagulant dose in laboratory bloom studies (Villacorte, 2014, Salinas-Rodriguez, 2015 and Tabatabai, 2014 as cited in Boerlage *et al*, 2017b). For example, the fouling potential of seawater during an algal bloom was reduced following coagulation and ultrafiltration pretreatment by 94%, 93%, and 88% for 100, 50, and 10 kDa MFI-UF test membranes, respectively (Salinas-Rodriguez, 2015). In MFI-UF laboratory experiments conducted by Tabatabai *et al.* (2014) to optimise coagulant dose, a larger 150 kDa MWCO test membrane was used for seawater solutions containing algal organic matter (0.5 mg/L as biopolymers). MFI-UF measurements showed the addition of 5 mg/L Fe reduced the fouling potential seven fold in the seawater with no measureable reduction when the coagulant dose was doubled.

6.4.5 Biological fouling potential

Multiple parameters and methods have been proposed to measure the biofouling potential in RO feedwater (see Chapter 5 for more detailed description of the methods). Water sample-based methods such as adenosine triphosphate (ATP), assimilable organic carbon (AOC) and biodegradable dissolved organic carbon (BDOC) has been used (Vrouwenvelder and van der Kooij, 2001; Amy *et al.*, 2011). Considering the potential impact of algal blooms, Liberman and Berman (2006) proposed to measure a set of parameters to determine the microbial support capacity RO feedwater, namely chlorophyll-a, TEP, bacterial activity, total bacterial count, inverted microscope observations of settled water samples, biological oxygen demand (BOD), total phosphorous and total nitrogen. Alternatively, inline monitors such as a biofilm monitor and a membrane fouling simulator (MFS) can be used to measure biofilm formation rate in RO (Vrouwenvelder and van der Kooij, 2001; Vrouwenvelder *et al.*, 2006). Most of the above-mentioned parameters and methods have been routinely used for non-saline waters including some RO plants.

Currently established AOC and ATP methods are not directly applicable to seawater due to high salt concentration (Amy *et al.*, 2011). However, newer AOC and ATP bioassays has been developed to overcome these challenges and can be routinely used to assess the biofouling potential of seawater during algal blooms and the efficiency of pretreatment to reduce it (Schneider *et al.*, 2012; Weinrich *et al.* 2011; Abushaban *et al.*, 2017). Currently, data showing correlation between either ATP or AOC and biofouling in seawater RO during algal blooms is still limited, due to the relatively recent developments of appropriate bioassays for seawater studies. Without using a bioassay, MFS can be used to measure the impact of AOM on the biofilm accumulation in RO spacers. Recent studies have demonstrated using MFS that the presence of AOM in the feedwater can indeed accelerate biofilm growth in RO (Villacorte *et al.*, 2017a; Dhakal, 2017).

6.5 OPERATIONAL & PRETREATMENT STRATEGIES

In addition to establishing a water monitoring programme to anticipate operational challenges in a RO plant during algal blooms, it is equally important that the operational strategies and pretreatment design implemented can cope with such challenges. A reliable pretreatment system is one that can continuously produce high quality RO feedwater while maintaining relatively stable hydraulic operation in terms of flow and pressure.

6.5.1 Toxin risk management in RO plants

WHO advocates a risk management approach to water quality where hazards to water quality such as toxins are identified, multiple barriers to hazards are developed and critical control points determined to ensure the hazards are controlled to reduce the residual risk to a negligible level. While pretreatment processes of desalination plants may provide multiple barriers for the removal of intracellular toxins through cell removal, the main barrier for removal of toxins extra- or intra-cellular is the RO membrane. Therefore, the RO would be defined as a critical control point (CCP) for algal toxin removal and the integrity of the membrane would be continuously monitored using conductivity (as a surrogate for salt rejection) to ensure toxin removal (Boerlage and Nada 2014). Alert and critical limits based on conductivity would be defined for the RO system and corrective actions undertaken to identify and eliminate the cause of the deviation in conductivity to bring the CCP back under control.

In practice, RO rejection efficiency losses can occur due to (1) chlorination and oxidation of the membranes; (2) accidental overdose of acid to below pH 3 for an extended period of time; or (3) an abundance of rolled permeate seals in the pressure vessels. In each case, the allowable permeate TDS would be exceeded, causing plant alarms for high conductivity in the first and second pass permeate. A salt passage increase (e.g., due to membrane ageing or oxidation) in the RO process unit would occur far sooner than any increase in product water toxin concentration. For this reason, a major increase in permeate TDS could be used to detect an integrity breach that could later lead to an increase in permeate toxin concentration. In a hypothetical study by Dixon *et al.* (2015), a set of theoretical RO projections were undertaken to understand the failure mode of how damage to the RO membrane may affect the permeate saxitoxin concentration during a typical bloom. For this hypothetical modelling study, RO inlet saxitoxin concentration was 10 μg/L (see Figure 9). To exceed a hypothetical local saxitoxin guideline value (1 μg/L) after the first pass, the plant would need to experience a first pass NaCl rejection reduction from 99.7 to 99.0%. Given a full two pass system, the second pass permeate would be approximately 0.3 μg/L even with damaged first pass elements. Given a partial split system with only 25% of water sent to second pass, this would still produce a combined permeate saxitoxin concentration under the 1 μg/L guideline limit for the above hypothetical scenario. In desalination plants susceptible to HABs that produce saxitoxin, a prudent monitoring program would include toxin monitoring of the feed seawater, the first pass permeate and the second pass permeate to confirm total removal of saxitoxin.

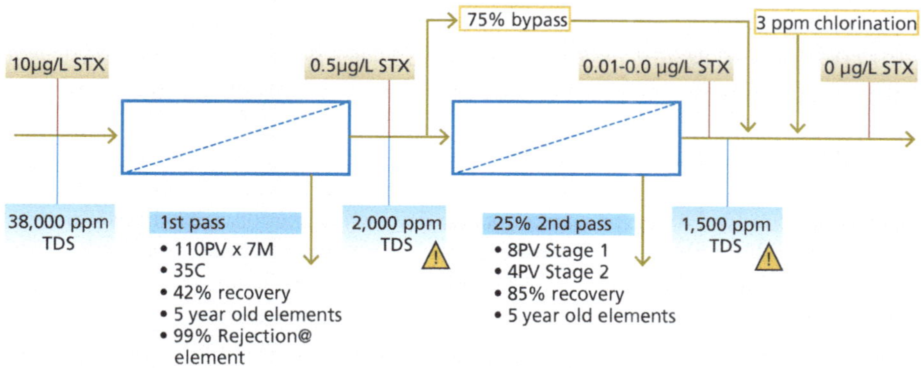

Figure 9 A summary showing a hypothetical scenario for saxitoxin removal through a typical partial two pass RO system. The figure illustrates that alarms will be generated for 1st pass and 2nd pass TDS before saxitoxin reaches a hypothetical local guideline concentration of 1 μg/L. Adopted from Dixon *et al.* (2015).

6.5.2 Seawater intake design considerations

Macroalgal (seaweeds) blooms may impact the performance of conventional surface intake structures due to clogging of intake screens. Such screens are typically installed in surface intakes to mitigate the impingement and entrainment of marine life larger than 1 mm.

Microalgal blooms generally do not impact surface intake structures but can substantially impact downstream processes in RO plants. A careful selection of the location and depth of a surface intake is an important consideration in areas prone to algal blooms. Installing a deep-water open ocean intake may prevent entry of micro-algae, but some algal species are motile and can migrate vertically within the water column. In such case, the RO plant could operate according to a predefined schedule, albeit with reduced production capacity. For instance, operating with deep intake during the day when algae are more likely to be found at the surface or by operating shallow intake during the night when algae tend to migrate to the lower water column (Anderson, 2017; Boerlage *et al.*, 2017a). Nevertheless, the distribution of extracellular AOM may not reflect the distribution of algal cells, so it can remain a challenge to the downstream processes (Boerlage *et al.*, 2017a).

RO plants with sub-surface intakes, especially vertical beach wells (see Chapter 3), are less vulnerable to algal blooms since such intakes can serve as a natural (slow) sand filters with relatively long retention times, that can substantially enhance removal of algae, bacteria and AOM from seawater entering seawater RO plants (Missimer *et al.*, 2013). Sub-surface intake structures have been reported to virtually remove algal cells and a significant fraction of bacteria (90-99%), biopolymers (>70%) and TEPs (34-92%) from seawater (Rachman *et al.*, 2014; Boerlage *et al.*, 2017a). Consequently, less-extensive pretreatment processes are needed to maintain stable operation in the plant. Unfortunately, sub-surface intakes are not always possible in some coastal locations where the geology (e.g., high mud content sediments, low permeability rocks) makes it unfeasible to install such structures due to high energy costs. So, depending on the local hydrogeology and the concentration of algae and duration of an algal bloom event, it is likely that most of the subsurface intake systems may allow a RO plant to operate continuously during a bloom without interruption (Boerlage *et al.*, 2017a). However, documented operational experience and treatment performance of RO plants with subsurface intake operating during a bloom is still rather scarce.

6.5.3 Chlorination and de-chlorination

Most RO plants practice intermittent chlorination/dechlorination at the intake with doses of up to 10 mg/L added for up to two hours on a daily, weekly or biweekly basis (Boerlage *et al.*, 2017a). Chlorination is now most commonly used on a periodic basis rather than continuously because it is known to cause biofouling of the downstream membranes (Winters, 1997). During algal blooms, chlorination at the intake can lyse algal cells which may complicate downstream processes if not managed correctly. Shock chlorination leads to more aggressive lysis of algal cells and subsequent coagulation pretreatment processes may not remove all AOM (see Figure 10). These AOM will eventually reach the RO membrane potentially causing organic and biological fouling. A strategy for avoiding cell lysis is to avoid shock chlorination during an algal bloom. A low continuous dose (0.1-0.2 mg/L) of hypochlorite may be a better approach to minimize the lysis of algal cells, while releasing some AOM to assist coagulation.

Since polyamide RO membranes are susceptible to oxidative degradation from chlorine, dechlorination of the RO feedwater (after main pretreatment) upstream of the RO membranes is always necessary when chlorination is implemented. This is achieved by adding a reducing agent - typically sodium metabilsulfite (SMBS). In theory, 1.34 mg of SMBS will remove 1.0 mg of free chlorine. In practice, however, 3.0 mg of SMBS is normally used to ensure complete dechlorination of 1.0 mg of chlorine (Dow Water and Process

Solutions, 2015). SMBS might lyse algal cells, but if pretreatment is operated efficiently, very few cells will be exposed to it prior to entering the RO train. Given SMBS is routinely used to preserve RO elements for long term storage, it may prevent biofouling in RO membranes to some degree.

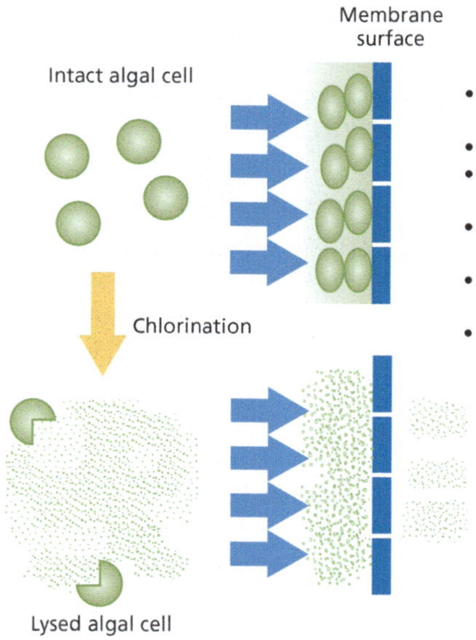

- Intact algal cells prevent IOM from passing through the UF membrane pores.
- Mucilage surrounding the algal cells is retained.
- Rapid transmembrane pressure increase due to compressible fouling layer.
- Released IOM can now transmit through the membrane pores.
- Mucilage now attaches to the membrane surface, preventing efficient cleaning.
- Reduced short-term fouling

Figure 10 Possible effects of algal cell lysis due to chlorination on UF membrane fouling and rejection. Adopted from Resosudarmo *et al.* (2017).

6.5.4 Dissolved air flotation

DAF is a clarification process suitable for removal of low-density particles that can float such as, algal cells, oil and grease, which are not effectively removed by just sedimentation or filtration (see also Chapter 3). Incorporating DAF prior to GMF has been recommended particularly for RO plants susceptible to the impact of algal blooms (Anderson and McCarthy, 2012). DAF as such, can reduce the concentration of algal cells to a large extent, protecting media filters from rapid clogging, reduced capacity, and breakthrough. A coagulant dose of 1 - 7 Fe^{3+}/L is usually required to render the process effective (Tabatabai, 2014). Additional coagulant might be dosed just before feeding the DAF effluent to downstream granular media filters to ensure an acceptable SDI in the RO feedwater.

Prior to being considered for RO pretreatment, several water treatment plants in the Netherlands and Great Britain were primarily using DAF for treatment of algal-laden surface water sources (van Puffelen *et al.*, 1995; Longhurst and Graham, 1987; Gregory, 1997). For such applications, DAF can remove 96% to 99.9% of algae when pretreatment and DAF are optimized (Henderson *et al.*, 2008a).

In RO pretreatment, DAF prior to GMF is usually recommended to enhance the robustness of the pretreatment scheme during algal bloom events, or in case of high coagulant concentrations are required during turbidity spikes (Rovel, 2003). DAF coupled with coagulation prior to GMF has been demonstrated to produce RO feedwater with $SDI_{15} < 4$ and algae removal of more than 99% when treating seawater containing various algae, including HAB species (Sanz *et al.*, 2005). After the severe HAB of 2008-2009, DAF is now being regularly considered/incorporated in new RO plants in the Gulf region upstream of GMF or MF/UF. In the Al-Dur plant in Bahrain, more than 99% removal of algal cells was reported during pilot testing of DAF combined with coagulation prior to GMF (Le Gallou *et al.*, 2011). The Al-Shuwaikh desalination plant in Kuwait equipped with DAF/UF as pretreatment consistently provided SDI < 2.5 for good quality feedwater and < 3.5 for deteriorated conditions during a HAB event (Park *et al.*, 2013).

Since algal blooms are a seasonal occurrence, a RO plant may not require DAF to operate all year round. If the DAF is only operational periodically, operators could consider bringing the DAF online while cell counts are low so that the plant is fully operational when counts increase. This argues for plankton monitoring near the plant intake and within the plant so that effective actions can be taken sufficiently early to minimize clogging and fouling in the downstream units (Dixon *et al.*, 2017).

6.5.5 Granular media filtration

Pretreatment using coagulation followed by granular media filtration (GMF) is currently one of the most commonly used pretreatment scheme for RO (see Chapter 3). GMF applied in RO pretreatment are typically rapid dual-media filters (DMF) in a single-stage configuration. However, in some cases where the source water contains high levels of organics (TOC > 6 mg/L) and suspended solids (turbidity > 20 NTU), two-stage filtration systems are applied to achieve desired SDI levels (Dixon *et al.*, 2017). Under this configuration, the first filtration stage is mainly designed to remove macroalgae, solids, and organics that are present in suspended form. Often when a plant is subject to algal blooms, coagulation is employed in the first stage filtration. The second-stage filters are configured to retain fine solids (including algal cells) and silt, and to remove a portion (20 to 50 %) of the soluble organics contained in the seawater by biofiltration (Dixon *et al.*, 2017).

During algal blooms, the effluent quality of GMF can be highly variable over time, with reported algae and biopolymer (algal-released organic macromolecules) removal efficiencies in the range of 48-90% and 17-47%, respectively (Plantier *et al.*, 2012; Salinas Rodriguez *et al.*, 2009). The deteriorating quality of the GMF effluent can be mitigated by increasing the coagulant dosed inline prior to the process. As discussed in Section 6.3.2.1, a high load of algae biomass in the raw water can lead to clogging of GMF and that increasing the coagulant dose may result in even higher clogging rates. To effectively control clogging in GMF and to ensure high quality RO feedwater and stable hydraulic operation during algal blooms, it is recommended to be preceded by coagulation/flocculation and a clarification step using sedimentation or flotation. The intermediate clarification step reduces the particulate/colloidal matter (including coagulated flocs) on the media filters (Anderson and McCarthy, 2012; Villacorte *et al.*, 2015b).

6.5.6 Microfiltration and ultrafiltration

The application of MF or UF in RO pre-treatment is considered a more reliable alternative to GMF (with or without coagulation), as MF/UF membranes are generally more effective in removing particulate and colloidal matter from water. As such, they are expected to be more reliable in producing low SDI RO feedwater even during an algal bloom. UF operation could be stabilised during algal blooms when preceded by in-line coagulation without flocculation or clarification. Other operational measures such as decreasing membrane flux, increasing backwash flux/frequency and applying a forward flushing and/or air scouring (when possible) may also improve the performance of UF during severe algal bloom situations. However, these measures often result in lower water productivity. Adaptive operation control measures such as described in Section 6.5.7.3 can optimize these hydraulic cleaning measures to control fouling without reducing net water production. Finally, installing a micro-screen with openings of 50 -150 μm upstream of UF membranes can potentially eliminate capillary plugging, particularly for inside-out capillary membranes.

In a UF-RO pilot seawater desalination facility in the Netherlands, the high concentration of AOM present during a severe algal bloom were reported to impair the operation of UF membranes, resulting in CEBs as frequent as once in 6 hours (Schurer *et al.*, 2012; 2013). Under such conditions, coagulant was dosed to control hydraulic performance of UF membranes. With optimized inline coagulation, operation was stabilized at relatively low doses of ferric chloride (<1 mg Fe/L) during the bloom period. Pre-coating of UF membranes with a layer of pre-formed flocs of ferric hydroxide at the start of each filtration cycle, intermittent coating and intermittent coagulation are other promising alternatives for controlling UF hydraulic performance during an algal bloom periods while further lowering chemical consumption (Tabatabai, 2014).

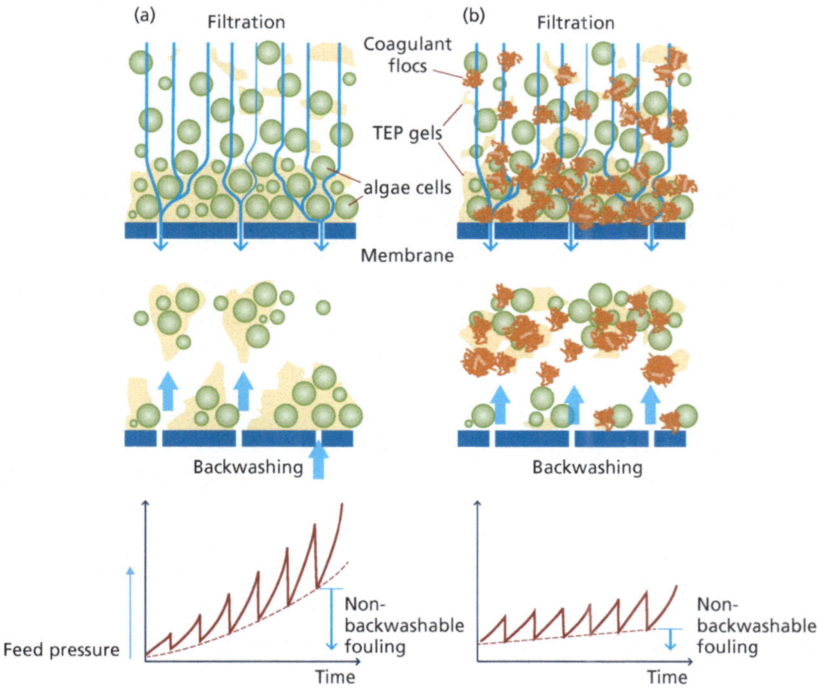

Figure 11 Graphical presentation of membrane fouling in UF system (a) operated during severe algal blooms and (b) fouling mitigation with optimised inline coagulation (Villacorte *et al.*, 2015a).

In general, coagulation can reduce the adverse effects of AOM on UF operation by reducing the fouling potential and compressibility of AOM layers on the membrane surface (Figure 11). This is mainly achieved through partial complexation of algal biopolymers and formation of colloidal Fe-biopolymer complexes at low coagulant dose (<1 mg Fe/L) and adsorption of algal biopolymers onto and enmeshment in iron hydroxide precipitates forming Fe-biopolymer aggregates at coagulant dose of 1 mg Fe/L and higher (Tabatabai *et al.*, 2014). However, if not optimized, coagulation may deteriorate the long-term UF operation. Unreacted iron species (monomers, dimers, trimers, etc.), ferrous iron and manganese – present in low-grade coagulants – can foul UF membranes by adsorbing on the membrane surface or within the pores, resulting in gradual irreversible fouling of UF membranes that will require chelation with cleaning solutions based on e.g., ascorbic and oxalic acids to release fouling (Tabatabai, 2014).

In terms of treatment performance, reported algae removals by UF membranes are consistently higher than 99% while biopolymer and TEP removals are typically higher than 40% (Villacorte *et al.*, 2013). New generation of UF membranes with low molecular weight cut-off (e.g., 10 kDa) can further improve RO feedwater due to higher removal of biopolymers (Villacorte *et al.*, 2015a; Dhakal, 2017).

6.5.7 Emerging pretreatment solutions

6.5.7.1 Ultrasonic algae control at the water intake

The use of ultrasonic technology to control algae in open water sources was recently developed by LGSonic (www.lgsonic.com). The ultrasonic unit comprise transmitters emitting specific ultrasound waves which can travel up to hundreds of meters through water targeting algae cells. Typically, algae and cyanobacteria can migrate vertically through the water column using their cellular gas vesicles to utilize sunlight near the surface and nutrients near the bottom. The ultrasonic sound waves create an ultrasonic pressure on the top layer of the water, which effects the buoyancy regulation of algal cells and prevents them from rising to the surface for photosynthesis (Figure 12). Such mobility limitation substantially hampers growth and multiplication of algae. The algae will be inactivated while the cell wall remains intact, potentially preventing the release of toxins from the algae into the water. The cells will eventually sink to the bottom of the water column which are then degraded by bacteria.

Ultrasonic algae control has been successfully implemented in freshwater sources, including ponds and drinking water reservoirs. Schneider *et al.* (2015) reported 22% reduction of coagulant consumption and 83% increase in filter run volumes in a water treatment plant after ultrasonic buoys were installed in its raw water reservoir. So far, application of such technology on or near the seawater RO plant intakes has not been reported. Some studies suggest that although ultrasonic waves are quite effective for cyanobacteria control, it may not be that effective in suppressing other bloom-forming algal species (Lürling and Tolman, 2014; Ahn *et al.*, 2007).

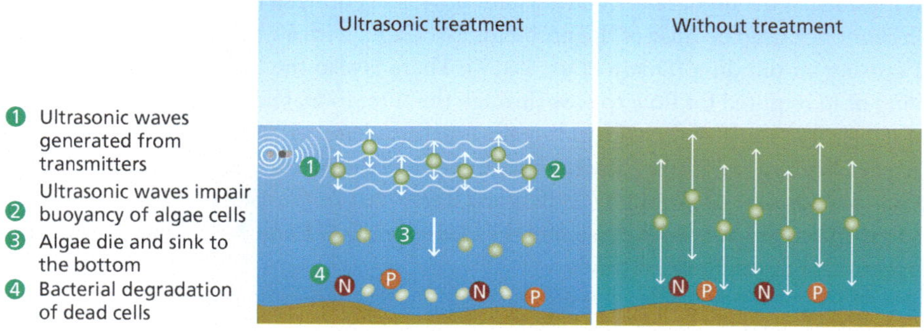

Figure 12 Left figure illustrates application of ultrasonic technology for algae control. Right figure illustrates normal migration movement of algae without ultrasonic treatment. Source: LGSonic (2019).

6.5.7.2 Integrated flotation-filtration pretreatment

Media filtration (GMF) or MF/UF preceded by DAF is considered as a robust pretreatment against algal blooms (Dixon *et al.*, 2017). However, DAF can take up a large footprint and additional complexity to the operation of the existing pretreatment system. To address these issues, integrated DAF-filtration pretreatment systems have been recently introduced. DAF integrated with GMF has been implemented in proprietary systems such as CoCo™ and Enflo-Filt™ or generic types called stack DAF, in-filter DAF or DAFF (Figure 13a). A hybrid DAF-MF/UF system has also been implemented in the proprietary AkvoFloat™ process (Figure 13b).

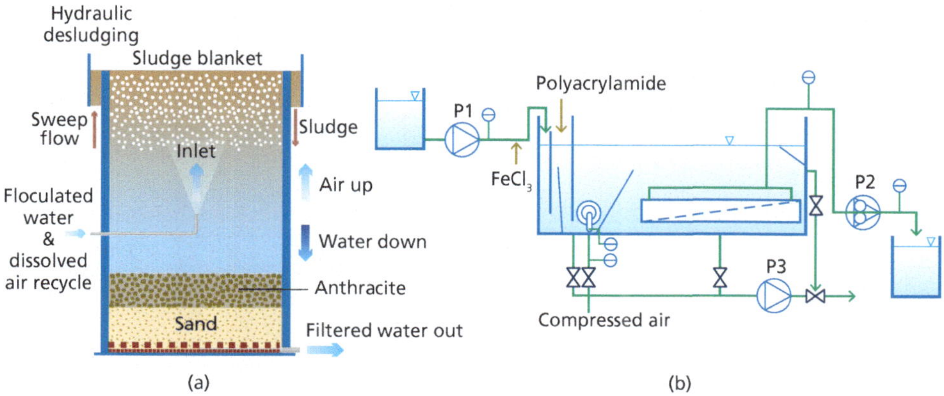

Figure 13 Schematics of (a) Enflo-Filt™ DAFF - combined DAF and granular media filter (Amato, 2014) and (b) AkvoFloat™ DAF-UF hybrid system (Ludwig and Beery, 2017).

Hybrid flotation-filtration systems offer the end-user the advantage of space savings and combined operational control. However, the operation of the DAF in terms of loading rate is restricted by the limits placed on the filter (GMF or MF/UF) and the physical property of the air bubble. Air bubbles with average diameters of 40-60 μm would have a rise rate in the range of 3-7 m/h, respectively (with large bubbles of 100 μm reaching rise rates of 20 m/h).

This means that the higher net flotation rates now being utilized in high-rate DAF of 30-50 m/h cannot be used because of the problems associated with air being drawn into the filter bed causing air blinding (Dixon *et al.*, 2017). There is also the lack of available hydraulic driving head required for flow to pass through the filter to match the higher DAF rates. For DAF-GMF, the whole system (including the flotation chamber) will need to be taken offline for 15 to 20 minutes during a filter backwash event. In the AkvoFloat™ system, backwashes typically take less than 2 minutes (Ludwig and Beery, 2017). Outside the algal bloom season or when oil and grease are not present in the feedwater, DAFF and AkvoFloat™ systems can be operated in direct filtration mode.

6.5.7.3 Auto-adaptive operation of MF/UF pretreatment

There are two main operational strategies to control fouling in MF/UF, namely: (1) setpoint control and (2) adaptive control (see Figure 14). The specific fouling control measures adopted during MF/UF operation relies heavily on the experience of the plant operator and their ability to respond quickly to upset in operational conditions typical in an algal bloom. Setpoint control means the operational settings of the plant are kept constant regardless of its performance. Such fixed settings are typically based on pilot testing and laboratory analysis, or based on recommendations from technical consultants, system builder or membrane supplier. Although it is rather easy to implement, it does not guarantee smooth operation of the plant, especially during extreme and unpredictable events like an algal bloom. On the other hand, adaptive control means operating the MF/UF plant with variable settings based on its observed performance. This strategy is more complicated to implement, but it is promising for MF/UF plants prone to sudden deterioration of water quality during algal blooms. It also facilitates optimization of chemicals used for coagulation and cleaning, as well as minimizing chemical residuals passing through MF/UF membrane to the downstream RO.

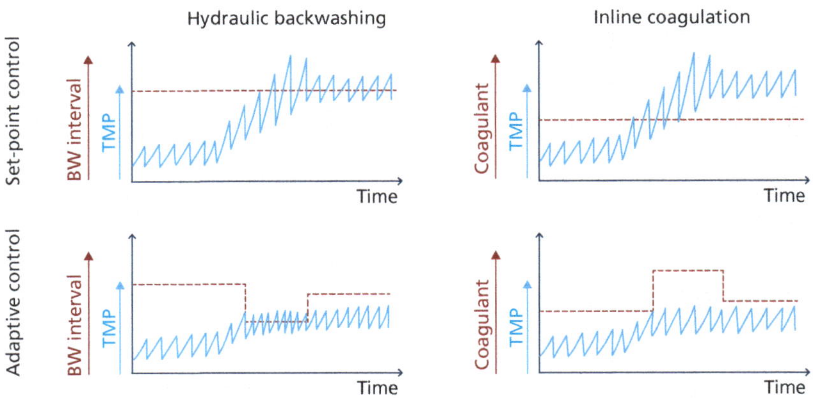

Figure 14 Simplified graphical comparison between setpoint and adaptive control in MF/UF. Adopted from Villacorte *et al.* (2018).

Set-point control is considered not robust against variations in feedwater quality. Typically, the mitigation measure during algal blooms is frequent chemical cleaning (e.g., CEB, CIP) which entails system downtime. Plant operators may choose to implement adaptive control based on variation of selected water quality parameters or hydraulic performance. The success of this control strategy can greatly vary from operator to operator. Moreover, there is no single water quality parameter or fouling index that can consistently predict the UF fouling rate in every plant location. To mitigate such challenges, auto-adaptive UF control algorithms have been developed using existing standard sensors in the plant (e.g., flow, pressure) to stabilise operation of the plant without requiring the operator to continuously adjust the plant settings (Gao *et al.*, 2017; Dominiak *et al.*, 2017; Cohen, 2017; Villacorte *et al.*, 2017c).

Grundfos recently developed the Smart Filtration Suite (SFS), which includes modules of auto-adaptive control algorithms designed to autonomously optimize in real time the hydraulic settings (see Figure 15) and chemical dosing in membrane systems (Pankratz, 2018; www.smartfiltrationsuite.com). These algorithms analyses standard parameters onsite (e.g., TMP, flow, power consumption) by extracting data from the pump motors and existing flow and pressure sensors, and automatically adjusts the set-points in the plant controller (e.g., PLC) and issue execution signals for the process (Dominiak *et al.*, 2014; Dominiak *et al.*, 2015). The way it works during operation is such: instead of specifying a process protocol (for instance 'filter for 30 minutes, then backwash for 60 seconds, repeat until TMP reaches 1.2 bar'), the algorithm analyses data in real time and dynamically issues execution signals such as 'make a backwash now', 'stop the backwash now', or 'increase filtrate flow to x'. This approach not only increases the overall process efficiency and system hydraulic capacity, but also increases the robustness of the process against disturbances and makes for quick reactions to dynamic conditions, beyond the capabilities of a human operator (Dominiak *et al.*, 2017).

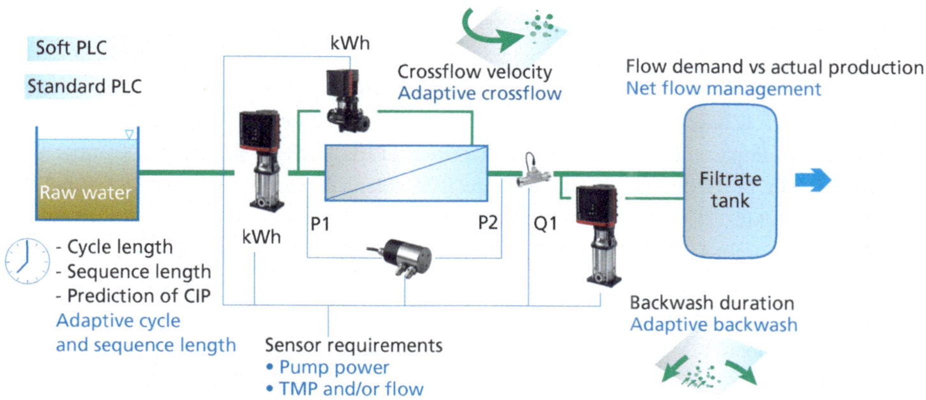

Figure 15 Control framework for auto-adaptive membrane filtration based on Grundfos SFS MF/UF algorithm. Adapted from Dominiak and Gissel (2017).

Bench scale tests with model seawater spiked with variable concentrations of algae showed that auto-adaptive control algorithms can effectively stabilise UF hydraulic performance during algal blooms with lower coagulant consumption and higher net water production (Dominiak *et al.*, 2018; Villacorte *et al.*, 2018). The algorithm is still to be validated in a full-scale MF/UF plant suffering from natural algal blooms, but its robustness has already been demonstrated with a full-scale MBR system and pilot-scale UF treatment of harbour seawater with high suspended solids loading (Dominiak *et al.*, 2017).

6.6 REFERENCES

Abushaban, A., Mangal, M.N., Salinas-Rodriguez, S.G., Nnebuo, C., Mondal, S., Goueli, S.A., Schippers, J.C., Kennedy, M.D., 2017. Direct measurement of atp in seawater and application of ATP to monitor bacterial growth potential in SWRO pre-treatment systems. Desalin. Water Treat. 99, 91-101.

Ahn C-Y., Joung S-H., Choi A., Kim H-S., Jang K-Y. & Oh H-M. (2007) Selective Control of Cyanobacteria in Eutrophic Pond by a Combined Device of Ultrasonication and Water Pumps, Environmental Technology, 28:4, 371-379.

Al-Hadidi, A. M. M. 2011. Limitations, Improvements, Alternatives for the Silt Density Index, PhD Thesis University of Twente, Enschede, Gildeprint Drukkerijen.

Alldredge, A. L., Passow, U., and Logan, B. E. (1993) The abundance and significance of a class of large, transparent organic particles in the ocean. Deep Sea Research (Part I, Oceanographic Research Papers), 40(6), 1131-1140.

Amato. T (2014). Personal Communication.

Amy GL, Salinas-Rodriguez SG, Kennedy MD, Schippers JC, Rapenne S, Remize P-J, Barbe C, Manes CLDO, West NJ, Lebaron P, Kooij DVD, Veenendaal H, Schaule G, Petrowski K, Huber S, Sim LN, Ye Y, Chen V and Fane AG (2011). Water quality assessment tools. In: Drioli, E., Criscuoli, A. & Macedonio, F. (eds.) Membrane-Based Desalination - An Integrated Approach (MEDINA). IWA Publishing, New York, pp 3-32.

Anderson C.R., Sapiano M.R.P., Prasad M.B.K., Long W., Tango P.J., Brown C.W. and Murtugudde R. (2010) Predicting potentially toxigenic Pseudo-nitzschia blooms in the Chesapeake Bay. Journal of Marine Systems, 83: 127 – 140.

Anderson D. M. (2017) Chapter 1. Harmful algal blooms. In: Anderson D. M., S. F. E. Boerlage, M. B. Dixon (Eds), Harmful Algal Blooms (HABs) and Desalination: A Guide to Impacts, Monitoring and Management. Paris, Intergovernmental Oceanographic Commission of UNESCO, 2017. 539 pp. (IOC Manuals and Guides No.78.) (English.) (IOC/2017/MG/78).

Anderson D. M., S. F. E. Boerlage, M. B. Dixon (2017), Harmful Algal Blooms (HABs) and Desalination: A Guide to Impacts, Monitoring and Management. Paris, Intergovernmental Oceanographic Commission of UNESCO, 2017. 539 pp. (IOC Manuals and Guides No.78.) (English.) (IOC/2017/MG/78).

Anderson, D. M., Cembella, A. D., and Hallegraeff, G. M. (2012) Progress in understanding harmful algal blooms: paradigm shifts and new technologies for research, monitoring, and management. Annual Review of Marine Science 4, 143-176.

Anderson, D. M., Glibert, P. M., and Burkholder, J. M. (2002) Harmful algal blooms and eutrophication: Nutrient sources, composition, and consequences. Estuaries 25, 704–726.

Anderson, D.M. (2014). HABs in a changing world: a perspective on harmful algal blooms, their impacts, and research and management in a dynamic era of climactic and environmental change. In: Kim, H.G., B. Reguera, G. Hallegraeff, C.K Lee, M.S. Han and J.K Choi (eds). Harmful Algae 2012, Proceedings of the 15th International Conference on Harmful Algae. International Society for the Study of Harmful Algae 2014, ISBN 978-87-990827-4-2, 16 pp.

Anderson, D.M., McCarthy, S., (2012). Red tides and harmful algal blooms: Impacts on desalination operations. Middle East Desalination Research Center, Muscat, Oman.

Arruda-Fatibello, S. H. S., Henriques-Vieira, A. A. and Fatibello-Filho, O. (2004) A rapid spectrophotometric method for the determination of transparent exopolymer particles (TEP) in freshwater. Talanta 62(1), 81-85.

ASTM D8002-15e1, Standard Test Method for Modified Fouling Index (MFI-0.45) of Water, ASTM International, West Conshohocken, PA, 2015, www.astm.org

Bar-Zeev E., Berman-Frank I., Girshevitz O., and Berman T. (2012) Revised paradigm of aquatic biofilm formation facilitated by microgel transparent exopolymer particles. PNAS 109 (23), 9119-9124.

Bellona, C., Drewes, J. E., Xu, P., and Amy, G. 2004. Factor affecting the rejection of during NF/RO treatment - a literature review. Water Research 38, 2795-2809.

Berge G (1962) Discoloration of the sea due to Coccolithus huxleyi bloom. Sarsia 6, 27–40.

Berktay, A. (2011). Environmental approach and influence of red tide to desalination process in the middle-east region. International Journal of Chemical and Environmental Engineering 2 (3), 183-188.

Berman, T., and Holenberg, M. (2005) Don't fall foul of biofilm through high TEP levels. Filtration & Separation 42(4), 30-32.

Bhaskar, P. V., & Bhosle, N. B. (2005). Microbial extracellular polymeric substances in marine biogeochemical processes. Current Science, 88(1), 45-53.

Boerlage S.F.E., Dixon M.B. and Anderson D.M. (2017c) Chapter 11: Case histories for harmful algal blooms in desalination. In: Anderson D. M., S. F. E. Boerlage, M. B. Dixon (Eds), Harmful Algal Blooms (HABs) and Desalination: A Guide to Impacts, Monitoring and Management. Paris, Intergovernmental Oceanographic Commission of UNESCO, 2017. 539 pp. (IOC Manuals and Guides No.78.) (English.) (IOC/2017/MG/78).

Boerlage S.F.E., Missimer T.M., Pankratz T.M., and Anderson D.M. (2017a) Chapter 6: Seawater intake considerations to mitigate harmful algal bloom impacts. In: Anderson D. M., S. F. E. Boerlage, M. B. Dixon (Eds), Harmful Algal Blooms (HABs) and Desalination: A Guide to Impacts, Monitoring and Management. Paris, Intergovernmental Oceanographic Commission of UNESCO, 2017. 539 pp. (IOC Manuals and Guides No.78.) (English.) (IOC/2017/MG/78).

Boerlage S.F.E., Villacorte L.O., Weinrich L., Tabatabai S.A.A., Maria D. Kennedy, and Schippers J.C. (2017b) Chapter 5: Harmful algal bloom-related water quality monitoring for desalination design and operation. In: Anderson D. M., S.F.E. Boerlage, M. B. Dixon (Eds), Harmful Algal Blooms (HABs) and Desalination: A Guide to Impacts, Monitoring and Management. Paris, Intergovernmental Oceanographic Commission of UNESCO, 2017. 539 pp. (IOC Manuals and Guides No.78.) (English.) (IOC/2017/MG/78).

Boerlage, S. F. E. 2008 Understanding the Silt Density Index and Modified Fouling Indices (MFI0.45 and MFI-UF). Desalination and Water Reuse Quarterly, May –June, 12-21.

Boerlage, S. F. E., and Nada, N. 2014. Algal toxin removal in seawater desalination processes, In: Proceedings of European Desalination Society, Cyprus.

Boerlage, S. F., Kennedy, M. D., Dickson, M. R., El-Hodali, D. E. and Schippers, J. C. 2002. The modified fouling index using ultrafiltration membranes (MFI-UF): characterisation, filtration mechanisms and proposed reference membrane. Journal of Membrane Science, 197(1), pp.1-21

Booth, B.C; Larouche, P; Bélanger, S; Klein, B; Amiel, D; Mei, Z.-P (2002). "Dynamics of Chaetoceros socialis blooms in the North Water". Deep Sea Research Part II: Topical Studies in Oceanography. 49 (22–23): 5003–25.

Caron D.A., Garneau, M.E., Seubert, E., Howard M.D.A.,, Darjany L., Schnetzer A., Cetinic, I., Filteau, G., Lauri, P., Jones, B. and Trussell, S. (2010) Harmful algae and their potential impacts on desalination operations off southern California. Water Research 44, 385-416.

Claquin P., Probert I., Lefebvre S., Veron B. (2008) Effects of temperature on photosynthetic parameters and TEP production in eight species of marine microalgae. Aquat Microb Ecol 51, 1-11.

Cohen Y. (2017) Self-adaptive RO desalination: Advances and challenges. In Proc. of 3rd International Conference on Desalination using Membrane Technology, Gran Canaria, Spain, 3-5 April 2017.

Decho, A.W. (1990). Microbial exopolymer secretions in ocean environments: their role(s) in food webs and marine processes Oceanogr. Mar. Biol. Ann. Rev. 28, 73-153.

Deeds, J. R., Mazzaccaro, A. P., Terlizzi, D. E., and Place, A. R. (2004). Treatment options for the control of an ichthyotoxic dinoflagellate in an estuarine aquaculture facility: A case study. Hall, S., Anderson, D., Kleindinst, J., Zhu, M., and Zou, Y. (Eds.), Proceedings of the Second International Conference on Harmful Algae Management and Mitigation, Asia-Pacific Economic Cooperation, APEC Publ. #204-MR-04.2, Singapore, pp. 177-181.

Desormeaux, E. D., Meyerhofer, P. F., and Luckenbach, H. 2009. Results from nine investigations assessing Pacific Ocean seawater desalination in Santa Cruz, California. Proceedings of the IDA World Congress, Dubai, UAE, November 2009.

Dhakal, N. (2017) Controlling Biofouling in Seawater Reverse Osmosis Membrane Systems, PhD thesis UNESCO-IHE/TU Delft, Delft.

Dixon M. B., Ho, L., Chow, C., Newcombe, G., Croue, J-P., Buisson, H., Cigana, J., and Treuger, R. (2012) Water Research Foundation Report #4016: Evaluation of integrated membranes for taste and odour and algal toxin control, Published by Water Research Foundation, Denver, Colorado, USA.

Dixon M.B., Boerlage S.F.E., Voutchkov N., Henderson R., Wilf M., Zhu I., Tabatabai S.A.A., Amato T., Resosudarmo A., Pearce G.K., Kennedy M., Schippers J.C., and Winters H. (2017) Chapter 9: Algal biomass pretreatment in seawater reverse osmosis. In: Anderson D. M., S. F. E. Boerlage, M. B. Dixon (Eds), Harmful Algal Blooms (HABs) and Desalination: A Guide to Impacts, Monitoring and Management. Paris, Intergovernmental Oceanographic Commission of UNESCO, 2017. 539 pp. (IOC Manuals and Guides No.78.) (English.) (IOC/2017/MG/78).

Dixon M.B., Richard Y., Ho L., Chow C.W.K., O'Neill B.K. and Newcombe G. (2011) A coagulationpowdered activated carbon-ultrafiltration - multiple barrier approach for removing toxins from two australian cyanobacterial blooms. Journal of Hazardous Materials, 186(2-3), 1553-1559.

Dixon, M. B. 2014. Removal of toxin and taste & odor compounds using membranes and associated processes. Proceedings of the Middle East Desalination Research Centre's HABs and Desalination workshop, Muscat, Oman, April 2014.

Dixon, M. B., Churman, H., and Henthorne, L. (2015) Harmful algae blooms and desalination: a cells journey from sea to SWRO. Proceedings of the IDA World Congress, San Diego, California, Sept 2015.

Dixon, M. B., Falconet, C., Ho, L., Chow, C. W. K., O'Neil, B., and Newcombe, G. (2010). Nanofiltration for the removal of algal metabolites and the effects of fouling. Water Science and Technology 61, 1189–1199.

Dominiak D., Gissel R.E. (2017) Autonomous process control algorithms for optimum filtration. In proceedings of IWA Specialized Conference on Instrumentation, Control & Automation. Québec, Canada. pp. 175-179.

Dominiak D.M., Yangali Quintanilla V., Villacorte L.O., Gissel R.E. (2018) Auto-adaptive fouling control in UF system during algal blooms using smart pumps. In Proc. of AWWA and AMTA Membrane Technology – Conference and Exposition, West Palm Beach, USA, March 2018.

Dominiak Dominik, Gissel Rikke, Svendsen Jacob, Zheng Xing, (Grundfos Holding A/S) (2015) Filterverfahren zum Filtern einer Flussigkeit und Filtervorrichtung zum Filtern einer Flussigkeit, Patent EP3187247A1.

Dominiak Dominik, Svendsen Jacob, Rasmussen Christian, Højsholt Rune (Grundfos Holding A/S), Control method for a filtration system (2014), Patent EP2985069 (B1).

Dow Water and Process Solutions (2015). Reverse Osmosis Technical Manual, http://msdssearch.dow.com/PublishedLiteratureDOWCOM/dh_08db/0901b803808db77d.pdf?filepath=liquidseps/pdfs/noreg/609-00071.pdf&fromPage=GetDoc, Accessed online, November 2015.

Edler L. and Elbrächter M. (2010) The Utermöhl method for quantitative phytoplankton analysis. In: Karlson, B., Cusack, C. and Bresnan, E. (editors). Microscopic and molecular methods for quantitative phytoplankton analysis. Paris, UNESCO. (IOC Manuals and Guides, no. 55.) (IOC/2010/MG/55) 110 pages.

Edzwald, J.K. (2010) Dissolved air flotation and me. Water Research, 44, 2077-2106.

Field C.B., Behrenfeld M.J., Randerson J.T., and Falkowski P. (1998) Primary Production of the Biosphere: Integrating Terrestrial and Oceanic Components. Science 281, 237-240.

Fogg, G. E. (1983) The ecological significance of extracellular products of phytoplankton photosynthesis. Bot. Mar., 26, 3–14.

Fonda-Umani S., Beran A., Parlato S., Virgilio D., Zollet T., De Olazabal A., Lazzarini B., and Cabrini M. (2004) Noctiluca scintillans Macartney in the Northern Adriatic Sea: long-term dynamics, relationships with temperature and eutrophication, and role in the food web, J. Plankton Res. 26(5), 545-561.

Gao L.X., Gu H., Rahardianto A., Christofides P.D., Cohen Y. (2017) Self-adaptive cycle-to-cycle control of in-line coagulant dosing in ultrafiltration for pre-treatment of reverse osmosis feed water. Desalination 401, 22–31.

Gledhill M. and Buck K.N. (2012) The organic complexation of iron in the marine environment: a review. Front. Microbio. 3:69.

Gotsis-Skretas, O. (1995). Mucilage appearances in greek waters during 1982-1994. Science of the Total Environment, 165, 229-230.

Gregory, R. (1997). Summary of General Developments in DAF for Water Treatment since 1976. Proceedings Dissolved Air Flotation Conference. The Chartered Institution of Water and Environmental Management, London, 1-8.

Henderson, R., Parsons, S.A., Jefferson, B. (2008a). The impact of algal properties and pre-oxidation on solid-liquid separation of algae. Water Research 42(8-9), 1827-1845.

Henderson, R.K., Baker, A., Parsons, S.A. and Jefferson, B. (2008b). Characterisation of algogenic organic matter extracted from cyanobacteria, green algae and diatoms. Water Research 42, 3435-3445.

Herman, P.H., Bredee, H.L. (1936) Principles of the mathematical treatment of constant pressure filtration. Journal of the Society of Chemical Industry 1-4.

Hess P., Villacorte L.O., Dixon M.B., Boerlage S.F.E., Anderson D.M., Kennedy M.D. and Schippers J.C. (2017) Chapter 2. Algal issues in seawater desalination. In: Anderson D. M., S. F. E. Boerlage, M. B. Dixon (Eds), Harmful Algal Blooms (HABs) and Desalination: A Guide to Impacts, Monitoring and Management. Paris, Intergovernmental Oceanographic Commission of UNESCO, 2017. 539 pp. (IOC Manuals and Guides No.78.) (English.) (IOC/2017/MG/78).

Huang, Y., Ortiz, L., Garcia, J., Aguirre, P., Mujeriego, R., and Bayona, J. M. 2004. Use of headspace solid-phase microextraction to characterize odour compounds in subsurface flow constructed wetland for wastewater treatment. Water Science and Technology 49, 89-98.

Huber S.A., Balz A., Abert M. and Pronk W. (2011) Characterisation of aquatic humic and non-humic matter with size-exclusion chromatography – organic carbon detection – organic nitrogen detection (LC-OCD-OND). Water Research 45:879-885.

Janse, I., Van Rijssel, M., Gottschal, J.C., Lancelot, C., Gieskes, W.W.C. (1996) Carbohydrates in the North Sea during spring blooms of Phaeocystis: A specific fingerprint. Aquat. Microb. Ecol. 10, 97–103.

Karlson B., Anderson C.R., Coyne K.J., Sellner K.G., and Anderson D.M. (2017) Chapter 3. Designing an observing system for early detection of harmful algal blooms. In: Anderson D. M., S. F. E. Boerlage, M. B. Dixon (Eds), Harmful Algal Blooms (HABs) and Desalination: A Guide to Impacts, Monitoring and Management. Paris, Intergovernmental Oceanographic Commission of UNESCO, 2017. 539 pp. (IOC Manuals and Guides No.78.) (English.) (IOC/2017/MG/78).

Karlson, B., Cusack, C., and Bresnan, E. (Eds.) 2010. Microscopic and molecular methods for quantitative phytoplankton analysis. Paris, UNESCO. (IOC Manuals and Guides, no. 55.). 110 pages. http://hab.iocunesco.org/index.php?option=com_oe&task=viewDocumentRecord&docID=5440

Kim H.-G. (2010) An Overview on the Occurrences of Harmful Algal Blooms (HABs) and Mitigation Strategies in Korean Coastal Waters In: A. Ishimatsu and H.-J. Lie (eds.) Coastal Environmental and Ecosystem Issues of the East China Sea, pp. 121–131.

Krasner, S. W., Hwang, C. J., and McGuire, M. J. 1983. A standard method for quantification of earthy-musty odorants in water, sediments, and algal cultures. Water Science and Technology 15, 127-138.

Lancelot, C. (1995). The mucilage phenomenon in the continental coastal waters of the north sea. Science of the Total Environment, 165, 83-102.

Laycock, M. V., Anderson, D. M., Naar, J., Goodman, A., Easy, D. J., Donovan, M. A., Li, A., Quilliam, M.A., Al Jamali, E., Alshihi, R., Alshihi, R. (2012) Laboratory desalination experiments with some algal toxins. Desalination, 293, 1-6.

Le Gallou, S., Bertrand, S., Madan, K.H. (2011). Full coagulation and dissolved air flotation: a SWRO key pretreatment step for heavy fouling seawater. In: Proceedings of International Desalination Association World Congress, Perth, Australia.

Leblanc K., Quéguiner B., Diaz F., Cornet V., Michel-Rodriguez M., Durrieu de Madron X., Bowler C., Malviya S., *et al.* (2018) Nanoplanktonic diatoms are globally overlooked but play a role in spring blooms and carbon export. Nature Communications 9, 953.

Leenheer J.A. and Croué J.P. (2003) Characterizing Aquatic Dissolved Organic Matter: Understanding the unknown structures is key to better treatment of drinking water. Environmental Science & Technology 37 (1), 18A-26A.

Leppard, G.G., 1993. In: Rao, S.S. (Ed.), Particulate Matter and Aquatic Contaminants. Lewis, Chelsea, MI, pp. 169– 195.

LGSonic (2019). Web accessed 2019-08-28: https://lg-sonic-lgsound1.netdna-ssl.com/wp-content/uploads/MPC-Buoy-brochure.pdf

Liberman, B., and Berman, T. (2006) Analysis and monitoring: MSC - a biologically oriented approach. Filtration & Separation, 43(4), 39-40.

Longhurst, S.J., Graham, N.J.D. (1987). Dissolved air flotation for potable water treatment: a survey of operational units in Great Britain. The Public Health Engineer 14(6), 71-76.

Ludwig J. and Beery M. (2017) Akvola: an integrated DAF-UF pilot. In: Anderson D. M., S. F. E. Boerlage, M. B. Dixon (Eds), Harmful Algal Blooms (HABs) and Desalination: A Guide to Impacts, Monitoring and Management. Paris, Intergovernmental Oceanographic Commission of UNESCO, 2017. 539 pp. (IOC Manuals and Guides No.78.) (English.) (IOC/2017/MG/78).

Lürling M., Tolman Y. (2014) Beating the blues: Is there any music in fighting cyanobacteria with ultrasound? Water Research 66, 361-373.

Maier, G., Glegg, G. A., Tappin, A. D., & Worsfold, P. J. (2012). A high resolution temporal study of phytoplankton bloom dynamics in the eutrophic taw estuary (SW England). Science of the Total Environment, 434, 228-239.

Maniyar M. (2018) Algae bloom driving up desalination cost in Oman. Times of Oman, January 20, 2018. Web: https://timesofoman.com/article/126445. Accessed 2019-08-19.

Mathiesen K. (2013) China's largest algal bloom turns the Yellow Sea green. The Guardian July 4, 2013. Web: https://www.theguardian.com/environment/2013/jul/04/china-algal-bloom-yellow-sea-green. Accessed 2019-08-19.

McGregor, G.B., Stewart, I., Sendall, B.C., Sadler, R., Reardon, K., Carter, S., Wruck, D., Wickramasinghe, W. (2012). First Report of a Toxic Nodularia spumigena (Nostocales/ Cyanobacteria) Bloom in Sub-Tropical Australia. I. Phycological and Public Health Investigations. Int. J. Environ. Res. Public Health 9, 2396-2411.

Mingazzini, M., and Thake, B. (1995). Summary and conclusions of the workshop on marine mucilages in the adriatic sea and elsewhere. Science of the Total Environment, 165, 9-14.

Missimer, T.M., Ghaffour, N., Dehwah, H.A., Rachman, R., Maliva, R.G., Amy, G. (2013). Subsurface intakes for seawater reverse osmosis facilities: Capacity limitation, water quality improvement, and economics. Desalination 322, 37–51.

Mopper, K., Zhou, J., Sri Ramana, K., Passow, U., Dam, H. G., & Drapeau, D. T. (1995). The role of surface-active carbohydrates in the flocculation of a diatom bloom in a mesocosm. Deep-Sea Research Part II, 42(1), 47-73.

Myklestad, S.M., (1995). Release of extracellular products by phytoplankton with special emphasis on polysaccharides. Science of the Total Environment 165, 155-164.

Okaichi T (1989). Red tide problems in the Seto Inland Sea, Japan. In: Okaichi T, Anderson DM, Nemoto T (eds) Red tides. Biology, environmental science and toxicology. Elsevier, New York, pp 137–144.

Olenina, I., Hajdu, S., Edler, L., Andersson, A., Wasmund, N., Busch, S., Göbel, J., Gromisz, S., Huseby, S., Huttunen, M., Jaanus, A., Kokkonen, P., Ledaine, I. and Niemkiewicz, E. (2006). Biovolumes and size-classes of phytoplankton in the Baltic Sea, HELCOM Balt. Sea Environ. Proc. 106, ISSN 0357-2994.

Olson RJ, Sosik HM (2007) A submersible imaging-in-flow instrument to analyze nano- and microplankton: Imaging FlowCytobot. Limnol. Oceanogr. Methods, 5: 195-208.

Orlova T.Y., Konovalova G.V., Stonik I.V., Selina M.S., Morozova T.V. and Shevchenko O.G. (2002). Harmful algal blooms on the eastern coast of Russia. In: Taylor, F.J.R. and Trainer, V.L. (Eds.). Harmful Algal Blooms in the PICES Region of the North Pacific. PICES Sci. Rep. No. 23, 152 pp.

Pankratz T. (2018) Grundfos' autonomous process controller. Water Desalination Report 54 (33), pp. 3-4.

Pankratz, T. (2008). Red Tides Close Desal Plants, Water Desalination Report, 44 (44).

Park, K.S., Mitra, S.S., Yim, W.K., Lim, S.W. (2013) Algal bloom - critical to designing SWRO pretreatment and pretreatment as built in Shuwaikh, Kuwait SWRO by Doosan. Desalination and Water Treatment 51(31-33), 1-12.

Passow, U. (2000). Formation of transparent exopolymer particles (TEP) from dissolved precursor material. Marine Ecology Progress Series 192, 1-11.

Passow, U. (2002). Transparent exopolymer particles (TEP) in aquatic environments. Progress in Oceanography 55(3), 287-333.

Passow, U., Alldredge, A.L. (1994). Distribution, size, and bacterial colonization of transparent exopolymer particles (TEP) in the ocean. Marine Ecology Progress Series 113, 185–198.

Passow, U., Alldredge, A.L. (1995). A dye-binding assay for the spectrophotometric measurement of transparent exopolymer particles (TEP). Limnol. Oceanogr 40(7), 1326-1335.

Petry, M., Sanz, M. A., Langlais, C., Bonnelye, V., Durand, J.-P., Guevara, D., Nardes, W.M., Saemi, C.H. (2007). The el coloso (chile) reverse osmosis plant. Desalination,203(1-3), 141-152.

Plantier, S., Castaing, J.B., Sabiri, N.E., Massé, A., Jaouen, P., Pontié, M. (2012). Performance of a sand filter in removal of algal bloom for SWRO pretreatment. Desalination and Water Treatment 51(7-9), 1838-1846.

Polak, E. H. and Provasi, J. 1992. Odor sensitivity to geosmin enantiomers. Chemical Senses 17, 23-26.

Qu, F., Liang, H., He, J., Ma, J., Wang, Z., Yu, H., Li, G. (2012a). Characterization of dissolved extracellular organic matter (dEOM) and bound extracellular organic matter (bEOM) of microcystis aeruginosa and their impacts on UF membrane fouling. Water Research 46(9), 2881-2890.

Rachman, R. M., Li, S., & Missimer, T. M. (2014). SWRO feedwater quality improvement using subsurface intakes in oman, spain, turks and caicos islands, and saudi arabia. Desalination, 351, 88-100.

Resosudarmo, A., Nappa, L., Ye, Y., Le-Clech, P., and Chen, V. 2017. Effect of physical and chemical stress on ultrafiltration membrane performance during marine algal blooms. Separation Science and Technology 52, 364 – 373.

Richlen, M.L., Morton, S.L., Jamali, E.A., Rajan, A., Anderson, D.M., (2010). The Catastrophic 2008-2009 red tide in the Arabian Gulf region, with observations on the identification and phylogeny of the fish-killing dinoflagellate cochlodinium polykrikoides. Harmful Algae 9(2), 163-172.

Rinaldi, A., Vollenweider, R. A., Montanari, G., Ferrari, C. R., & Ghetti, A. (1995). Mucilages in italian seas: The adriatic and tyrrhenian seas, 1988-1991. Science of the Total Environment, 165, 165-183.

Rovel, J.M. (2003). Why a SWRO in Taweelah - pilot plant results demonstrating feasibility and performance of SWRO on Gulf water? In: Proceedings of International Desalination Association World Congress, Nassau, Bahamas.

Salinas Rodríguez, S.G., Kennedy, M.D., Schippers, J.C., Amy, G.L. (2009). Organic foulants in estuarine and bay sources for seawater reverse osmosis - Comparing pretreatment processes with respect to foulant reduction. Desalination and Water Treatment 9 (1-3), 155-164.

Salinas-Rodriguez, S. G., Amy, G. L., Schippers, J. C., and Kennedy, M. D. (2015) The Modified Fouling Index Ultrafiltration constant flux for assessing particulate/colloidal fouling of RO systems. Desalination 365, 79-91.

Sanz, M.A., Guevara, D., Beltrán, F., Trauman, E. (2005). 4 Stages pretreatment reverse osmosis for South-Pacific seawater: El Coloso plant (Chile). In: Proceedings of International Desalination Association World Congress, Singapore.

Sasaki, T., Okabe J., Masahiro H., Hayashi H., Iida Y (2013) Cesium (Cs) and strontium (Sr) removal as model materials in radioactive water by advanced reverse osmosis membrane. Desalination Water Treatement 51, 1672-1677.

Schippers J.C. (2012). personal communications.

Schippers, J. C., and Verdouw, J. 1980. The Modified Fouling Index, a method of determining the fouling characteristics of water. Desalination 32, 137-148.

Schneider O.D., Weinrich L.A., Brezinski S. (2015) Ultrasonic Treatment of Algae in a New Jersey Reservoir. Journal AWWA 107 (10), E533-E542.

Schneider, O. D., Giraldo, E., Weinrich, L., Salinas, S., and Kennedy, M. (2012) Investigation of Organic Matter Removal in Saline Waters by Pretreament. Water Research Foundation: Denver, CO.

Schoonenberg Kegel, F., Rietman, B. M., and Verliefde, A. R. D. (2010) Reverse osmosis followed by activated carbon filtration for efficient removal of organic micropollutants from river bank filtrate. Water Science and Technology 61(10), 2603-2610.

Schurer, R., Janssen, A., Villacorte, L., Kennedy, M.D. (2012). Performance of ultrafiltration and coagulation in an UF-RO seawater desalination demonstration plant. Desalination and Water Treatment 42(1-3), 57-64.

Schurer, R., Tabatabai, A., Villacorte, L., Schippers, J.C., Kennedy, M.D. (2013). Three years operational experience with ultrafiltration as SWRO pretreatment during algal bloom. Desalination and Water Treatment 51 (4-6), 1034-1042.

Selina, M. S., Konovalova, G. V., Morozova, T. V., & Orlova, T. Y. (2006). Genus alexandrium halim, 1960 (dinophyta) from the pacific coast of russia: Species composition, distribution, and dynamics. Russian Journal of Marine Biology, 32(6), 321-332.

Sellner, K.G.,Doucette, G.J., Kirkpatrick, G.J., (2003). Harmful algal blooms: causes, impacts and detection. Journal of Ind. Microbiol. Biotechnol. 30, 383-406.

Seubert, E. L., Trussell, S., Eagleton, J., Schnetzer, A., Cetinic, I., Lauri, P., Jones, B. H., and Caron, D. A. 2012. Algal toxins and reverse osmosis desalination operations: laboratory bench testing and field monitoring of domoic acid, saxitoxin, brevetoxin and okadaic acid. Water Research 46(19), 6563-6573.

Shikata T., Nagasoe S., Matsubara T., Yoshikawa S., Yamasaki Y., Shimasaki Y., Oshima Y., Jenkinson I.R., Honjo T. (2008). Factors influencing the initiation of blooms of the raphidophyte Heterosigma akashiwo and the diatom Skeletoma costatum in a port in Japan. Limnology & Oceanography 53, 2503-2518.

Sim L.N., Suwarno S.R., Lee D.Y.S., Cornelissen E.R., Fane A.G., Chong T.H. (2019) Online monitoring of transparent exopolymer particles (TEP) by a novel membrane-based spectrophotometric method, Chemosphere 220, 107-115.

Smayda, T. J. (1997). Harmful algal blooms: their ecophysiology and general relevance to phytoplankton blooms in the sea. Limnol. Oceanogr., 42, 1137–1153.

Smetacek, V., & Zingone, A. (2013). Green and golden seaweed tides on the rise. Nature, 504 (7478), 84-88.

Smith J.C., Cormier R., Worms J., Bird C.J., Quilliam M.A., Pocklington R., Angus R., Hanic L. (1990). Toxic blooms of the domoic acid containing diatom Nitzschia pungens in the Cardigan River, Prince Edward Island. In: Graneli E, Sundström B., Edler L., Anderson D.M. (eds) Toxic marine phytoplankton. Elsevier, New York, pp 227–232.

Stumpf RP, Tomlinson MC, Calkins JA, Kirkpatrick B, Fisher K, Nierenberg, K., Currier, R., Wynne, T.T (2009). Skill assessment for an operational algal bloom forecast system. Journal of Marine Systems, 76(1-2), 151-161.

Suurnäkki, S., Gomez-Saez, G. V., Rantala-Ylinen, A., Jokela, J., Fewer, D. P., and Sivonen, K. (2015) Identification of geosmin and 2-methylisoborneol in cyanobacteria and molecular detection methods for the producers of these compounds. Water Research 68, 56-66.

Tabatabai, S. A. A. (2014) Coagulation and Ultrafiltration in Seawater Reverse Osmosis Pretreatment UNESCO-IHE/TU Delft, Delft.

Tabatabai, S. A. A., Schippers, J. C., and Kennedy, M. D. (2014) Effect of coagulation on fouling potential and removal of algal organic matter in ultrafiltration pretreatment to seawater reverse osmosis. Water Research 59, 283-294.

Tas, S., & Okus, E. (2011). A review on the Bloom Dynamics of a Harmful Dinoflagellate Prorocentrum minimum in the Golden Horn Estuary. Turkish Journal of Fisheries and Aquatic Sciences, 11(4), 523-531.

Tester P.A., Wiles K., Varnam S.M., Velez Ortega G., Dubois A.M., and Arenas Fuentes V. (2004). Harmful Algal Blooms in the Western Gulf of Mexico: Karenia brevis Is Messin' with Texas and Mexico! pp. 41-43. In: Steidinger, K. A., J. H. Landsberg, C. R. Tomas, and G. A. Vargo (Eds.) Harmful Algae 2002. Florida Fish and Wildlife Conservation Commission, Florida Institute of Oceanography, and Intergovernmental Oceanographic Commission of UNESCO.

Thornton, D.C.O., Fejes, E.M., DiMarco, S.F., Clancy, K.M. (2007). Measurement of acid polysaccharides (APS) in marine and freshwater samples using alcian blue. Limnol. Oceanogr 5, 73-87.

Thuy, N., Huang, CP. & Lin, JL. Visualization and quantification of transparent exopolymer particles (TEP) in freshwater using an auto-imaging approach. Environ Sci Pollut Res (2017) 24 (21): 17358–17372.

Trainer, V. L., Adams, N. G., Bill, B. D., Anulacion, B. F. and Wekell, J. C. (1998). Concentration and dispersal of a Pseudo-nitzschia bloom in Penn Cove, Washington, USA. Nat. Toxins 6, 113–125.

van der Hoeven (1984) Observations of surface water temperature in the Netherlands: series from KNMIRWS (in Dutch). Scientific report W.R. 84-5, ISSN 0169-1651.

van Puffelen, J., Buijs, P.J., Nuhn, P.N.A.M., Hijen, W.A.M. (1995). Dissolved air flotation in potable water treatment: the Dutch experience. Water Science and Technology, 31(3-4), 149-157.

Verliefde, A. R. D., Cornelissen, E. R., Heijman, S. G. J., Verberk, J. Q. J. C. and Amy, G. L. (2009) Construction and validation of a full-scale model for rejection of organic micropollutants by NF membranes. Journal of Membrane Science 339(1), 10-20.

Villacorte L.O., Dominiak D.M., Gissel R.E., van de Ven W., Yangali-Quintanilla V.A. (2017c) Auto-adaptive fouling control in UF system during algal blooms using smart pumps. In Proc. of the IWA Membrane Technology Conference, Singapore, 5-9 September 2017.

Villacorte L.O., Ekowati Y., Calix-Ponce H.N., Kisielius V., Kleijn J.M., Vrouwenvelder J.S., Schippers J.C. and Kennedy M.D. (2017a) Biofouling in capillary and spiral wound membranes facilitated by marine algal bloom. Desalination 424, 74-84.

Villacorte L.O., Gissel R.E., Dominiak D.M. (2018) Auto-adaptive fouling control in UF during algal blooms. In Proc. of EDS conference on Desalination for the Environment: Clean Water and Energy, Athens, Greece, 3-6 September 2018.

Villacorte L.O., Schippers J.C., and Kennedy M.D (2017b) Appendix 3. Methods for measuring transparent exopolymer particles and their precursors in seawater. In: Anderson D. M., S. F. E. Boerlage, M. B. Dixon (Eds), Harmful Algal Blooms (HABs) and Desalination: A Guide to Impacts, Monitoring and Management. Paris, Intergovernmental Oceanographic Commission of UNESCO, 2017. 539 pp. (IOC Manuals and Guides No.78.) (English.) (IOC/2017/MG/78).

Villacorte L.O., Tabatabai S.A.A., Anderson D.M., , Schippers J.C., Kennedy M.D. (2015a) Seawater reverse osmosis desalination and (harmful) algal blooms. Desalination 360, 61-80.

Villacorte, L. O. (2014) Algal blooms and membrane-based desalination technology. PhD Thesis UNESCO-IHE/TUDelft, ISBN 978-1-138-02626-1, CRC Press/Balkema, Leiden.

Villacorte, L. O., Ekowati, Y., Neu, T. R., Kleijn, J. M., Winters, H., Amy, G., Schippers J. C., and Kennedy M. D. (2015b). Characterisation of algal organic matter produced by bloom forming marine and freshwater algae. Water Research 73, 216–230.

Villacorte, L. O., Ekowati, Y., Winters, H., Amy, G., Schippers, J. C., and Kennedy, M. D. (2015d). MF/UF rejection and fouling potential of algal organic matter from bloom forming marine and freshwater algae. Desalination 367, 1-10.

Villacorte, L.O., Ekowati, Y., Calix-Ponce, H.N., Amy, G.L., Schippers, J.C., Kennedy, M.D. (2015c). Improved method for measuring transparent exopolymer particles (TEP) and their precursors in fresh and saline water. Water Research 70 (1), 300–312.

Villacorte, L.O., Ekowati, Y., Winters, H., Amy, G.L., Schippers, J.C., Kennedy, M.D. (2013). Characterisation of transparent exopolymer particles (TEP) produced during algal bloom: a membrane treatment perspective. Desalination and Water Treatment 51 (4-6), 1021-1033.

Villacorte, L.O., Kennedy, M.D., Amy, G.L., Schippers, J.C. (2009). The fate of transparent exopolymer particles (TEP) in integrated membrane systems: removal through pretreatment processes and deposition on reverse osmosis membranes. Water Research 43(20), 5039-5052.

Voutchkov, N. (2010). Considerations for selection of seawater filtration pretreatment system. Desalination, 261 (3), 354-364.

Vrouwenvelder, J.S., van der Kooij, D. (2001). Diagnosis, prediction and prevention of biofouling of NF and RO membranes. Desalination 139(1-3), 65-71.

Vrouwenvelder, J.S., van Paassen, J.A.M., Wessels, L.P., van Dam, A.F., Bakker, S.M. (2006). The membrane fouling simulator: A practical tool for fouling prediction and control. Journal of Membrane Science 281(1-2), 316-324.

Weinrich L.A., Schneider O.D. and LeChevallier M.W. (2011) Bioluminescence-Based Method for Measuring Assimilable Organic Carbon in Pretreatment Water for Reverse Osmosis Membrane Desalination. Appl. Environ. Microbiol 77 (3), 148-1150.

Wetsteyn, L. P. M. J., Peeters, J. C. H., Duin, R. N. M., Vegter, F., & de Visscher, P. R. M. (1990). Phytoplankton primary production and nutrients in the oosterschelde (the netherlands) during the pre-barrier period 1980-1984.Hydrobiologia, 195(1), 163-177.

Winters, H. (1997) Twenty years experience in seawater reverse osmosis and how chemicals in pretreatment affect fouling of membranes. Desalination 110(1-2), 93-98.

Winters, H., Isquith, I., Arthur, W. A., & Mindler, A. (1983). Control of biological fouling in seawater reverse osmosis desalination. Desalination, 47(1-3), 233-238.

Winters, H., Isquith, I.R. (1979). In-plant microfouling in desalination. Desalination 30(1), 387-399.

Wynne, T. T., Stumpf, R. P., Tomlinson, M. C., Schwab, D. J., Watabayashi, G. Y., and Christensen, J. D. (2011) Estimating cyanobacterial bloom transport by coupling remotely sensed imagery and a hydrodynamic model. Ecological Applications 2, 2709-2721.

Zendong, Z., Abadie, E., Mazzeo, A., Hervé, F., Herrenknecht, C., Amzil, Z., Dell'Aversano, C., and Hess, P. 2015. Determination of the concentration of dissolved lipophilic algal toxins in seawater using pre-concentration with HP-20 resin and LC-MS/MS detection, in: MacKenzie, L. (Ed.), 16th International Conference on Harmful Algae 27th-31st October 2014. Cawthron Institute, Nelson, New Zealand and International Society for the Study of Harmful Algae, Wellington, New Zealand.

Zhou, J., Mopper, K., and Passow, U. 1998. The role of surface-active carbohydrates in the formation of Transparent Exopolymer Particles by bubble adsorption of seawater. Limnology and Oceanography 43(8), 1860-18.

Chapter 7

Inorganic fouling

Jan C. Schippers

The main learning objectives of this chapter are the following:

- Understand the origin of iron and manganese in groundwater and beach wells
- Apply solutions to avoid and control fouling due to iron and manganese in membrane systems

7.1 INTRODUCTION

Membrane fouling due to iron and manganese primarily happens in ground water, artificial recharge and beach/shore well water. Iron fouling is frequently observed and is causing:

- Loss in permeability of the membranes, resulting in the need for higher feed pressure;
- Increase in pressure drop across the feed/concentrate channel of spiral wound elements and fibre bundle of hollow fibre membrane elements.
- Increase in salt passage due to enhanced concentration polarization in the foul layer.

Several sources of iron fouling have been identified e.g.,

- Anaerobe ground water containing ferrous iron.
- Corrosion products from pipe materials and equipment.
- Hydroxide flocs from coagulation processes.
- Complexes with natural organic matter.

The ferrous iron (II) ion by itself will not cause membrane fouling because it is very well soluble. However, if the water contains dissolved oxygen, it will be oxidized to ferric iron (III) and form deposits on the membrane surface and spacers.

Manganese fouling is less common than iron fouling and is mainly observed in groundwater, infiltration and beach/shore well water. However, some manganese is observed in surface water e.g., river water, lakes due to manganese containing runoff, water as well.
Manganese is usually present as manganese (II), which is very well soluble. When the water contains oxygen, it will be oxidized slowly to manganese (IV) to form insoluble oxides, which result in membrane fouling.

 Seawater Reverse Osmosis Desalination
Editors: Sergio Salinas, Jan Schippers, Gary Amy, In Kim, Maria Kennedy
doi: 10.2166/9781780409863_0187

The focus of this chapter will be on fouling due to iron and manganese origination from ground water and water abstracted through artificial recharge and beach wells.

7.2 ORIGIN OF IRON AND MANGANESE

Ground waters, similar to rivers and lakes, are part of the hydrological cycle and are characterized by steady flow velocities. Average flow velocities observed in aquifers range from 3 mm to 30 km per year and residence time vary between couples of weeks to 50,000 years. Rain water (directly or indirectly through infiltration via rivers, ditches, etc.) is the source of most ground waters and does not contain iron or manganese.

These inorganic compounds occur naturally in soils, rocks and minerals. In the aquifer the water comes in contact with these solid materials dissolving them. An aquifer is an underground layer of water – bearing permeable rock, rock fractures (karstic layers composed of limestone) or unconsolidated material (gravel, sand, silt). When iron and manganese are present, they are in the dissolved form, because undissolved forms (suspended) will be removed by attachment on soil material and settling during travelling.

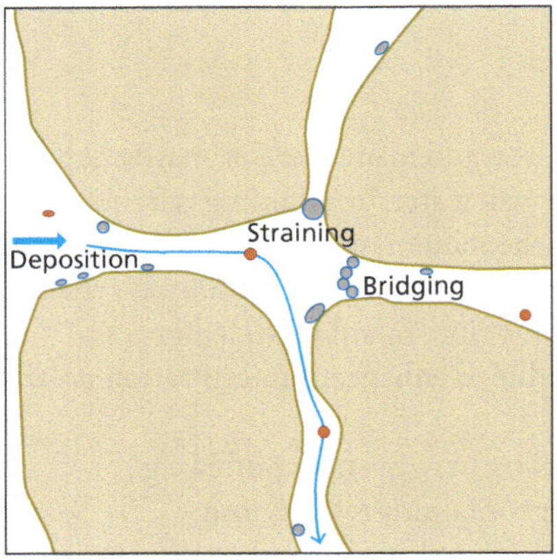

Figure 1 Removal suspended particles in the soil by depositing, straining and bridging (Adapted from Zwart, 2007)

The dissolved form of iron is in the reduced form (ferrous) namely Fe(II) or Fe^{2+}. Manganese is present in the reduced form as well Mn(II) or Mn^{2+}. However, most iron and manganese containing materials are in the oxidized form namely Fe(III) or Fe^{3+} and Mn(IV) or Mn^{4+}, having extremely low solubility's at pH values occurring in natural waters.
Some common iron oxide and manganese oxides comprise:

- Hematite (α-Fe_2O_3);
- Goethite (α-FeOOH);
- Maghemite (Υ-Fe_2O_3);
- Lepidocrocite (Υ-FeOOH);

- Freshly precipitated hydrous oxide or ferric hydrite ($Fe[OH]_3$ n H_2O), and
- Pyrolusite (MnO_2)
- Birnessite ($MnO_{1.9}$)

7.2.1 Anaerobic conditions

Iron (III) and manganese (IV) present in these oxides are reduced to iron(II) and manganese(II) and consequently dissolve under anaerobic conditions only, which means that no oxygen is present. Infiltrating rainwater use to be saturated with oxygen, consequently anaerobic condition can occur only if oxygen is consumed in the soil (aquifer). Oxygen consumption occurs in many soils due to bacteria which oxidize organic matter, which is commonly present. This organic matter originates from decaying remaining's of trees and plants together with gravel, sand and silt which have been deposited to form (large) layers of sediments. Ground waters are mostly abstracted from these layers.

Organic matter in soils consists mainly of humic substances, which are the result of bacterial activities (under anaerobic conditions) in the past. A large variety of humic substances exists. $C_{10}H_{18}O_{10}$ represents a simplified formula for these substances. Bacterial activities in the soil are responsible for the formation of ammonium (NH_4^+) and methane (CH_4) as well as from organic matter.

Bacteria consume oxygen to oxidize humic substances, ammonium and methane.

Humic substances: $2\,C_{10}H_{18}O_{10} + 19\,O_2 \rightarrow 20\,CO_2 + 18\,H_2O$
Ammonium: $2\,NH_4^+ + 4\,O_2 \rightarrow 2\,NO_3^- + 4\,H^+ + 2\,H_2O$
Methane: $CH_4 + 2\,O_2 \rightarrow CO_2 + 2\,H_2O$

Iron and manganese are dissolving under anaerobic condition which means oxygen depletion due to bacterial activities. This process is illustrated in the reaction equations below:

$$4\,Fe(OH)_3 + 8\,H^+ \rightarrow 4\,Fe^{3+} + 4\,OH^- + 8H_2O$$
$$4\,Fe^{3+} + 4\,OH^- \rightarrow 4\,Fe^{2+} + O_2 + H_2O$$

$$4\,Fe(OH)_3 + 8\,H^+ \rightarrow 4\,Fe^{2+} + O_2 + H_2O$$

and,

$$6\,MnO_2 + 12\,H^+ \rightarrow 6\,Mn^{2+} + 3\,O_2 + 6\,H_2O$$

Oxygen in these equations is not really formed, when the reaction goes → direction. It is a hypothetically formed intermediate illustrating that this oxygen is taken by bacteria to oxidize humic substances, ammonium or methane.
In reality oxygen is not being transferred but electrons (*e*). This mechanism is illustrated below;

$2H_2O = O_2 + 4\,H^+ + 4\,e$ oxidation
$4\,Fe^{3+} + 4\,e = 4\,Fe^{2+}$ reduction

$2H_2O + 4\,Fe^{3+} = O_2 + 4\,H^+ + 4\,Fe^{2+}$ **redox reaction**

Beside reduction of iron(III) and manganese(IV) nitrate and sulfate can be reduced as well to form nitrogen (denitrification) and hydrogen sulfide.

$$4\,NO_3^- + 4H^+ + \text{bacteria} \rightarrow 2\,N_2 + 5\,O_2 + 2\,H_2O$$
$$SO_4^{2-} + 2H^+ + \text{bacteria} \leftrightarrow H_2S + 2\,O_2$$

Oxygen is in these equations a hypothetically formed intermediate in oxidizing organic matter.

7.2.2 Aerobic conditions

Pyrite (FeS_2) might be present under anaerobic conditions in aquifers as well. When exposed to water containing oxygen, it will be oxidized to Fe (II) and sulfate.

$$2\,FeS_2 + 7\,O_2 + 2\,H_2O \rightarrow 2\,Fe^{2+} + 4\,SO_4^{2-} + 4\,H^+$$

In this way iron (II) will be introduced into the water and the pH will drop due to the formation of H^+.

7.2.3 Degree of anaerobia

The classification aerobic and anaerobic conditions, simplifies what really is happening somewhat too much, to explain the sequence of occurrence of the discussed reactions. Actually, the degree of aerobia and anaerobia is governing the sequence of occurrence. This degree is expressed as redox potential (E_h expressed in mV) or negative logarithm of electron concentration ($-\log[e^-] = pe$) and can be measured. Relation is $pe = 16.9 \times E_h$. A positive redox potential indicates that the water is in an aerobic condition and a negative redox potential indicates that it is in an anaerobic condition. Figure 2 illustrates the sequence of occurrence of redox reaction as a function of redox (pe).

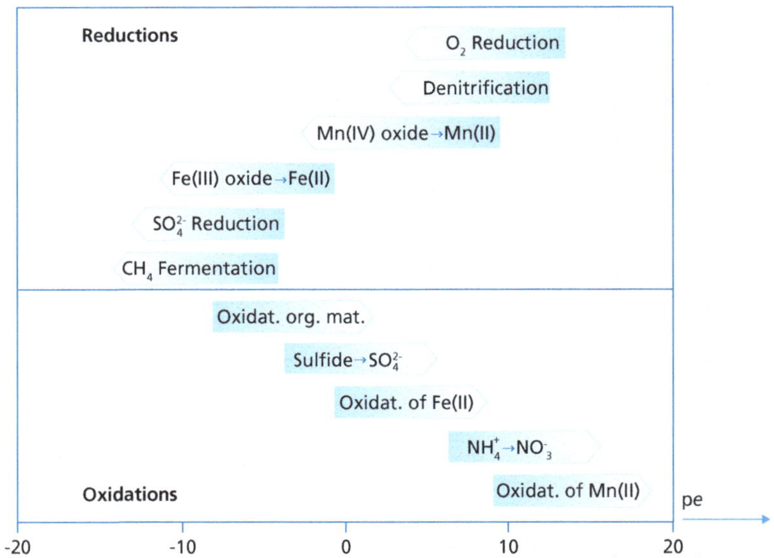

Figure 2 Sequences of important redox processes at pH =7 in natural waters (Adapted from Appelo and Postma, 2005)

7.3 COMPOSITION OF GROUNDWATER AND BEACH WELLS

Some ground waters exhibit:

- very low turbidity (< 0.1 NTU).
- very low SDI (< 1).
- very low concentrations of iron (< 0.05 mg/L) and manganese (< 0.01 mg/L).

This type of ground waters is usually not causing any fouling in reverse osmosis and nanofiltration plants.

Many ground waters and water abstracted with beach wells contain iron, resulting in turbidity and some manganese. Usually the turbidity appears a couple of minutes after aeration, when oxygen is introduced. This oxygen will oxidize Fe(II), which is very well soluble into the Fe(III) form. It hydrolyses into ferric hydroxide ($Fe[OH]_3$) which is insoluble and precipitates. The precipitate results into high turbidity and high SDI. This phenomenon is illustrated in Figure 3.

Manganese present in the form of Mn(II) will not oxidize after aeration due to a very low reaction rate. It will only oxidize when a catalyst is present.

$$4\,Fe^{2+} + O_2 + 10\,H_2O \rightarrow 4\,Fe(OH)_3\downarrow + 8H^+$$

$$6\,Mn^{2+} + 3\,O_2 + 6\,H_2O \rightarrow 6\,MnO_2\downarrow + 12\,H^+$$

Figure 3 Groundwater samples immediately after sampling (left) and a couple of minutes after aeration (right). Location: Baq'a, Amman, Jordan (Jan C. Schippers).

When rain water is travelling, through a soil containing organic material, into deeper layers it contains initially oxygen. Gradually the oxygen is consumed and the water arrives in an anaerobic state, resulting in the release of manganese (first) and iron (later). Figure 4 illustrates this situation in the soil in a very simplified way.

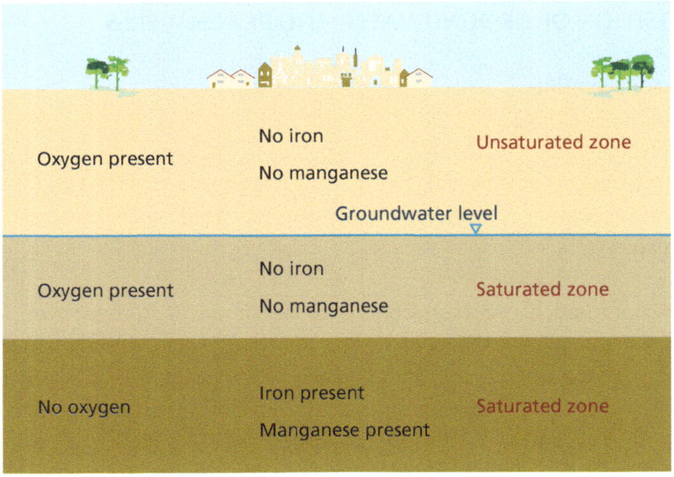

Figure 4 Unsaturated and saturated zone with and without oxygen and absence/presence manganese and iron.

Three zones are supposed to occur namely:

- Unsaturated zone. In this zone the soil is moist, but the pores are not filled with water. Water cannot be abstracted from this zone.
- Saturated zone in which the pores are filled with water and oxygen is still present. Consequently, manganese and iron cannot dissolve, so they are absent in the water.
- Saturated zone in which the water does not contain oxygen anymore as a result manganese and iron can dissolve and the water may contain the constituents.

In the soil more processes than just release of manganese and iron. These processes and their sequence of occurring are illustrated in Figure 5.

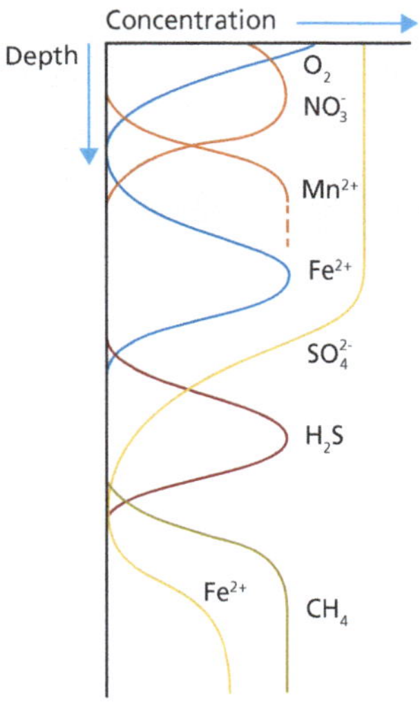

Figure 5 Process and sequence occurring in the soil (Adapted from Appelo and Postma, 2005)

In the aerobe zone, any released ammonium (from organic matter) or introduced with rain water in agriculture aeries will be oxidized to nitrate. In the anaerobe zone it will be reduced to nitrogen (N_2).

Figure 4 illustrates a very simplified situation, in reality the soil is much more complicated. In Figure 6 a more realistic picture is given. However, at many locations the situation is even more complicated.

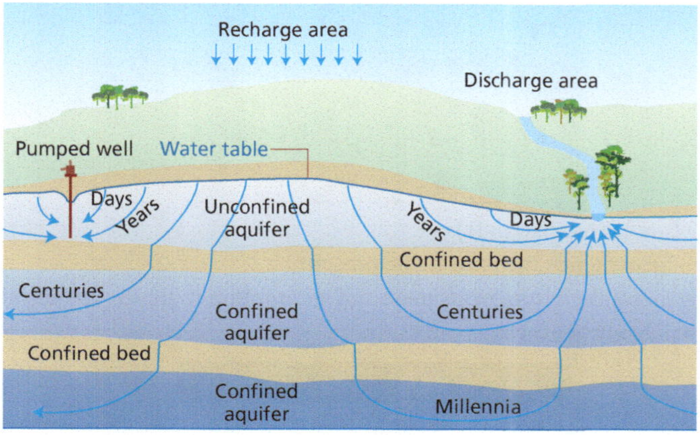

Figure 6 Soil with different aquifers separate by confining beds, which are poorly permeable (Adapted from USGS, 2020)

Oxygen, ammonium, nitrate, manganese and iron are key parameters in ground water. Based on these parameters ground waters can be categorized in three types. This finding is illustrated with the composition of well water from three different locations namely; Fuerteventura (Spain), Gouda (Netherlands) and Tarfaya (Morocco).

Example 1 – Fuerteventura (Spain)

Does this ground water contain oxygen and/or ammonium? Explain your answer.

Answer: The absence of iron and manganese and presence of nitrate indicates that oxygen will be present and ammonium will be absent.

Ion	mg/L	Ion	mg/L
Ca^{2+}	210	HCO_3^-	342
Mg^{2+}	250	SO_4^{2-}	743
Na^+	1,185	Cl^-	2,118
K^+	44	NO_3^-	49
Mn	0.0	pH	7,5
Fe	0.0	NH_4^+	?

Example 2 – Gouda (Netherlands)

Will oxygen be present? Will nitrate be present?

Answer: The presence of iron, manganese and ammonium indicates that oxygen is absent and nitrate will likely be absent

Ion	mg/L	Ion	mg/L
Ca^{2+}	94	HCO_3^-	205
Mg^{2+}	12	SO_4^{2-}	71
Na^+	80	Cl^-	159
K^+	4	NO_3^-	?
Mn	0.7	pH	7,5
Fe	2.8	O_2	**?**
NH_4^+	0.7		

Example 3 – Tarfaya (Morocco)

Does this water contain oxygen? Explain your answer.

Answer: The presence of iron and manganese indicates that oxygen is absent. However, the presence of nitrate and absence of ammonium indicates that oxygen is presence. This inconsistency can be explained when both aerobic and anaerobic water are abstracted at the same time.

Ion	mg/L	Ion	mg/L
Ca^{2+}	149	HCO_3^-	191
Mg^{2+}	131	Cl^-	2,049
Na^+	1,130	SO_4^{2-}	325
K^+	56	NO_3^-	62
Fe	0.1	F	5.4
Mn	0.03	pH	7.3
NH_4^+	0.0	O_2	**?**

7.3.1 Beach/shore wells

Water abstracted through beach wells originates for a main part or fully from the sea. Another part (small) comes from groundwater. Sea water itself does not contain iron or manganese. Frequently the seawater travelling through the soil becomes anaerobe due to the presence of organic matter. As a result, manganese and iron might release from the soil. At some locations, oxygen is not fully consumed by bacteria, the concentrations of manganese and iron are very low and no pre-treatment for reverse osmosis is required.

Other locations show rather high and increasing concentrations and pre-treatment is necessary. Increasing concentrations originate from deeper layers in the soil and or groundwater abstracted from anaerobe layers. Reducing the abstraction rate – at the expense of capacity – usually alleviates the problems.

Figure 7 illustrates the beach well principle. The geological situation might be more complicated than shown or different e.g., at some locations the soil is karstic (lime stone) e.g., in Ghar Lapsi, Malta. At this location the first large scale seawater RO plant in Europe has been put into operation in 1985 and is abstracting seawater through shore wells.

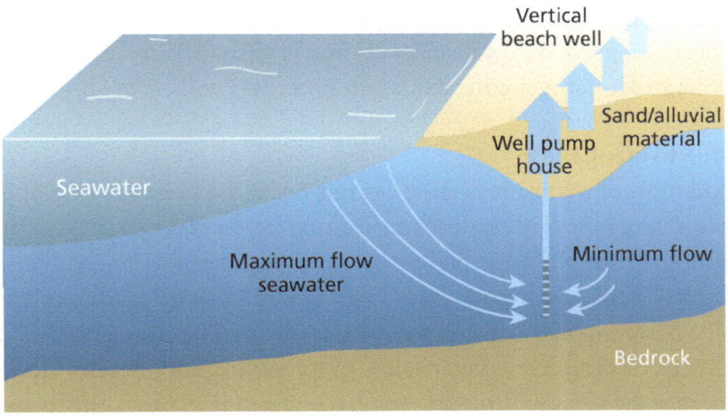

Figure 7 Well intake system located along a shoreline. This is truly a "beach well" system that promotes direct recharge from the sea and minimizes capture of landward water resources. Minimal flow should come from the shoreline direction to avoid aquifer impacts and entry of poor quality (Adapted from Missimer, *et al.*, 2013)

7.4 MEMBRANE FOULING DUE TO IRON AND MANGANESE

7.4.1 Fouling due to iron

When iron is present in ground water it is in the reduced state and very well soluble. As soon as oxygen enters the water, iron will be oxidized according the equation:

$$4\,Fe^{2+} + O_2 + 10\,H_2O \rightarrow 4\,Fe(OH)_3 \downarrow + 8\,H^+$$

Ferric hydroxide, which is formed by a homogeneous reaction in the water, will precipitate because it is very insoluble. It forms colloidal and suspended particles. Oxygen might be introduced through different pathways, e.g.,

- Mixing anaerobe water with aerobe water in the wells
- Storage tank(s) in the transport system to convey the ground water to the reverse osmosis plant
- Leakages in the sealing of pumps
- Leakages in the valves

These colloidal and suspended particles will deposit on the membrane surface and spacers. Elevated levels of SDI and $MFI_{0.45}$ will occur and ferric hydroxide will stain the membrane filter, used in these tests, brownish yellow. These tests are very useful tools in detecting introduction of oxygen in the system.

Homogeneous oxidation of iron (II) to iron (III) is likely not completed when the water enters the reverse osmosis system and gives rise to a second oxidation mechanism namely heterogeneous oxidation. In this process Fe(II) is adsorbed on the membrane surface and spacers, and subsequently very fast oxidized to Fe (III) to form (adsorbed on the membrane) $Fe(OH)_3$. This ferric hydroxide has a much higher adsorption capacity than the membrane surface and spacers itself and forms a dense layer on the membrane surface. Consequently, the oxidation process is speeded up substantially and is termed "auto-catalytic".
Both homogeneous and heterogeneous oxidation processes will occur simultaneously.

7.4.2 Fouling due to manganese

Homogenous oxidation of manganese (II) does not occur, because the rate of oxidation at ambient pH levels is very low. Only heterogeneous oxidation takes place. Similar to iron (II) oxidation, manganese is initially adsorbed on the membrane surface and spacers itself or on the previously formed ferric hydroxide layer. The adsorbed Mn(II) will oxidize rather fast and formed MnO_2 has a high adsorption capacity resulting in a fast, autocatalytic process. Usually the build-up of this dense layer takes quite some time, because the adsorption capacity of the membrane is rather limited. But once a layer is in place, the formation of the dense black layer of MnO_2 is fast. If iron hydroxide is present this process will be enhanced.

Iron fouling can be easily recognized, because it stains the membrane surface and spared yellow/brown. Manganese deposits results in a black coloration. See Figure 8.

$$6\,Mn^{2+} + 3\,O_2 + 6\,H_2O \rightarrow 6\,MnO_2\downarrow + 12\,H^+$$

Figure 8 Membranes fouled with iron (left) and manganese (right) (DuPont, 2020)

7.5 RATE OF OXIDATION IRON (II) AND MANGANESE (II)

The rate of homogeneous oxidation of iron and manganese depends on the oxygen concentration, the pH (- log [H+]) and concentration of iron (II) and manganese (II).

$$d[Fe^{2+}]/dt = -k_{Fe}[Fe^{2+}][O_2]/[H^+]^2$$
$$d[Mn^{2+}]/dt = -k_{Mn}[Mn^{2+}][O_2]/[H^+]^2$$

In Figure 9 the effects of the concentration of oxygen and pH on the rate of oxidation of iron (II) are illustrated.

Rate of oxidation of iron (II) by oxygen depends on: pH and oxygen concentration, as follows:

- The lower the pH, the lower the rate.
- The higher the pH, the higher the rate.
- The lower the oxygen concentration the lower the rate of oxidation.

The rate oxidation manganese oxidation is negligible at pH values below 9.
A catalyst, in the form of MnO_2 and/or Mn_3O_4, is needed to speed up the rate of oxidation. This is an autocatalytic process. As consequence fouling starts slowly and speeds up gradually.
Remark: Below pH 6.9 the rate of oxidation (even with a catalyst) is very low.

Oxidation of manganese is much slower than iron and almost negligible ate ambient pH levels. Figure 10 illustrates the difference in rate of oxidation between iron (II) and Mn(II). In this figure:

- $[Fe(II)]_0$ = concentration Fe(II) at time t =0 (at the start of the oxidation)
- $[Fe(II)]_t$ = concentration Fe(II) at time t = *t* (*t* minutes after the start of the oxidation)

Examples:

- At time t = 0, Log $[Fe(II)]_t/[Fe(II)]_0 = 0$ because $[Fe(II)]_t = [Fe(II)]_0$
- Log $[Fe(II)]_t/[Fe(II)]_0 = -1$ means $[Fe(II)]_t = 0.1$ [Fe(II)]0 or 90% of Fe(II) has been oxidized.

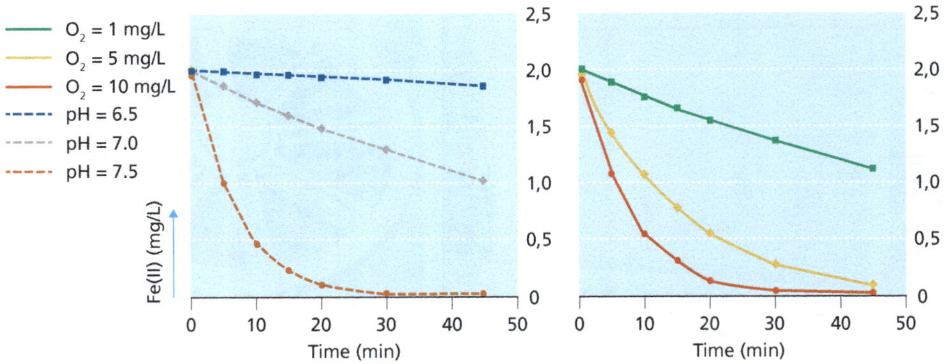

Figure 9 Effect of pH (left) and oxygen concentration (right) on the rate of homogeneous oxidation of iron (II) (Adapted from Stumm and Morgan, 1996)

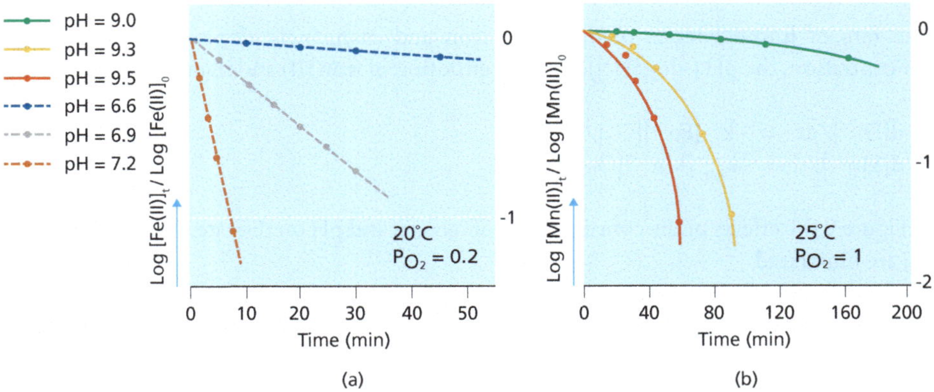

Figure 10 Rate of oxidation a) Fe (II) and b) Mn (II) (Adapted from Stumm and Morgan, 1996)

In heterogeneous oxidation the present adsorbed ferric hydroxide and manganese dioxide play a dominant role. In particular the rate of oxidation of Mn(II) is largely increased by the autocatalytic mechanism.

Remark: Ground water abstracted from layers with and without oxygen will cause sever well clogging as well. On and near the well screen similar fouling processes as on the membranes will occur.

7.6 HOW TO AVOID FOULING DUE TO IRON (II) AND MANGANESE (II)

Four options are identified namely:

1. Abstract water that does not contain any iron or manganese. Unfortunately, this raw water source is not frequently available.
2. Abstract water that does not contain any oxygen and exclude oxygen. Several plants apply this option successfully.

3. When inevitably oxygen enters the system. Lowering the pH will be very useful, to such a level that the rate of oxidation is low. This approach might require substantial amounts of acid.
4. Abstract water (does not matter whether oxygen is present or not). Treat the water by e.g., aeration followed by rapid (green) sand filtration.

7.6.1 Controlling membrane fouling due to iron and manganese

Feed water abstracted from layers with oxygen and consequently no iron and manganese will be present, will not cause membrane fouling. Usually SDI and $MFI_{0.45}$ will be very low as well.

Feed water abstracted from layers without oxygen and iron and manganese present, will not cause membrane fouling.

- Condition: Oxygen must be excluded completely from entering the feed water in the well and reverse osmosis plant. Because iron and manganese need very little oxygen to oxidize namely.
  1 mg Fe(II) needs 0.14 mg O_2
  1 mg Mn(II) needs 0.29 mg O_2

Example 4 – How much oxygen is dissolved in water when is saturated?

100% air saturation is the equilibrium point for gases in water. According to Henry's Law, the dissolved oxygen content of water is proportional to the percent of oxygen (partial pressure) in the air above it

At constant temperature, the amount of gas absorbed by a given volume of liquid is proportional to the pressure in atmospheres that the gas exerts.

$[gas] = K_H \cdot \rho_{gas}$ where: K_H (Henry's constant) is a solubility factor, varying from gas to gas.

O_2 in atmosphere ~20.3% = 0.203 atm.

$K_{H@20\,^{\circ}C} = 1.39$ (mmol O_2 / (kg H_2O×atm)) (for pure water)

Thus, the amount of O_2 that will dissolve in water at 20 °C:

$$[O_2]=(0.203\,atm)\left(\frac{1.39\,mmol\,O_2}{kf\cdot atm}\right)=0.282\frac{mmol\,O_2}{kg}=9.03\frac{mg\,O_2}{kg}$$

Feed waters abstracted from layers with and without oxygen, in one well, will cause severe membrane fouling (and well clogging). Because water with oxygen and no iron and manganese, will mix in the well with water without oxygen and with iron and manganese. The same situation occurs when water is abstracted without any oxygen but oxygen is introduced e.g., in a storage tank, leaking valves and leaking seals of pumps.

Dosing acid can effectively control the rate of fouling, because the rate of oxidation can be minimized due to the lower pH. Several plants apply this strategy successfully. The Pembroke plant in Malta applies acid dosing already for several decades. Figure 11 gives the process scheme of this plant.

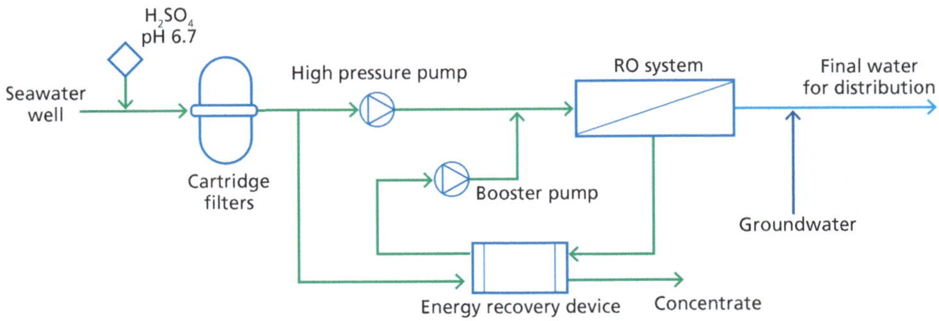

Figure 11 Process scheme Pembroke plant in Malta. Acid is dosed to control iron and manganese fouling (Adapted from Lagartos, *et al.*, 2019)

7.6.2 Removal of Iron and Manganese

7.6.2.1 Aeration followed by sand filtration

Aeration followed by rapid sand and green sand filtration is commonly and successfully applied to remove iron and manganese from fresh ground water for many decades. Pressurized steel filters and concrete open, gravity filters are applied. Figure 12 gives the principle of a pressurized filter and Figure 13 is a picture of such filters in practice.

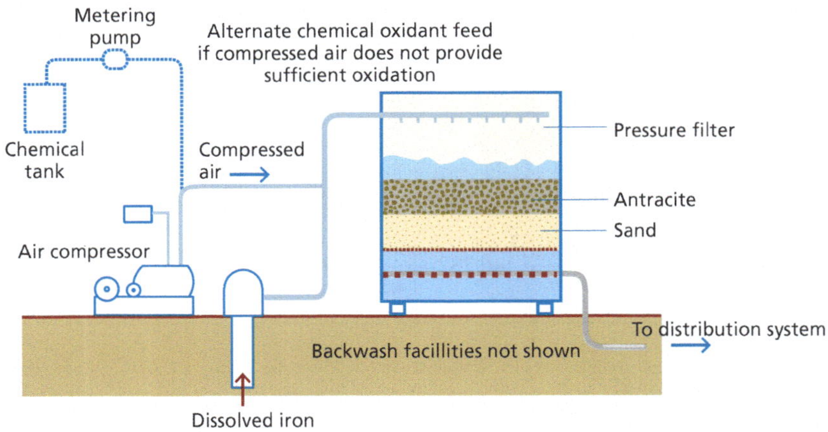

Figure 12 Principle of filter used for iron and manganese removal from groundwater (Adapted from AWWA, 1995)

7.6.2.2 Iron removal

Two physical mechanisms are involved in the removal of iron namely:

- **Oxidation-floc formation mechanism.**
  Homogeneous oxidation in the supernatant and subsequent formation of ferric hydroxide flocs occur. This process continues in the filter bed. Flocs are removed by the (depth) filtration process.

Figure 13 Rapid sand filters in groundwater treatment (Buamah, 2009)

- **Adsorption-oxidation mechanism.**
  Heterogeneous oxidation takes place. This after that iron (II) has been adsorbed at the dense layer of ferric hydroxide, which is present on the sand grains. The oxidation of the adsorbed iron (II) is very fast and a new layer of ferric hydroxide is being formed. This process "Catalytic iron removal" as well.

These mechanisms are illustrated in Figure 14.

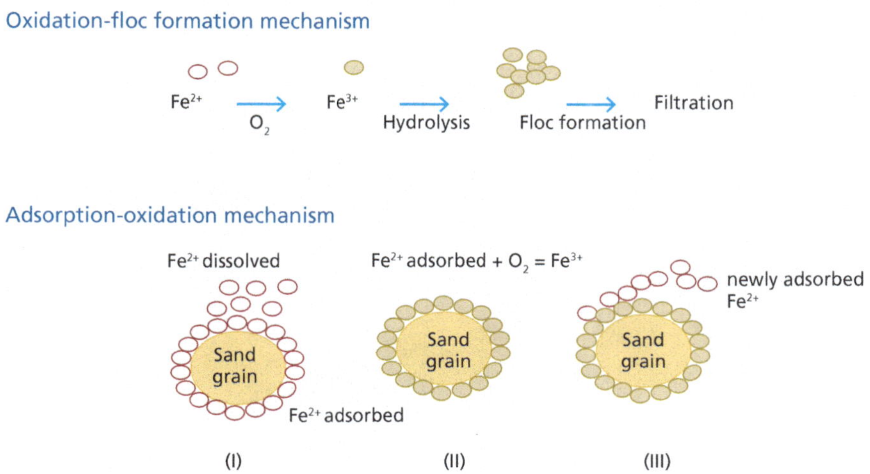

Figure 14 Physicochemical iron removal mechanisms (Adopted from Sharma, 2001)

Several plants treating fresh (low salinity) ground water apply **pre-chlorination** to enhance the rate of oxidation of Fe(II) and Mn(II). Intermittent dosing **potassium permanganate** is applied as well to enhancing the oxidation of Mn(II) that is adsorbed on the surface of filter media e.g., sand or green sand.

Dosing of these chemicals is not recommended when sand filtration is applied as pre-treatment for reverse osmosis, because overdosing will damage and/or foul the membranes. Moreover, chlorination results in formation of assimilable (biodegradable) organic matter from natural organic matter (humic acids). These are oxidized to smaller organic compounds, which are easily assimilable for bacteria. Biofouling of the membrane elements will occur.

In literature frequently biological iron removal has been reported. There is no doubt that in several plants, bacteria play and paramount role. Bacteria are able to catalyse the rate of oxidation of iron (II). It is not unlikely that both mechanisms namely, the physical chemical and the biochemical mechanism play a role in many/several plants.

7.6.2.3 Manganese removal

In sand filters homogeneous oxidation does not take place because the rate is too low. Autocatalytic process is responsible for effective removal of manganese. In this process the catalyst plays an essential role because Mn(II) is only oxidized if adsorbed on MnO_2 or more precise $(Na^{+}_{0.7}Ca^{2+}_{0.3})\ Mn_7O_{14}$, named **birnessite.**

This mineral has a very high specific surface area due to its amorphous structure, when formed in filters. Figure 15 illustrates this very high specific surface area.
In turned out that manganese removal in rapid sand filters at pH < 6.9, becomes problematic, due to low rate of oxidation of the adsorbed Mn(II).

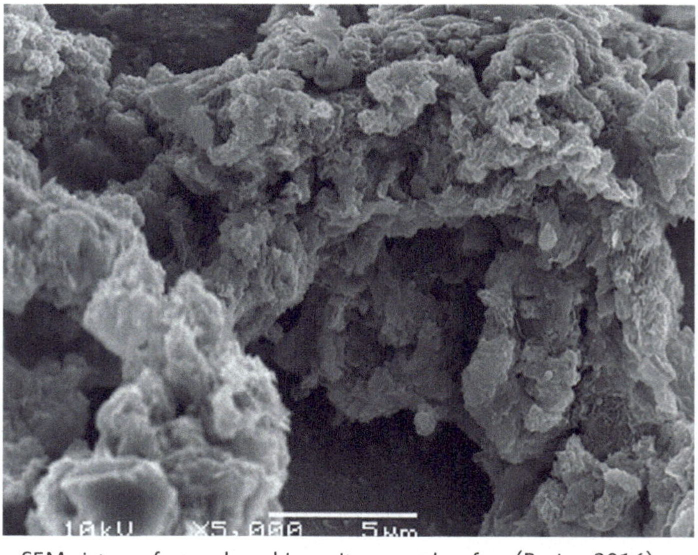

Figure 15 SEM picture of amorphous birnessite on sand surface (Bruins, 2016)

7.6.2.4 Polishing with cartridge filtration

Commonly cartridge filtration is applied prior to the reverse osmosis plant. Justifications for this polishing step are:

- Removing suspended and colloidal particles escaping from the sand filters.
- Main pre-treatment in groundwater, when iron and manganese concentrations are very low.
- Protection of the high-pressure pumps against sand. Originating from e.g., wells and rapid sand filters (damaged filter nozzles).

Cartridges with pores ranging from 100 μm down to 1 μm are applied. In practice mainly 5 – 20 μm cartridges are used. Replacement frequencies vary from "once per week to once per year" depending on the water quality.

Figure 16 shows cartridges used after a sand filter in a reverse osmosis plant treating brackish groundwater (likely partly infiltrated seawater) Gran Canaria, Spain. See Figure 17.

Figure 16 Used cartridge filters

Figure 17 Sand filter. Gran Canaria, Spain.

Figure 18 Reverse osmosis unit producing irrigation water for a green house in Gran Canaria, Spain

Example 5

Where are the cartridge filters located?

What is the capacity of this reverse osmosis unit?

Answer:

a) the figure 18, the cartridge filters are located in the vertical metallic vessel in front of the pump to safeguard the integrity of the RO membranes and of the pump.

b) The RO unit has two stages (3 pressure vessels in the first stage and 2 pressure vessels in the second stage). Assuming 6 RO elements (40 m^2 each) placed per pressure vessel, and an average flux of 20 $L/m^2/h$.

We have:

Capacity = Flux · membrane area

Capacity first stage = $20\ L/m^2/h \cdot [40\ m^2 \cdot (3 \cdot 6)] = 14400\ L/h$

Capacity second stage is half of the first stage = 7200 L/h.

Total capacity of the RO installation is 21.6 m^3/h.

7.7 SUMMARIZING

- Many ground waters contain iron (II) and manganese (II).
- Beach/shore wells and water abstracted from artificial recharge areas may contain these natural contaminants as well.
- Some ground waters do not contain these contaminants.
- Iron (II) and manganese (II) appear mainly under anaerobic conditions.
- Membrane fouling can be controlled by:
    a. Abstracting anaerobic (ground) water and keeping it strictly anaerobic. If excluding oxygen is problematic lowering the pH by dosing acid is an option;
    b. Removing iron (II) and manganese (II) by aeration followed by rapid (green) sand filtration. Pre-Chlorination is applied sometimes but not recommended

These two options are illustrated in Figure 19 and Figure 20.

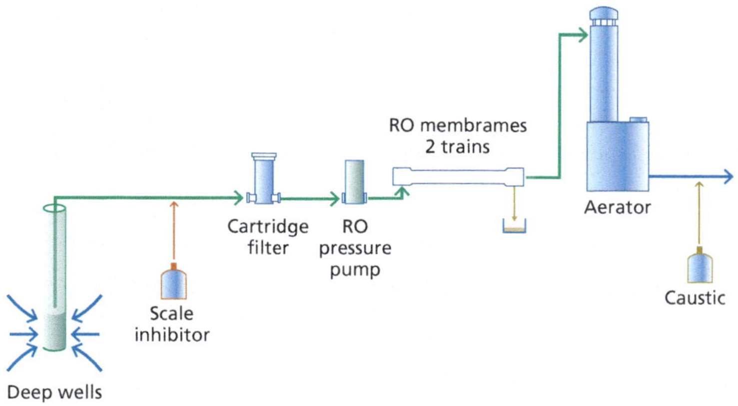

Figure 19 Process scheme for anaerobic groundwater treatment (Adapted from Nemeth-Harn, 2018)

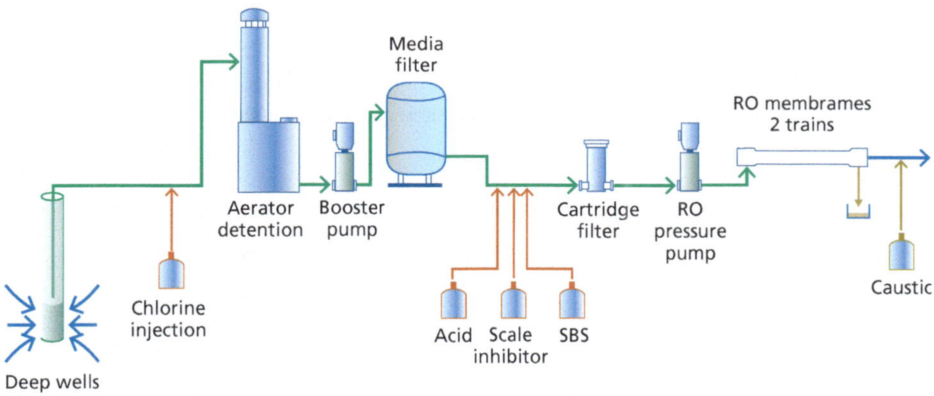

Figure 20 Process scheme with sand filtration for anaerobic groundwater treatment (Adapted from Nemeth-Harn, 2018)

7.8 REFERENCES

Appelo CAJ, Postma D (2005) Geochemistry, Groundwater and Pollution CRC Press, Amsterdam

AWWA (1995) Water Treatment. Principles and practivces of water supply operations, 2nd edn AWWA, Denver, CO

Bruins J (2016) Manganese removal from groundwater : role of biological and physico-chemical autocatalytic processes. CRC Press/Balkema

Buamah R (2009) Adsorptive Removal of Manganese, Arsenic and Iron from Groundwater CRC Press/ Balkema, Delft

DuPont (2020) FILMTEC™ Reverse Osmosis Membranes Technical Manual. In: Water solutions (ed), pp. 207.

Lagartos A, Rozenbaoum E, Oruc M, Hyung H, Armas JCd, Sacco D (2019) Long-term boron rejection of thin-film nanocomposite membrane at Pembroke Desalination Plant in Malta: a case study. Desalination and Water Treatment 157: 274-280

Missimer TM, Ghaffour N, Dehwah AHA, Rachman R, Maliva RG, Amy G (2013) Subsurface intakes for seawater reverse osmosis facilities: Capacity limitation, water quality improvement, and economics. Desalination 322: 37-51 DOI https://doi.org/10.1016/j.desal.2013.04.021

Nemeth-Harn J (2018) Direct Membrane Treatment of Anaerobic High Iron and Manganese Groundwater. In: Fox Engineering (ed), March 23 edn, pp. https://foxeng.com/direct-membrane-treatment-anaerobic-high-iron-manganese-groundwaters/.

Sharma SK (2001) Adsorptive iron removal from groundwater Swets & Zeitlinger B.V., Lisse

Stumm W, Morgan JJ (1996) Aquatic chemistry: chemical equilibria and rates in natural waters, 3rd edn Wiley interscience publication, New York

USGS (2020) Groundwater and aquifers, Online, pp. https://www.usgs.gov/special-topic/water-science-school/science/aquifers-and-groundwater?qt-science_center_objects=0#qt-science_center_objects.

Zwart AH-d (2007) Investigation of clogging processes in unconsolidated aquifers near water supply wells. PhD Dissertation, Ponsen & Looyen BV

Chapter 8

Scaling

M. Nasir Mangal, Sergio G. Salinas-Rodríguez,

Victor A. Yangali-Quintanilla,

Maria D. Kennedy, Jan C. Schippers

The main learning objectives of this chapter are the following:

- Understand what is scaling, mention scaling species
- Understand factors affecting scaling
- Calculate the scaling potential of a water
- Propose solutions to control scaling in RO systems

8.1 MEMBRANE SCALING

Scaling is the formation of hard mineral layer due to the crystallization/precipitation of supersaturated sparingly soluble salts (e.g., calcium carbonate, calcium sulphate, barium sulphate, etc.) onto the membrane surface/feed spacer as illustrated in Figure 1. Scale forms a dense layer having a high hydraulic resistance, resulting in significant reduction in permeability of the membrane.

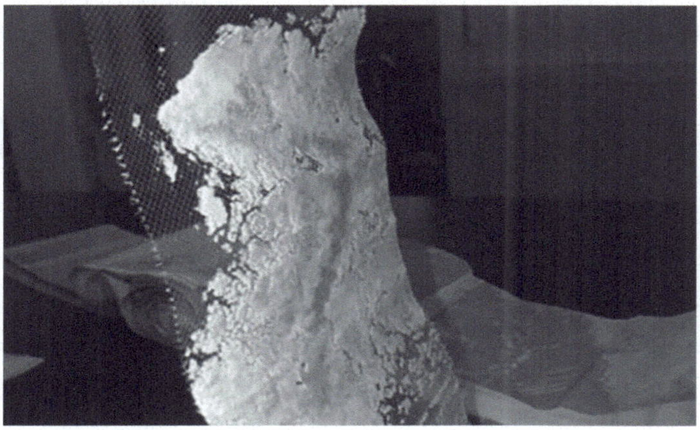

Figure 1 Scaling in RO systems (Kucera, 2011)

 Seawater Reverse Osmosis Desalination
Editors: Sergio Salinas, Jan Schippers, Gary Amy, In Kim, Maria Kennedy
doi: 10.2166/9781780409863_0207

Scaling adversely affects the performance of the RO such as:

- Lowering the permeate production, due to decrease in membrane permeability
- Increasing operational costs, due to higher operating pressure, cleaning costs, etc.
- Deteriorating the permeate water, due to increasing salt passage
- Shortening the life of membranes, due to frequent cleanings needed for restoring membrane permeability

Scaling of RO membranes is a challenging problem both in seawater (in seawater calcium carbonate plays a role only) and brackish water desalination. However, in treating brackish water, scaling is the main reason for operating RO systems at low recoveries, which leads to:

- High specific energy consumption (kWh/m^3),
- Less production of permeate water,
- More production of concentrate (waste),
- High chemical costs due to pre- and post-treatment.

As scaling is a concentration phenomenon, it starts in the last stage (tail elements) where the concentration of sparingly soluble salts is the highest. The high concentration of sparingly soluble salts exceeds their solubility limits, which as a result, triggers the formation of crystals onto the membrane surface.

Before discussing in more detail, the scaling of RO, it is necessary that a brief overview is given about the fundamental concepts of crystallization/scaling such as solubility, supersaturation, and mechanisms of scale formation.

8.1.1 Solubility of salts and supersaturation

Solubility of a salt in water is the ability of that salt to dissolve in water. It is a chemical property which is affected due to various parameters such as temperature, pressure, pH, ionic strength, etc. For instance, for some salts (e.g., KNO_3, $NaNO_3$, $BaSO_4$, etc.), solubility increases with increase in temperate, while for salts (e.g., $CaCO_3$, $CaSO_4$, etc.), solubility decreases when temperature is increased. Solubility of salts are generally expressed as mol of salt per litre of water (mol/L), gram of salt per litre or mL of water (g/L or g/mL), and gram of salt per gram of water (g/g), etc. In Table 1, the solubilities of salts in pure water are presented. The cells highlighted in grey represent the salts with very low solubilities.

Table 1 Solubilities of salts in pure water (18 °C) in g/L

	Na	Ca	Mg	K	Ba	Sr
Cl	360	730	560	330	370	510
SO_4	170	2	350	110	0.002	0.11
NO_3	840	1,220	740	300	90	70
CO_3	190	0.013	1	1,080	0.02	0.011
F	45	0.016	0.076	930	1.6	0.1

Determining whether a compound is saturated, under saturated, or super saturated is straight forward for silica (SiO_2); however, complicated for e.g., calcium sulphate and other salts. Simply adding together, the calcium concentration and sulphate concentration and comparing with solubility of $CaSO_4$ is not correct, because: the concentration of calcium and sulphate are in general not matching, or calcium is in excess or sulphate is in excess; solubility depends on temperature; and solubility depends on presence of other ions (salinity). Consequently, we cannot use the Table 1.

Furthermore, for sparingly soluble salts, another term which is used to grasp information about the solubility is the solubility product (K_{SP}). K_{SP} is the equilibrium constant of salts which represents the level a salt that may dissociate to its ionic species. Salts with low solubility in water have small K_{SP} values and vice versa. It is calculated as the mathematical multiplication of the molar concentrations of the dissociated ions raised to the power of their stoichiometric coefficients as described in Equation 8.2 for a dissolution reaction of

$$A_mB_{n(S)} \leftrightarrow mA_{(aq)} + nB_{(aq)} \quad \text{Eq. 8.1}$$

$$K_{SP} = [A]^m[B]^n \quad \text{Eq. 8.2}$$

Example 1 – What is the K_{SP} expression for $CaCO_3$?

Answers:

The equilibrium reaction is

$$CaSO_{3(S)} \leftrightarrow Ca^{2+}{}_{(aq)} + CO_3{}^{2-}{}_{(aq)}$$

So, the corresponding equilibrium constant is

$$K_{SP} = [Ca^{2+}][CO_3{}^{2-}]$$

As K_{SP} is dependent on temperature, the value should be always mentioned with the corresponding temperature at which they were measured/calculated. In Table 2, K_{SP} values of some scaling species at 25 °C are presented. Among the various scaling species shown in Table 2, calcium sulphate has the highest solubility, while calcium phosphate has the lowest solubility. Furthermore, two polymorphs of calcium carbonate are presented in Table 3, where polymorphs refer to crystalline compounds which have the same chemical composition but can exist in two or more crystalline shapes due to different arrangement of ions in the crystal lattice (Le Pevelen and Tranter, 2017). As shown, calcite is less soluble than aragonite.

Table 2 K_{SP} values of sparingly soluble salts at 25 °C

Compound	Formula	K_{SP}
Calcium carbonate (calcite)	$CaCO_3$	3.36×10^{-9}
Calcium carbonate (aragonite)	$CaCO_3$	6.0×10^{-9}
Calcium fluoride	CaF_2	5.3×10^{-9}
Calcium sulphate	$CaSO_4$	9.1×10^{-6}
Barium sulphate	$BaSO_4$	1.1×10^{-10}
Calcium phosphate	$Ca_3(PO_4)_2$	2.0×10^{-29}
Iron(II) carbonate	$FeCO_3$	3.2×10^{-11}
Strontium sulphate	$SrSO_4$	3.2×10^{-7}

For a scaling salt with a known K_{SP}, the molar solubility of the salt can be calculated as explained in Example 2. Similarly, for a compound with a known molar solubility, K_{SP} can be calculated.

Example 2 – The K_{SP} of $Ca_3(PO_4)_2$ is 2×10^{-29} as shown in Table 2

a) What is the molar solubility of $Ca_3(PO_4)_2$?
b) What is the solubility of $Ca_3(PO_4)_2$ in mg/L.

Solution:
The equilibrium reaction is

$$Ca_3(PO_4)_{2(S)} \leftrightarrow 3Ca^{2+}{}_{(aq)} + 2\,PO_4{}^{3-}{}_{(aq)}$$

So, the corresponding equilibrium constant is

$$K_{SP} = [3Ca^{2+}]^3[2PO_4{}^{3-}]^2$$

Let us denote the solubility of $Ca_3(PO_4)_2$ as S in mol/L. Then, for a saturated solution we have: $[Ca^{2+}] = 3S$, $2[PO_4{}^{3-}] = 2S$

Substituting this into the K_{SP} expression above,

$$K_{SP} = [3S]^3[2S]^2$$

$$2 \cdot 10^{-29} = 108 \cdot S^5$$

$$S = 7.1 \cdot 10^{-7}\ mol/L$$

a) Molar solubility of $Ca_3(PO_4)_2$ is 7.1×10^{-7} M.

b) Solubility of $Ca_3(PO_4)_2 = 7.1\cdot10^{-7}\dfrac{mol}{L}\cdot\dfrac{310.18\cdot10^3\,mg}{1\,mol} = 0.22mg/L$

Supersaturation is known as the driving force for the formation of crystals. Supersaturation develops when the concentrations of inorganic ions for a given scaling specie exceed the equilibrium concentration or the solubility limit. In other words, a solution is referred as supersaturated with respect to a given salt when the ion product (IP) of the salt exceeds the K_{SP}.

Based on the concentrations of scaling salt present in water, a water solution can be categorized as:

- **Saturated**: water is in equilibrium with a salt; not more can dissolve.
- **Undersaturated**: water can dissolve more salt than present in the water.
- **Supersaturated**: water contains more salts than can dissolve; precipitation may occur.

Theoretically compounds will precipitate when the solubility is exceeded. However, it has been demonstrated that some compounds in particular $BaSO_4$ show 'stable' super saturated solutions.
It should be noted, that the crystallization process not only involves the supersaturated conditions, but also depends on the precipitation kinetics (nucleation and crystal growth) (Koutsoukos, 2010).

8.1.2 Precipitation kinetics

8.1.2.1 Nucleation

When a solution exceeds a critical supersaturation level, nucleation process begins. The term nucleation refers to the formation of nuclei or clusters by association of lattice ions (Boerlage *et al.*, 2002). The formed nuclei can grow to the critical size under supersaturated conditions where they remain stable and can further grow to macroscopic crystallite as shown in Figure 2 (Dalmolen, 2005). If the nuclei do not reach to the critical size, they are unstable and therefore they will re-dissolve in the solution.

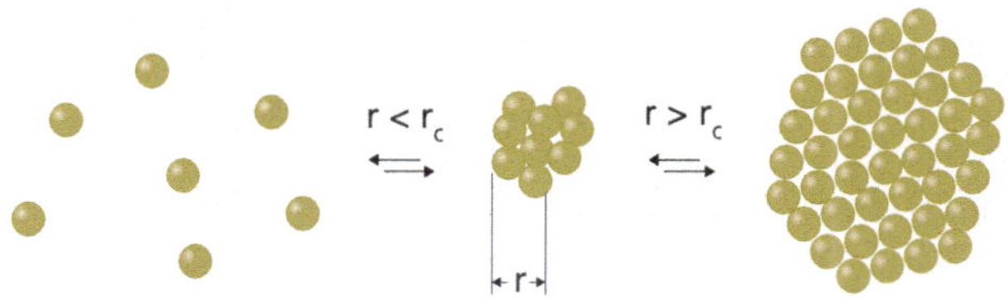

Figure 2 Process of nucleation (Dalmolen, 2005)

Nucleation process is categorized into two groups; primary nucleation and secondary nucleation as shown in Figure 3 (Mullin, 2001). The primary nucleation refers to the formation of clusters in the absence of other crystalline substances, while the secondary nucleation is the development of clusters in the presence of another crystalline matter (Mullin, 2001).

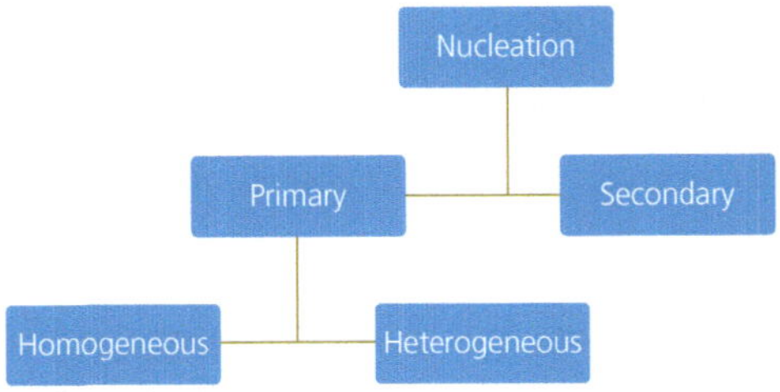

Figure 3 Nucleation types

Primary nucleation is further classified into two groups such as homogenous nucleation and heterogeneous nucleation. Homogenous nucleation is expected to prevail at high supersaturated conditions and occurs in the bulk solution, whereas heterogeneous nucleation is anticipated to be dominant at lower supersaturation levels and requires a surface as illustrated in Figure 4 (Boerlage *et al.*, 2000). The difference between the heterogeneous and secondary nucleation is that for the secondary nucleation, the surface should be the surface of the crystals, while for heterogeneous nucleation, the surface can be wall of a reactor, surface of the membrane, other particles, etc.

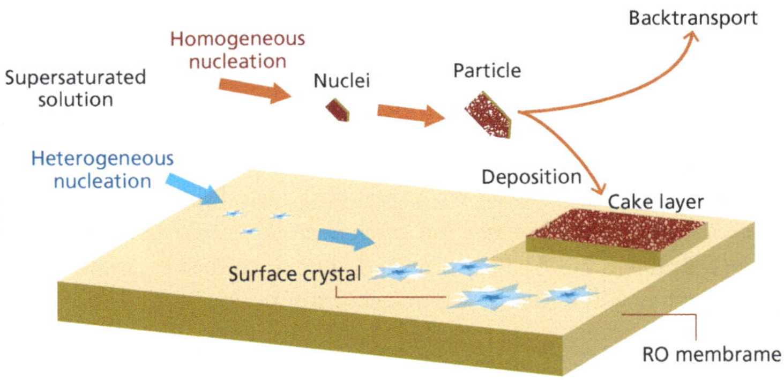

Figure 4 Nucleation mechanisms (Adopted from Oh *et al.*, 2009)

8.1.2.2 Crystal growth

After the nuclei reaches to the critical size, they start to grow into visible crystals Mullin, 2001. The process of the crystal growth is complex which can include the following steps (Cubillas and Anderson, 2010, Elwell and Scheel, 1975, Stumm and Morgan, 1981):

- bulk diffusion of ions to the crystal surface
- surface adsorption of ions
- surface diffusion of ions or ion pairs, and
- integration of molecules into the crystal lattice

8.1.2.3 Concept of induction time

Induction time 'tind' is referred to the time between the development of supersaturated conditions and formation of critical nucleus or detectable crystals (Chien *et al.*, 2007), (Boerlage *et al.*, 2000). It is composed of three time periods such as relaxation time (t_r), nucleation time (t_n), and growth time (t_g) (Kashchiev, 2000). The time needed to initiate nucleation from time zero to steady state condition is called the relaxation time (Guan, 2009). Nucleation time is defined as the time needed to form a stable nucleus and the period in which detectable crystal are formed from the stable nucleus is known as growth time (Kashchiev, 2000).

Induction time depends on the supersaturation level of a water solution, but it is mainly dependent on the precipitation kinetics. Some researchers have found a linear relationship between log tind and ($1/\log^2 S_r$) for various scaling species such as $CaCO_3$, $BaSO_4$, S_rSO_4 and $BaCrO_4$ (Söhnel and Mullin, 1988, Verdoes *et al.*, 1992). Where, S_r is an index for the determination of scaling potential of sparingly soluble salts in water (refer to section 8.4.1.2 for detailed information).

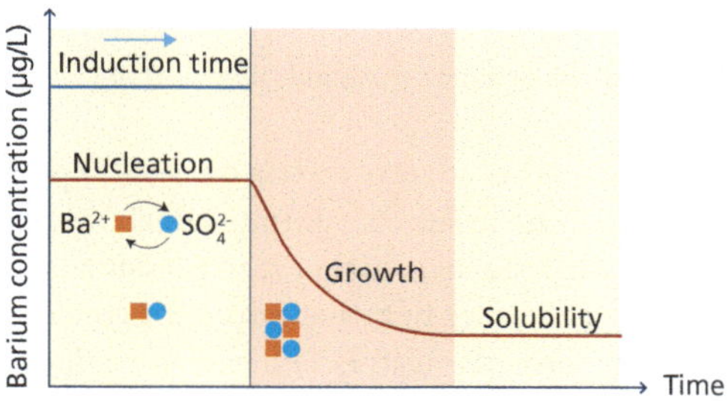

Figure 5 Example of an induction time test highlighting the three phases: nucleation, growth, and solubility. For $BaSO_4^{2-}$ the initial stable super saturation is very long. (Adapted from Boerlage, 2001)

In order to measure the induction time several methods have been developed, including but not limited to the pH method (Waly, 2011), the conductivity method (Söhnel and Mullin, 1978), turbidity or scattered light method (Shih *et al.*, 2006, Abdel-Aal *et al.*, 2004, Prisciandaro *et al.*, 2001), and the concentration of calcium (Verdoes *et al.*, 1992). Among these methods, the pH method was reported to be the most accurate one for the induction time measurement of $CaCO_3$ (Waly, 2011).

8.2 FACTORS AFFECTING SCALING

Scaling, more precisely the formation of crystals, can be affected by various factors such as pH, temperature, operating pressure, permeation rate, flow velocity, and the presence inorganic and organic substances, i.e., humic substances (Troup and Richardson, 1978, Antony *et al.*, 2011, Sheikholeslami, 2003, MacAdam and Parsons, 2004, Pastero *et al.*, 2004). Also, the effect of concentration polarization as illustrated in Figure 6 is significant on scaling in RO applications (Lee *et al.*, 1999, Chong and Sheikholeslami, 2001, Dydo *et al.*, 2003).

Concentration polarization is the presence of high concentrations of salts on the membrane surface (C_m) in comparison to the bulk solution (C_b) which happens when solutes/ions are largely rejected by the membrane (Lee *et al.*, 1999). Due to this phenomenon, sparingly soluble salts exceed their equilibrium (saturation) limit on the membrane surface which consequently result in the scale formation.

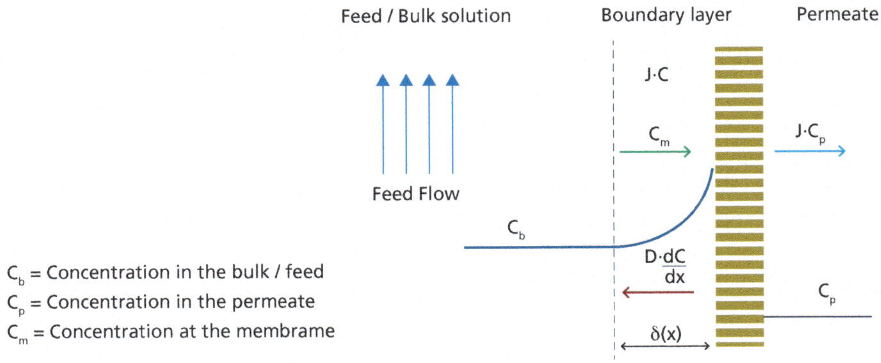

Figure 6 Principle of concentration polarisation (Kennedy *et al.*, 2014)

Concentration polarization has also other adverse effects on the RO performance such as; it leads to the increase in osmotic pressure in the concentrate which results in the reduction of net driving pressure (NDP). As a consequence, higher feed pressure is required to maintain the permeate flow rate constant. The other negative impact of concentration polarization is the increase in salinity of the permeate water. The concentration polarization can be expressed by Equation 8.3 (Brusilovsky *et al.*, 1992).

$$\frac{C_m}{C_b}=\frac{J\delta}{e^D} \qquad \text{Eq. 8.3}$$

Where, J is the permeation flux, δ is the characteristic boundary layer thickness and D is the diffusivity of solutes.

The degree of the concentration polarization in RO application is related to the operating conditions such as flux and element recovery, water chemistry, temperature, membrane properties, and module geometry (Antony *et al.*, 2011).

8.2.1 pH in RO concentrate and in RO permeate

In RO processes, the pH of the concentrate is different from the pH in the feed water. Consequently, we have to calculate or measure the pH of the concentrate.
The pH in the feed water is:

$$pH_f = 6.4 + \log\frac{\left[HCO_3^-\right]}{\left[CO_2\right]_f} \qquad \text{Eq. 8.4}$$

In the concentrate the $[HCO_3^-]$ is higher than in the feed namely, times the CF (concentration factor = 1 / 1-R, assuming, that the membrane salt rejection is about 100%). Remark: CO_2 is not increasing in the concentrate because, carbon dioxide is passing the membranes.
The pH in the RO concentrate stream is higher than in the RO feed water and equals:

$$pH_C = 6.4 + \log\frac{\left[HCO_3^-\right]_f}{\left[CO_2\right]_f} \qquad \text{Eq. 8.5}$$

(Remark: this formula is applicable only up to pH 8)

Substituting: $[HCO_3^-]_c = [HCO_3^-]_f \times 1/(1-R) = [HCO_3^-]_f \times CF$

$[CO_2]_{conc} = [CO_2]_{feed}$

results in: $pH_c = pH_f + \log(1/1-R) = pH_f + \log CF$

$$\mathbf{pH_c = pH_f + \log CF} \qquad \text{Eq. 8.6}$$

Remark: For recoveries of 0.50, 0.75 and 0.90, the pH in the concentrate will be 0.3, 0.6 and 1.0 higher, than in the feed water, respectively. This increase in pH value can be observed in the example presented in Figure 7.

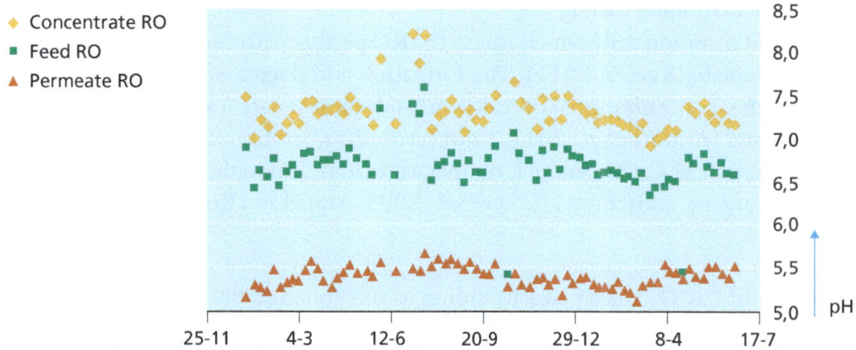

Figure 7 pH in feed water and concentrate water of an RO plant in Klazienaveen, the Netherlands, recovery = 75 %.

pH in the concentrate of RO systems is higher and in permeate lower than in feed water.

Why is this?

The pH in the RO concentrate will be higher because:

$pH_c = pH_f + \log(1/1-R) = pH_f + \log CF$

For R = 75 % (CF=4), the Log CF = log 4 = 0.6, thus, the pH_c is 0.6 higher than in the RO feed water.

The pH in the RO permeate will be lower because hydrogen carbonate is mainly rejected by the membrane.

$$pH_p = 6.4 + \log \frac{[HCO_3^-]_f}{[CO_2]_f} \qquad \text{Eq. 8.7}$$

For instance, assuming 90% rejection (f) for HCO^-_3

And substituting:

$$[HCO_3^-]_p = [HCO_3^-]_f \cdot (1 - f) = [HCO_3^-]_f \cdot 0.1$$

$$[CO_2]_p = [CO_2]_{feed}$$

$$pH_p = 6.4 + \log \frac{[HCO_3^-]_p}{[CO_2]_{c=f}} = 6.4 + \log \frac{[HCO_3^-]_f \cdot (1 - f)}{[CO_2]_{c=f}}$$

$$pH_p = pH_f - 1.0 \qquad \text{Eq. 8.8}$$

8.3 TYPES OF SCALE ENCOUNTERED IN RO

The most common types of scales encountered in RO applications are:

- calcium carbonate
- calcium sulphate
- silica/metal silicates
- barium sulphate
- calcium phosphate

8.3.1 Calcium carbonate scaling

One of the most common scale which affects the RO performance is due to the precipitation of calcium carbonate (Kucera, 2011). The formation and degree of $CaCO_3$ scaling mainly depends on the concentrations of calcium and bicarbonate/carbonate in the feed/concentrate water (Antony *et al.*, 2011, Tzotzi *et al.*, 2007). Other factors which have effect on the precipitation of $CaCO_3$ are pH, temperature, ionic strength, presence of impurities (inorganic and organic substances) (Chen *et al.*, 2005, Amjad and Koutsoukos, 2010, Waly, 2011).

The formation of $CaCO_3$ takes place according to the Equation 8.9 when a water solution becomes supersaturated with respect to $CaCO_3$.

$$Ca^{2+} (aq) + CO_3^- \leftrightarrow CaCO_3 (s) \qquad \text{Eq. 8.9}$$

The pH of the water has substantial effect on the formation $CaCO_3$ scale, since the concentrations of the various carbonate species (H_2CO_3, HCO_3^-, and CO_3^{2-}) mainly depend on the pH as shown in Figure 8. When the pH of water increases, the conversion of bicarbonate to carbonate increases as well, which therefore rises the potential of $CaCO_3$ precipitation. The equilibrium reactions of the carbonate system are presented below.

$$CO_2 (g) \leftrightarrow CO_2 (aq) \qquad \text{Eq. 8.10}$$
$$CO_2 (g) + H_2O \leftrightarrow H_2CO_3 \qquad \text{Eq. 8.11}$$
$$H_2CO_3 \leftrightarrow HCO_3^- + H^+ \qquad \text{Eq. 8.12}$$
$$HCO_3^- \leftrightarrow CO_3^- + H^+ \qquad \text{Eq. 8.13}$$

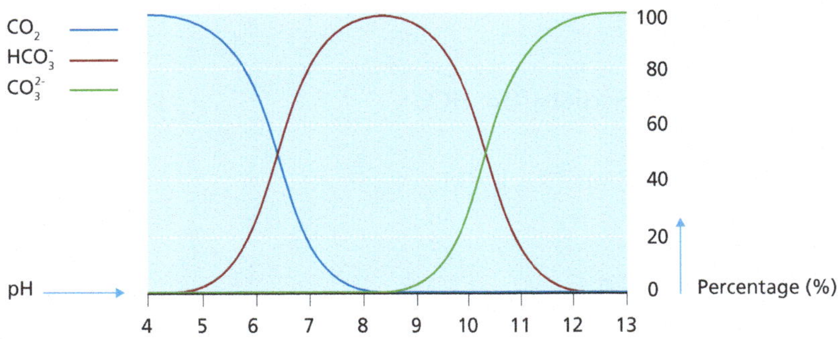

Figure 8 Theoretical carbonic species as a function of pH

In literature, six forms of $CaCO_3$ scale deposits are reported to exist depending on the pH, temperature and presence of foreign substances (impurities), such as (Chakraborty *et al.*, 1994, Brecevic and Nielsen, 1989, Coleyshaw *et al.*, 2003, Elfil and Roques, 2001):

- Three anhydrous forms (calcite, aragonite and vaterite),
- Two hydrated forms (calcium carbonate monohydrate and calcium carbonate hexahydrate),
- One amorphous calcium carbonate.

Among various forms, calcite (which exist in cubical shapes) is the most stable form. In Figure 9, SEM image of the RO membrane which was scaled with calcium carbonate is illustrated.

Figure 9 SEM of the RO membrane scaled with calcium carbonate

8.3.2 Calcium sulphate scaling

Calcium sulphate is from the group of non-alkaline scales encountered on the RO membrane surface (Antony *et al.*, 2011). Precipitation of $CaSO_4$ is reported to occur when the IP of the Ca^{2+} and SO_4^{2-} ions exceeds the K_{sp} according to the following reaction:

$$Ca^{2+} + SO_4^{2-} + xH_2O \rightarrow CaSO_4 \cdot xH_2O \downarrow \qquad \text{Eq. 8.14}$$

Where x can be 0, ½, or 2 based on different forms of calcium sulphate.

Calcium sulphate scale can occur in three different forms (Lee and Lee, 2000, Schausberger *et al.*, 2009):

- Gypsum ($CaSO_4 \cdot 2H_2O$),
- Hemihydrate ($CaSO_4 \cdot ½H_2O$)
- Anhydrite ($CaSO_4$)

Gypsum is the most common scale which exist at ambient temperature and generally in two different morphologies such as needles and platelets as demonstrated in Figure 10 (Shih *et al.*, 2005, Antony *et al.*, 2011, Seewoo *et al.*, 2004). The crystal morphology of the gypsum scale depends mainly on the concentrations of Ca^{2+} and SO_4^{2-} ions (Deckers *et al.*, 1984). It was demonstrated that gypsum had needle-like morphology at low concentrations (0.3 M $CaSO_4$), while high levels (0.725 M $CaSO_4$) favoured the platelet morphology with smoother surfaces.

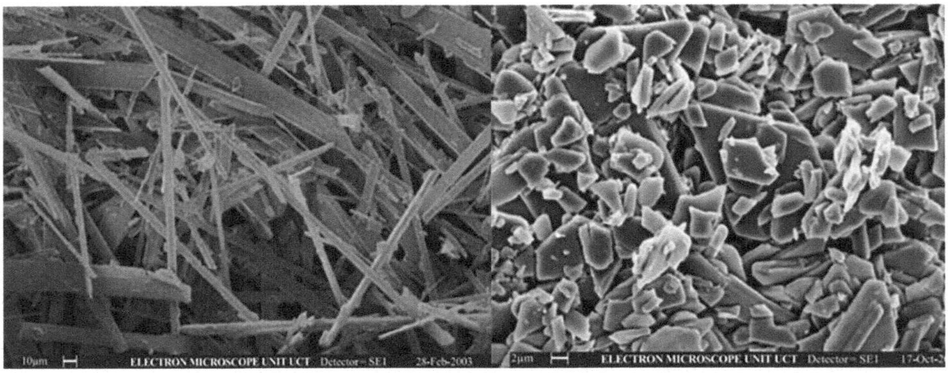

Figure 10 Gypsum scale deposits. a) needle-like morphology, b) platelets morphology (Seewoo *et al.*, 2004)

8.3.3 Silica/metal silicates

Silica can severely affect the membrane performance either by precipitating as colloidal silica or precipitating as metal silicates (Neofotistou and Demadis, 2004). The type and formation of silica deposits mainly depends on the pH and concentration of silica in the solution (Sahachaiyunta *et al.*, 2002). At and below neutral pH, silica is in the undissociated form as meta silicic acid $(H_2SiO_3)n$ which at high concentrations polymerizes into insoluble colloidal silica and results in silica scaling (Nishida *et al.*, 2009, Bremere *et al.*, 2000, Antony *et al.*, 2011).

The metal silicate precipitation occurs above neutral pH, as at high pH, the silicic acid dissociates and forms silicate anions $(SiO_3^{2-})_n$ which can react with metal ions such as calcium, magnesium, manganese, and aluminium (Antony *et al.*, 2011). In a study conducted by Gabelich *et al.* (2005), they observed kaolinite scale $(Al_2Si_2O_5(OH)_4)$ on the membrane surface when aluminium sulphate (alum) coagulation was used prior to reverse osmosis (RO) treatment.

8.3.4 Barium sulphate scaling

Barium sulphate precipitation can also adversely affect the performance of RO membranes. Overtime, barium sulphate deposits may lead to a very hard layer on the membrane surface which may not be easily removed with cleaning and therefore replacement of the RO membranes maybe needed (Boerlage *et al.*, 2000). The solubility of the barium sulphate is very low (1×10^{-5} mol/L or 2.3 mg/L in pure water) (Van der Leeden, 1991). Therefore, concentrate water at very low recoveries can be supersaturated with respect to barium sulphate. As mentioned earlier, precipitation is not only governed by the supersaturation but also depends on the precipitation kinetics which involve the formation of nuclei and further crystal growth. Boerlage *et al.* (2002) reported that $BaSO_4$ has a long stable phase prior to nucleation in the supersaturated state. In Table 3, $BaSO_4$ scaling risk at different saturation ratios (Sr) produced by Boerlage *et al.* (2002) is presented.

Table 3 The different levels of $BaSO_4$ supersaturation before scaling occurs (Boerlage *et al.*, 2000)

Supersaturation ratio (Sr) limits	Temperature (°C)				
	5	10	15	20	25
Risky supersaturation limit, i.e., high scaling risk at Sr > RSr (Induction = 5 h)	6	5.7	5.5	5.2	5
Low scaling risk at SSr > Sr > RSr	5.4-6	5.2-5.7	5-5.5	4.8-5.2	4.6-5
Safe supersaturation limit, i.e., no scaling risk at Sr < SSr (Induction time = 10 h)	5.4	5.2	5	4.8	4.6

8.3.5 Calcium phosphate scaling

Calcium phosphate scale can occur on the membrane surface when high concentration of calcium and orthophosphate ions, exceeding the solubility limit, are present in the concentrate. Phosphate can be present in different forms depending on the pH such as PO_4^{3-}, HPO_4^-, $H_2PO_4^-$, and H_3PO_4. Various compounds of calcium phosphate are reported to form under certain conditions of pH, temperature, ionic strength, molar ration of calcium to phosphate such as:

- Crystalline forms of calcium phosphate:
- Monocalcium phosphate monohydrate
- Monocalcium phosphate anhydrous
- Dicalcium phosphate dihydrate
- Dicalcium phosphate anhydrous
- Octacalcium phosphate
- Tricalcium phosphate
- Tetra calcium phosphate
- Hydroxyapatite
- Amorphous forms of calcium phosphate:
- Amorphous calcium phosphate

In RO processes, amorphous calcium phosphate is generally responsible for the flux decline instead of the crystalline phases of calcium phosphate. In Figure 11, SEM image of the amorphous calcium phosphate scaling on RO is presented.

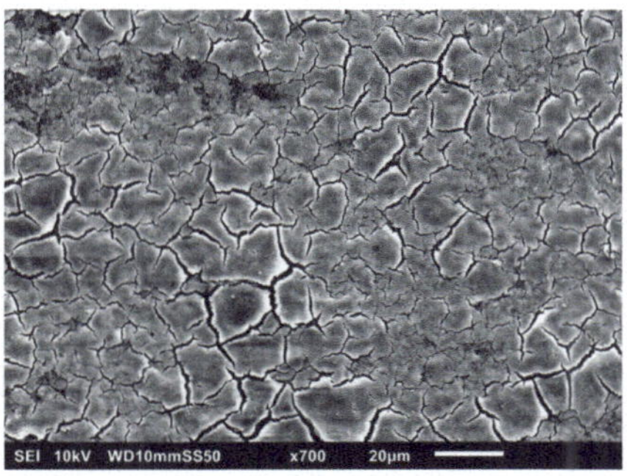

Figure 11 SEM of the RO membrane fouled with amorphous calcium phosphate

8.4 PREDICTION OF SCALING TENDENCY

8.4.1 Scaling indices

There are a number of indices available to measure the scaling tendency of the sparingly soluble salts in a water solution. The most commonly used in RO applications are:

- Saturation index (SI);
- Supersaturation ratio (Sr);

For $CaCO_3$ scaling, following indices are also used:

- Langelier saturation index (LSI);
- Stiff-Davis stability index (S&DSI);
- Calcium carbonate precipitation potential (CCPP)

8.4.1.1 Saturation index (SI)

SI is the logarithmic ratio between the ion activity product (*IAP*) and the thermodynamic solubility product (K_{sp}) of a sparingly soluble salt in water. For instance, when $CaCO_3$ is the scaling specie, SI can be calculated according to Equation 8.15.

$$SI = \log\frac{IAP}{K_{sp}} = \log\frac{\gamma_{Ca}\left[Ca^{2+}\right]\gamma_{CO_3}\left[CO_3^{2-}\right]}{K_{sp}\ of\ CaCO_3} \quad \text{Eq. 8.15}$$

For a water solution, when:

- SI = 0 the solution is just saturated or is in equilibrium.
- SI > 0 the solution is supersaturated, precipitation may occur.
- SI < 0 the solution is undersaturated, more salt can be dissolved.

In Equation 8.15, γ represents the activity coefficient which is used to determine the effective concentration of ions in a solution. The activity coefficient is dependent on the ionic strength, as it decreases with an increase in the ionic strength. The γ can be calculated by the Equation 8.16.

$$\log\gamma = \frac{0.5\cdot Z_i^2\cdot\sqrt{I}}{1+\sqrt{I}} \quad \text{Eq. 8.16}$$

Where,

Z_i = charge (oxidation number) of ion *i*,

I = Ionic strength

The term Ionic strength is defined as the total concentration of ions in a solution and is calculated by Equation 8.17.

$$I = \frac{1}{2}\sum i\sum i^2\cdot C_i \quad \text{Eq. 8.17}$$

Where,

C_i = molar concentration of ion *i*.

An empirical formula (Equation 8.18) is also sometimes used to roughly calculate the ionic strength from the total dissolve solids (TDS) concentration.

$$I\frac{(mol)}{L} \approx 2.5\cdot 10^{-5}\cdot TDS\frac{(mg)}{L} \quad \text{Eq. 8.18}$$

8.4.1.2 Supersaturation Ratio (Sr)

S_r is the square root of the ratio between the ion activity product (IAP) and the thermodynamic solubility product (K_{sp}) of a sparingly soluble salt in water. For instance, when $CaCO_3$ is the scaling specie, S_r can be calculated according to Equation 8.19.

$$S_r = \sqrt{\frac{\gamma_{Ca}\left[Ca^{2+}\right]\gamma CO_3\left[CO_3^{2-}\right]}{K_{sp}\ of\ CaCO_3}} \qquad \text{Eq. 8.19}$$

For a water solution, when:

Sr = 0 the solution is just saturated or is in equilibrium.

Sr > 1 the solution is supersaturated, precipitation may occur.

Sr < 1 the solution is undersaturated, more salt can be dissolved.

Example 3 – An RO unit is treating groundwater at 75 % recovery. The water analysis of the feed water revealed 200 mg/L of Ca^{2+}, 480 mg/L of SO_4^{2-} and 400 mg/L HCO_3? The pH of the feed water was found approximately 7.2. TDS concentration was approximately 1,200 mg/L.

Question: Is there any tendency of $CaSO_4$ scaling in the RO unit?

Given: K_{sp} of $CaSO_4 = 6.1 \cdot 10^{-5}$

Solution:

At 75 % recovery, $CF = \frac{1}{1-R} = \frac{1}{1-0.75} = 4$

In the concentrate:

$$\left[Ca^{2+}\right] = 200\frac{mg}{L}\cdot 4 = 800\frac{mg}{L} = 800\frac{mg}{L}\cdot\frac{1\,mol}{40{,}000\,mg} = 2\cdot 10^{-2}\frac{mol}{L}$$

$$\left[SO_4^{2-}\right] = 480\frac{mg}{L}\cdot 4 = 1{,}920\frac{mg}{L} = 1{,}920\frac{mg}{L}\cdot\frac{1\,mol}{96{,}000\,mg} = 2\cdot 10^{-2}\frac{mol}{L}$$

$$TDS = 1{,}200\frac{mg}{L}\cdot 4 = 4{,}800\frac{mg}{L}$$

$$I = 2.5\cdot 10^{-5}\cdot TDS = 2.5\cdot 10^{-5}\cdot 4{,}800 = 0.12\frac{mol}{L}$$

$$\log\gamma_{Ca^{2+}} = \log\gamma_{SO_4^{2-}} = \frac{0.5\cdot(+2)^2\cdot\sqrt{0.12}}{1+\sqrt{0.12}} = -0.51$$

$$\gamma_{Ca^{2+}} = \gamma_{SO_4^{2-}} = 0.31$$

$$SI = \log\frac{\gamma_{Ca}\left[Ca^{2+}\right]\gamma_{CO_3}\left[SO_4^{2-}\right]}{K_{sp}\ of\ CaSO_4} = \frac{0.31(0.02)(0.31)(0.02)}{6.1\cdot 10^{-5}} = 0.63$$

As SI > 0, therefore $CaSO_4$ scaling is likely to occur.

8.4.1.3 Langelier saturation index (LSI)

LSI is the most common method used for assessing the feed water potential for calcium carbonate scaling in RO applications and it is derived from theoretical concept of saturation (Sheikholeslami, 2005). According to ASTM method, LSI is applicable for water with total dissolved concentration (TDS) up to 10,000 mg/L (Singh, 2014). It is calculated by the equation 8.20.

$$LSI = pH - pH_S \qquad \text{Eq. 8.20}$$

Where, pH is the measured water pH and pHs is the pH at saturation in calcite or calcium carbonate and is calculated by Equation 8.21.

$$pH_S = (9.3 + A + B) - (C + D) \qquad \text{Eq. 8.21}$$

Where:

$$A = \frac{(\log_{10} TDS - 1)}{10} \qquad \text{Eq. 8.22}$$

$$B = -13.12 \cdot \log_{10}(^{\circ}C + 273) + 34.55$$
$$C = \log_{10}(Ca^{2+} \text{ as } CaCO_3) - 0.4$$
$$D = \log_{10}(\text{alkalinity as } CaCO_3)$$

For a water solution, when:

- LSI = 0 the solution is just saturated with $CaCO_3$.
- LSI > 0 the solution is supersaturated, $CaCO_3$ precipitation may occur.
- LSI < 0 the solution is undersaturated, more $CaCO_3$ salt can be dissolved.

In the concentrate, pH can be calculated using Equation 8.6 up to pH 14.

8.4.1.4 Stiff-Davis Stability Index (S&DSI)

The S&DSI is used to assess the scaling potential of calcium carbonate scaling for high saline water (TDS > 10,000 mg/L) (Singh, 2014). This method, similar to LSI, is based on the actual pH of the water solution and pH of the water solution saturated with respect to $CaCO_3$.

For a water solution, when:

- S&DSI = 0 the solution is just saturated with $CaCO_3$.
- S&DSI > 0 the solution is supersaturated, $CaCO_3$ precipitation may occur.
- S&DSI < 0 the solution is undersaturated, more $CaCO_3$ salt can be dissolved.

8.4.1.5 Calcium carbonate precipitation potential (CCPP)

CCPP measurements are used to determine the amount of calcium carbonate which will theoretically precipitate. The positive values of CCPP indicates that calcium carbonate will precipitate, whereas the negative values of CCPP represents the amount of calcium carbonate which will dissolve in a water solution. Generally, CCPP is measured using computer programs.

Example 4 – Refer to the information of Example 3.

Given: Temperature = 20 °C

Question: What is the LSI value in the concentrate? Will $CaSO_3$ scaling occur in the RO unit?

Solution:

At 75 % recovery, $CF = \frac{1}{1-R} = \frac{1}{1-0.75} = 4$

In the concentrate:

$$\left[Ca^{2+}\right] = 200\frac{mg}{L}\cdot 4 = 800\frac{mg}{L} = 800\frac{mg}{L}\cdot\frac{1\,mmol}{40\,mg}\cdot\frac{2\,mEq}{1\,mmol}\cdot\frac{100\,mg\,as\,CaCO_3}{2\,mEq}$$

$$\left[Ca^{2+}\right] = 2{,}000\frac{mg}{L}\,as\,CaCO_3$$

$$\left[HCO_3^-\right] = 480\frac{mg}{L}\cdot 4 = 1{,}920\frac{mg}{L} = 1{,}920\frac{mg}{L}\cdot\frac{1\,mmol}{61\,mg}\cdot\frac{1\,mEq}{1\,mmol}\cdot\frac{100\,mg\,as\,CaCO_3}{2\,mEq}$$

$$\left[HCO_3^-\right] = Alkalinity = 1{,}573.8\frac{mg}{L}\,as\,CaCO_3$$

$$TDS = 1{,}200\frac{mg}{L}\cdot 4 = 4{,}800\frac{mg}{L}$$

$$pH_c = 7.2 + \log\frac{1}{1-0.75} = 7.8$$

$$A = \frac{(\log_{10} TDS - 1)}{10} = \frac{(\log_{10} 4{,}800 - 1)}{10} = 0.27$$

$B = -13.12 \cdot \log_{10}(^\circ C + 273) + 34.55 = -13.12 \cdot \log_{10}(20 + 273) + 34.55\ 2.18$

$C = \log_{10}(Ca^{2+}\text{ as }CaCO_3) - 0.4 = \log_{10}(2000) - 0.4 = 2.9$

$D = \log_{10}(\text{alkalinity as }CaCO_3) = \log_{10}(1573.8) = 3.2$

$pH_s = (9.3 + A + B) - (C + D) = 9.3 + 0.27 + 2.18) - (2.9 + 3.2) = 5.65$

$LSI = pH - pH_s = 7.8 - 5.65 = 2.15$

As LSI = 2.15, therefore $CaCO_3$ scaling will to occur in the RO unit.

8.5 SCALING PREDICTIONS WITH COMPUTER SOFTWARE

8.5.1 Commercial Programs

A number of commercial programs are available which can be used to predict the scaling potential in RO. Most of these programs are developed by antiscalant suppliers and membrane manufacturers. The programs are:

- Genesys Membrane Master (MM5) – Genesys International
- Sokalan® RO-Xpert-BASF
- Hyd-RO-dose – French Creek Software
- Argo Analyzer (Winflow) – Suez
- Avista Advisor – Avista Technologies Inc.

- WAVE – DOW membrane projection software
- IMSDesign (Integrated Membrane Solutions Design) – Hydranautics membrane projection software

8.5.2 PHREEQC

PHREEQC is a computer program developed by U.S. Geological Survey (USGS). The program is written in the C and C++ programming languages Parkhurst and Appelo, 1999. The program calculates the scaling potential of various scaling species. The scaling potential results are expressed in terms of SI.

8.6 MONITORING SCALING IN RO

It is essential to monitor scaling in early stage to avoid the occurrence of severe scaling in the RO unit, to control scaling and to know when to clean the RO.

8.6.1 Sensors and data monitoring

The first condition to allow monitoring of scaling in RO systems is the availability of sensors. Sensors are part of the instrumentation in most of the RO systems, however the availability of standard sensors with signal transmitters (for data logging) in those systems is not ubiquitous. By standard sensors, it must be considered the minimal required to allow data normalization as specified by a guideline standard, specifications of designers/ consultants or information required by technical manuals or tools provided by membrane manufacturers. Standard sensors and the related water stream location may include feed pressure, concentrate pressure, permeate pressure, feed temperature, feed conductivity, feed pH, feed flow, permeate conductivity, permeate flow, and others like ORP when chlorine addition and removal is a required condition to control biofouling. Figure 12, simplifies the layout of a typical RO system, and shows the main equipment and sensors location and type needed to allow data monitoring.

RO (low-pressure, high-pressure) and NF (tight or loose) membrane are both influenced greatly by changes in temperature, conductivity, pH and operational conditions like recovery and constant/variable permeate flow operation. Therefore, flow and pressure readings as raw data are of little use if data normalization is not implemented. The way how readings/data is collected is also paramount to determine a correct operation of the system. Manual data collection is almost out-dated and mainly used in low-budget projects or systems. There is a trend that most of the new systems are moving to more modern forms of data collection and monitoring. SCADA (supervisory control and data acquisition) systems based on PLC (programmable logic controllers) have been around for almost 45 years in control processes for the automotive, steel, and nuclear power industries (Synchrony Inc, 2001). The adoption of SCADA control and data monitoring was less popular in the water industry and early adoption may be traced back to the early 90's. SCADAs experienced an increasing trend of use in water treatment systems in the last 20 years, which extended the use of SCADA from water distribution systems where the technology was adopted earlier. Digitalisation and data collection in cloud servers is not recent as may be interpreted by the increasing trend of IoT (Internet of Things). In fact, the water industry was aware of it from the beginning of the internet era and could be realised with communication implementation from SCADAS, so web-based access was feasible from multiple users to access all data and setpoints from remote locations (Synchrony Inc, 2001). The main difference from those early days to now is the increased use of standard technologies with added security that evolved into simplified communication to the cloud via router gates or from PLCs with robust security protocols such as OPC unified architecture (UA).

Data normalization is performed to allow an adequate comparison of membrane operation to an initial condition of the membrane and the system in time (e.g., initial permeate flow) or to an existing standard condition (e.g., temperature at 25 °C). RO/NF data normalization is a crucial implementation to allow monitoring of scaling. In the following section, some explanations and examples of data normalization are presented.

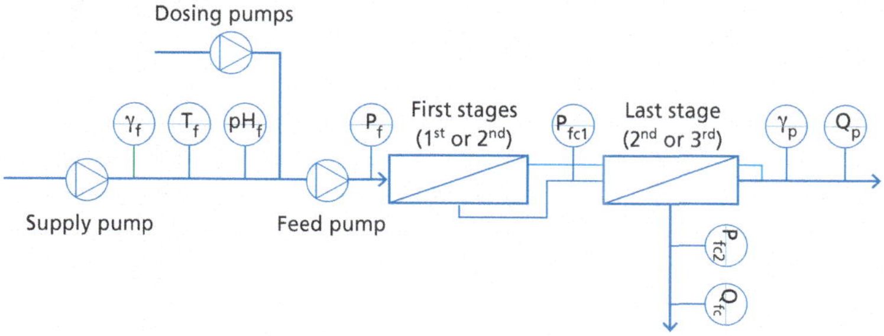

Figure 12 Sensors for data monitoring in RO system (γ conductivity, T temperature, P pressure, pH, Q flow, sub-indices f=feed, f_c=feed-concentrate, p=permeate)

8.6.2 Parameters used to monitor scaling in RO systems

In practice, some indicators such as pressure drop, net driving pressure (NDP), permeate flow, and salt passage are used.

Pressure drop is monitored in the last element and/or last stage. Pressure drop or the hydraulic differential pressure (ΔP) is the difference between the feed pressure entering the last stage/last element and the concentrate pressure leaving the last element. When the water temperature and flows are constant, ΔP should be constant unless something deposits on the membrane surface/feed spacer and hence causes blockage in the passage of the flow. Therefore, any increase in the ΔP can be attributed to the occurrence of scaling. It is worth mentioning that ΔP may not be much helpful in case of an amorphous scale with a very thin layer on the membrane surface since the layer may not cause significant blockage for the passage of the flow. In this case, permeability may decline considerably, but the increase in ΔP may not be noticeable.

$$NPD = \Delta P_{act} \cdot CF(Q) \cdot CF(T) \quad \text{Eq. 8.23}$$

Where: ΔP_{act} is actual pressure drop
$CF(Q)$ is correction factor for flow
$CF(T)$ is correction factor for temperature.

The pressure drop in one element is approximately 0.2 bar. Section 2.6 in chapter 2 describes further the empirical formula for normalizing pressure drop according to Schock and Miquel (1987).

Net driving pressure (NDP) is the actual net driving pressure in producing permeate water by passing the saline/concentrate water through the RO membrane. It is calculated as the average of the feed and concentrate pressure minus the osmotic pressure and the permeate pressure. When the water temperature, conductivity, flows are constant, any increase in NDP of the last element/stage can be attributed to the occurrence of scaling.
Net driving pressure is calculated with the following formula:

$$NPD = P_f - \frac{\Delta P_{fc}}{2} - P_p - \pi_{fc} + \pi_p \qquad \text{Eq. 8.24}$$

Where: NDP = net driving pressure; P_f = feed pressure; ΔP_{fc} = pressure drop, equal to $P_f - P_c$ or $P_{fc1} - P_{fc2}$; P_p = permeate pressure; π_{fc} = feed-concentrate osmotic pressure; π_p = permeate osmotic pressure; TCF = temperature correction factor.

π is osmotic pressure, a measure of the chemical potential of dissolved salts and other dissolved substances in the water. The chemical potential of pure water changes when it contains dissolved substances. The osmotic pressure reflects the changes of the activity coefficient of water. The activity coefficient of water is affected by all the ion-solvent interactions. There are different formulas to calculate osmotic pressure, one general formula is by using Equation 8.25.

$$\pi = R \cdot \Phi \cdot T \cdot \Sigma_i \, \alpha_i \, c_i \qquad \text{Eq. 8.25}$$

Where: R = universal gas constant; Φ = osmotic coefficient; π = osmotic pressure; T = temperature; α_i = activity coefficient of water for *i*; c_i = concentration of ionic specie *i*.

One example of a formula for feed concentrate osmotic pressure (π_{fc}) is:

$$\pi_{fc} = 2.654 \cdot T \cdot C_{fc}/(10^6 - C_{fc}) \qquad \text{Eq. 8.26}$$

C_{fc} is calculated from C_f. The concentration of salts can be derived from conductivity measurements with the following general formula:

$$C_f = A \cdot \gamma^3 + B^3 \cdot \gamma^2 + C \cdot y + D \qquad \text{Eq. 8.27}$$

Where: γ = conductivity; A, B, C, D = constants.
Constants are calculated from historical data of conductivity and total dissolved solids.
C_{fc} is feed-concentrate in the membrane surface, one example of an equation to account for up-concentration of salts from the feed to the feed-concentrate stream and then in the membrane surface is:

$$C_{fc} = -CP \cdot C_f \cdot R^{-1} \cdot \ln(1\text{-}R) \qquad \text{Eq. 8.28}$$

Where CP is the concentration polarization factor, as example, CP = 1.1 for low brackish groundwater applications. R is total recovery of the system, as example R could be 70 %. Recovery is also system and application dependent but in the range 30 to 90 %. CP will be water treatment application dependent and normally in the range 1 – 2.

One example for C_f calculation from conductivity in µS/cm (γ_f) is given as

$$C_f = 0.76 \cdot \gamma_f - 3.07 \qquad \text{Eq. 8.29}$$

For measured conductivity in the feed, γ = 1500 µS/cm, C_f = 1136.93, and replacing in Equation 8.28

$C_{fc} = - 1.1 \cdot 1136.93 \cdot 0.7^{-1} \cdot \ln 0.3$

C_{fc} = 2151 mg/L

Replacing C_{fc} in Equation (8.26) with feed temperature T_f= 15 °C = 288.15 K

$\pi_{fc} = 2.654 \cdot 288.15 \cdot 2151 / (10^6 - 2151)$

π_{fc} = 1.65 bar

The last term of Equation (8.24) is osmotic pressure of the permeate. In general, the concentration of salts in the permeate water is very low and can be calculated with a general formula of

$$C_p = E \cdot \gamma^2 + F \cdot \gamma + G \qquad \text{Eq. 8.30}$$

Where: γ = conductivity; E, F, G = constants.

As an example, the calculation with Equation

$$\pi_p = 7.49 \cdot 10^{-4} \cdot \gamma_p - 0.13 \cdot 10^{-3} \qquad \text{Eq. 8.31}$$

For measured conductivity in the permeate, γ_p = 13 µS/cm, gives

$\pi_p = 7.49 \cdot 10^{-4} \cdot 13 - 0.19 \cdot 10^{-3}$

π_p = 0.01 bar

Assuming the following measured data:

P_f is measured feed pressure, P_f = 9.41 bar

ΔP_{fc} is pressure drop along the last pressure vessel, $P_f - P_{fc}$ (both measured), ΔP_{fc} = 0.3 bar

P_p is measured permeate pressure, P_p = 0.03 bar

Now, by replacing all terms in Equation (8.24)

NDP = 9.41 – 0.3/2 – 0.03 – 1.65 + 0.01

NDP = 7.59 bar

As mentioned earlier, when the water temperature, conductivity, flows are constant, any increase in NDP can be attributed to the occurrence of scaling. To compensate for those, a better alternative to monitor scaling is corrected temperature net driving pressure (NDP_{tc}) The approach for calculating NDPtc is to convert the net driving pressure to a representative temperature reference, since standard conditions are determined to 25 °C, then the temperature correction factor (TCF) is represented by the general formula

$$TCF = 1.03^{(T-25)} \qquad \text{Eq. 8.32}$$

Another formula alternative for temperature correction factors are provided by membrane manufacturers, this information is accessible. The information can be stored in a database, and by this, temperature correction factors to 25 °C can be specified in the control program of a dosing pump for a membrane product name. One example of TCF equation is:

$$TCF = e^{2700\cdot(3.35\cdot 10^{-3} - Tf^{-1})}$$ Eq. 8.33

Replacing $T_f = 288.15$ in Eq. 8.33,
TCF = 0.734

Finally, the NDP_{tc} is calculated with Equation 8.34

$$NDP_{tc} = NDP \cdot TCF$$ Eq. 8.34

$NDP_{tc} = 7.59 \cdot 0.734 = 5.57$ bar

Permeate flow is also a useful parameter for monitoring scaling in the RO unit. When the NDP and temperature are constant, the decrease in permeate flow of the last element/stage can be an indication of scaling. As the permeate flow is related to the NDP and temperature, any increase/decrease in the permeate flow due to variations in NDP and temperature should be factored out. Therefore, the permeate flow should be normalized. For the normalization, commercial programs are available from the membrane manufacturers (e.g., DOW, Hydranautics, etc.). According to DOW, the normalization of permeate flow is executed based on Equation 8.35.

$$Q_N = Q_i \cdot \frac{NDP_r}{NDP_t} \cdot \frac{TCF_r}{TCF_t}$$ Eq. 8.35

Where,
Q_N = Normalized permeate flow at time t
Q_i = Actual permeate flow at time t
NDP_r = NDP at reference point
NDP_t = NDP at time t
TCF_r = Temperature correction factor at reference conditions
TCF_t = Temperature correction factor at time t

According to Hydranautics, TCF is calculated by Equation 8.36.

$$TCF = e^{2700\cdot\left[\frac{1}{(273+T)} - \frac{1}{298}\right]}$$ Eq. 8.36

Where T is temperature in degree Celsius (°C)

Salt passage is also used as a parameter to monitor the occurrence of scaling. At fixed recovery, when the feed conductivity and temperature are constant, an increase in salt passage could be attributed to scaling due to concentration polarization effect. The salt passage should be also normalized, for instance, to factor out any increase in salt passage due to increase in feed temperature. Based on Hydranautics, salt passage normalization can be done using Equation 8.37.

$$\%SP_N = \%SP_a \cdot \frac{EPF_a}{EPF_s} \cdot \frac{STCF_s}{STCF_a}$$ Eq. 8.37

Where,
$\% SP_n$ = Normalized salt passage in percentage to standard conditions
$\% SP_a$ = Actual salt passage in percentage
EPF_s = Permeate flow of the element at standard conditions
EPF_n = Permeate flow of the element at actual conditions
$STCF_s$ = TCF for salt transport at standard conditions
$STCF_s$ = TCF for salt transport at actual conditions

8.6.3 Monitoring systems

There are a number of monitoring systems available which can be used to continuously monitor scaling in RO applications. These monitoring devices are installed on the concentrate stream of the last stage in RO applications that provide an additional recovery to the overall recovery of the RO facility. Due to the provision of additional recovery, it is expected that scaling would generally occur first within the monitoring device before the actual membranes of the RO. These monitoring devices are:

- Mass balance
- Membrane Scale Guard
- Integrated scaling monitor
- Black Box – Avista Technologies Inc.

Mass balance of sparingly soluble compound(s) e.g., in the last stage of a plant.

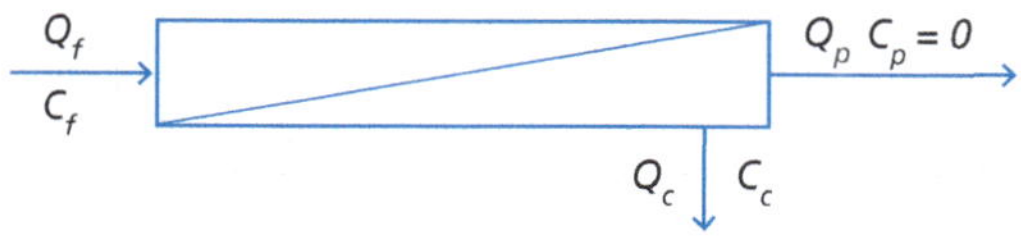

Figure 13 Flow and concentration balance in an RO system

Doing a mass balance, we have:

$$Q_f \times C_f - (Q_p \times C_p + Q_c \times C_c) = \text{Deposited material}$$

The disadvantages of following this approach are: *i*) inevitable inaccuracy in flow measurements and chemical analysis, are making the outcome not very accurate, *ii*) the method is laborious, *iii*) the method is costly.

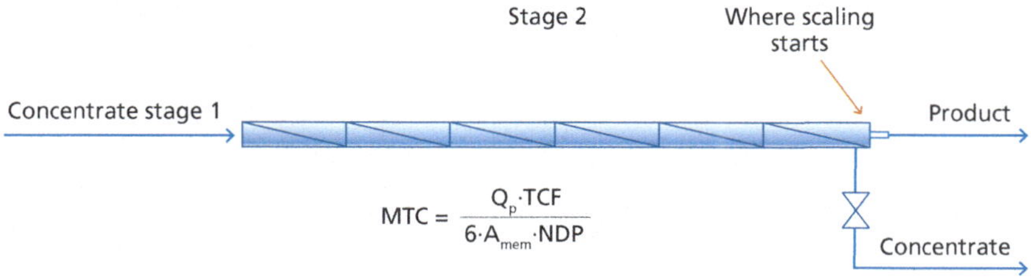

Figure 14 Monitoring mass transfer coefficient in the second stage of a BWRO system

Monitoring the permeability loss is rather inaccurate. Because, if 30 % MTC decrease in the last element occurs, this appears as a decrease of only 5 % on the MTC decrease of stage 2, as illustrated in Figure 14. The method is anyway useful, however, rather inaccurate and late in alarming.

Scale guard as shown in the Figure 15 can be fed with the concentrate of the last stage of pilot or full-scale RO application. Scaling guard can be operated at additional recovery in the range of 1 – 4 %. Scaling is monitored by observed decrease in the mass transferred coefficient of the scale guard.

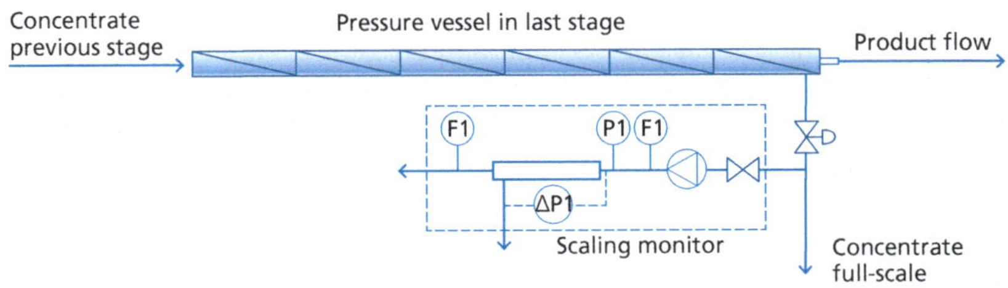

Figure 15 Scale guard (Adopted from van de Lisdonk *et al.*, 2000)

In the membrane scale guard, the flow conditions are the same as in last RO element of the last stage, there is an extra recovery in the element inside the Membrane ScaleGuard, and the MTC of the RO element in Membrane ScaleGuard is measured.

An example of application of the membrane scale guard is in the Netherlands, at Vitens Water Supply Company in an RO plant treating anoxic groundwater, with 5-3-2 staging, 4 elements per pressure vessel. The membrane scale guard was used to find the optimum process conditions. The settings of the pilot were recovery 75 % and 80 % with dose of an antiscalant. The membrane Scale Guard connected to membrane concentrate increases the conversion with about 2% where rapid detection of scaling is expected. Results are presented in Figure 16.

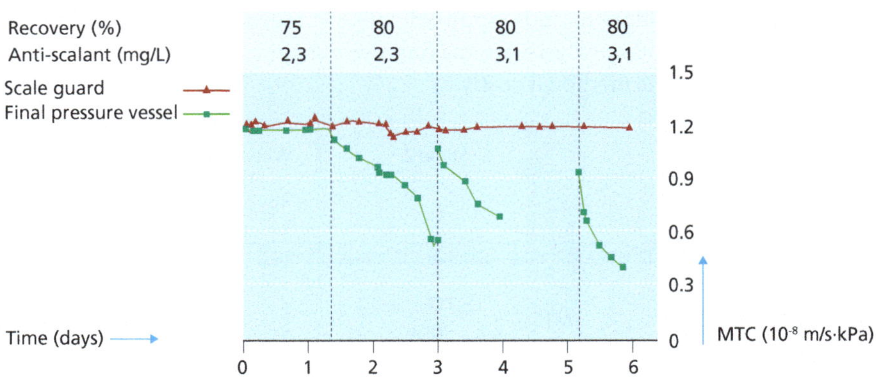

Figure 16 Results show early warning with the Membrane ScaleGuard at Vitens Pilot plant (Adopted from van de Lisdonk *et al.*, 2000)

No significant decrease in average MTC in stage 3 was observed. However, significant decrease in the individual elements measured. For elements 1, 2, 3 and 4, the MTC decrease was 4 %, 6.5%, 8.5 %, and 24 % respectively.

Internal Scale Monitor is based on monitoring accurately the normalized flux of the last element of the last stage, based on: permeate flow, pressure, osmotic pressure (conductivity) in concentrate and temperature. Because scaling usually starts in the very last element of the last stage. Integrated scaling monitors are installed in several full-scale plants in the Netherlands. The increase in differential pressure and decrease in permeate flow is attributed to scaling. In case, the internal scale monitor scales it can affect the operation of the RO unit which is considered as the main drawback of this type of scale guard.

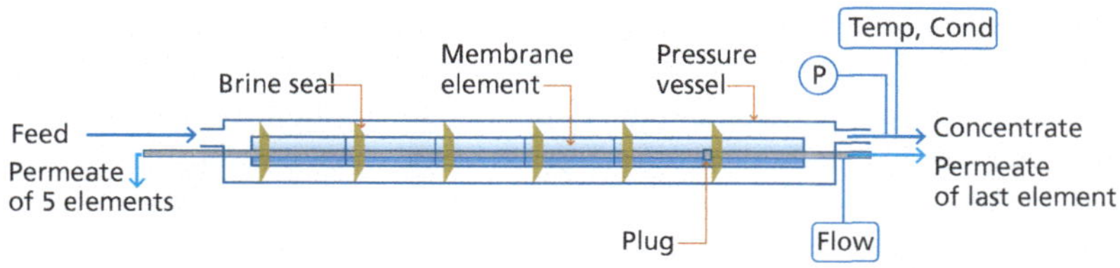

Figure 17 Schematic of the internal scale monitor in the last element of the last stage in RO systems

8.7 SCALING CONTROL AND ANTISCALANTS

There are several strategies to control scaling in RO systems. The first one is not exceeding the solubility of any compound. This approach likely limits the RO recovery to a large extent, which results in higher pre-treatment and energy cost and wastage of water. The second strategy is dosing acid to eliminate super saturation, but this approach is only applicable for calcium carbonate and calcium phosphate. The third strategy is dosing an antiscalant in the RO feed (with or without acid dosing). Antiscalants allow significant super saturation of specific sparingly soluble inorganic compounds.

To prevent scaling in RO applications, various chemical, physical and mechanical approaches have been proposed, which can be summarized into three groups (Antony *et al.*, 2011): I) altering the feed water characteristics, II) optimization of operating parameters and system design, and III) addition of scale inhibitors.

8.7.1 Altering feed water characteristics

The potential for scale formation can be minimized by changing the feed water quality in RO applications (Antony *et al.*, 2011). The alteration of feed water can be achieved by various techniques including but not limited to the acidification and ion-exchange softening of the feed water.

For $CaCO_3$ scaling, as discussed earlier, at high pH conditions CO_3^{2-} is the dominant specie in carbonic system which reacts with Ca^{2+} and leads to $CaCO_3$ scaling. To prevent $CaCO_3$ scaling, acidification of the feed water which is one of the earliest methods is used. In acidification, the pH of the feed water is lowered to 5-7 which raises the solubility of the calcium carbonate (Antony *et al.*, 2011). As can be seen from Equations 8.38 and 8.39, addition of acid shifts the equilibrium equations to the left side, as a result $CaCO_3$ scaling potential is reduced. This method is not much attractive, since huge amount of acid is required to reduce the water pH. It should be noted, with acid addition the permeate water may become very acidic which requires much more chemical post-treatment to meet the desired water quality parameters.

$$HCO_3^- \leftrightarrow CO_3^- + H^+ \qquad \text{Eq. 8.38}$$

$$wH_2CO_3 \leftrightarrow HCO_3^- + H^+ \qquad \text{Eq. 8.39}$$

Another technique for the mitigation of $CaCO_3$ scaling is the reduction of Ca^{2+} concentration in the feed water using ion exchange. In this method, the Ca^{2+} ions are replaced with the Na^+ ions and are adsorbed on the ion exchange resins according to the equation 8.40 (Antony *et al.*, 2011). The ion exchange method can eliminate the need for the acidification of the feed water. However, this method requires huge chemical consumption and involves high capital costs and also the difficulty with the brine regenerate discharge makes the method less attractive (Kelle Zeiher *et al.*, 2003).

$$Ca^{2+} + 2\,NaZ \rightarrow 2Na^+ + CaZ_2 \qquad \text{Eq. 8.40}$$

8.7.2 Optimization of operating parameters and system design

The scaling tendency of calcium carbonate can be minimized by changing the RO system design and operational parameters that keep the calcium carbonate salt in unsaturated conditions or slow down the precipitation kinetics (Antony *et al.*, 2011). It includes the following approaches:

1. Feed flow reversal
2. Limiting product recovery

The feed flow reversal technique was proposed by Lauer (1997) to achieve high recoveries in RO applications without or with minimum chemical addition. This technique is based on the reduction of the elapsed nucleation time by periodically changing the feed entrance and concentrate exit positions, or in other words, by reversing the unsaturated feed flow at time less than the induction time for the scale formation (Antony *et al.*, 2011).

Limiting product recovery is another way of preventing calcium carbonate scale. In this approach, the recovery of the RO application is lowered to reduce the supersaturation level of the concentrate water to undersaturated conditions. By reducing the recovery, the adverse effect of the concentration polarization is also reduced due to less solute concentration on the membrane surface, thus the potential for the scale formation is minimized. However, this approach is not very attractive and economical since it results in high specific energy consumption. In addition, the high amount of concentrate disposal is also an issue.

8.7.3 Addition of scale inhibitors/antiscalants

Antiscalant addition to the feed water is one of the most widely used and an effective technique to prevent scaling in RO applications (Lee *et al.*, 1999, Pervov, 1991Greenlee *et al.*, 2010). One reason which makes antiscalant addition more attractive is the low dose requirement to overcome scaling problem (Antony *et al.*, 2011).

It should be noted that the antiscalants do not completely eliminate the scaling formation, but they disrupt the stages of the crystallization process. More specifically, they delay nucleation phase of crystallization and/or retard the growth phase of crystallization (Amjad, 1996, Antony *et al.*, 2011). Additionally, with antiscalant addition, the solubility limits of the sparingly soluble salts is increased which makes it possible to achieve higher recovery in RO applications (Drak *et al.*, 2000). In general, the main mechanisms of the scale inhibitors in preventing scale formation can be grouped into three categories (Darton, 2000): threshold inhabitation, crystal modification, and dispersion.

Threshold inhibition is the ability of the antiscalant to prevent the formation of crystals in a supersaturated solution at the nucleation phase. More specifically, when the crystal formation starts to occur at submicroscopic level, the negative groups of the antiscalant attached to the cationic sites of the scale nuclei which then disrupts the electronic balance that is required to encourage the growth of the crystals (Avista, 2016).

Crystal modification is the ability of an antiscalant to cause distortion in the shape of the scaling crystals which also changes the properties of the crystals, making it soft and non-adherent. The modified crystals are generally reported to be more in oval in shape and less compact (Avista, 2016).

Dispersion is the property of an antiscalant to adsorb on the crystals and impart extra negative charge which then keeps the crystals separated in the solution, thus terminating any further crystal growth (Avista, 2016).

8.7.4 Antiscalants

A number of commercial antiscalants are available which are designed for specific types of scale. The commonly used antiscalants in RO applications can be categorized in three different groups based on their compositions and properties, which includes: Polyphosphates, Phosphonates/ organophosphates, Polycarboxylates and Biobased antiscalants (Antony *et al.*, 2011, van Engelen and Nolles, 2013).

Polyphosphate antiscalant such as sodium hexametaphosphate (SHMP) $(NaPO_3)6$ was the first antiscalant applied in RO applications to control scaling (Antony *et al.*, 2011Darton, 2000). Polyphosphate are formed either in linear or cyclic form from the condensation of ortho-phosphoric acid. It is reported that the scaling control performance of this type of antiscalant is due to $0–P–(0)_3$ bond (Ghani and Al-Deffeeri, 2010). Polyphosphates seems more attractive antiscalant, as at very low levels they can prevent $CaCO_3$ scaling at room temperatures (Rahman and Amjad, 2010). Hatch and Rice (1939) reported that 1 to 5 mg/L of SHMP was efficient in preventing the precipitation of $CaCO_3$. However, the major drawback of polyphosphate antiscalant is that at high temperature, it hydrolyses to orthophosphate which not only results in suppression of antiscalant efficiency but also leads to calcium phosphate scaling (Darton, 2000).

Phosphonate antiscalants, unlike polyphosphates, are less vulnerable to hydrolysis at high temperatures as they contain phosphonic groups in their structure. The phosphonic groups are connected with C-P bonds which are more stable than the P-O-P bonds (Ghani and Al-Deffeeri, 2010, Antony *et al.*, 2011). It is reported that inhibition ability of phosphonates is higher than polyphosphate especially for the inhibition $CaCO_3$, $Mg(OH)_2$, and $BaSO_4$ scaling (Sawada, 1997). The other advantage of phosphonate antiscalant is their high water-solubility. However, like polyphosphate, the phosphonate antiscalants have the same disadvantage which is their likelihood to form calcium phosphate scales (Gill, 1999, Butt *et al.*, 1995).

Polycarboxylates antiscalants are low molecular weight polyelectrolytes and anionic in nature. Due to their anionic nature, the scale formation is prevented due to dispersion of the crystals nuclei and also modification of the crystal shapes (Antony *et al.*, 2011). The common polycarboxylates antiscalants are polyacrylic acid (PAA), polymaleic acid (PMA), polyaspartic acid (PASP), and Polyepoxysuccinic acid (PESA). Among all, polyacrylates are the most widely used antiscalant because of their high efficiency in preventing scale formation (Yuchi *et al.*, 2007, Antony *et al.*, 2011).

Biobased antiscalant such as carboxymethyl inulin (CMI) is becoming more favourable in RO applications, since it considered to be more environmentally friendly in comparison to the other antiscalants. CMI is known as a threshold inhibitor and it functions in three ways to control scaling: complexing of metal ions, crystal growth inhibition, and dispersancy (van Engelen and Nolles, 2013).

8.8 DETERMINATION OF ANTISCALANT DOSE IN RO SYSTEMS

8.8.1 Dosage determination of scale inhibitor (antiscalant)

The current state-of-the-art is to "rely on recommendations of equipment and antiscalant manufacturers when determining appropriate antiscalant selection and doses necessary for a specific feed water and design recovery" (Crittenden *et al.*, 2012). Some antiscalants providers use their own software programs to recommend the dosage of antiscalant for the RO/NF system. Nevertheless, the practice of dosing a constant dose in the range 1-5 mg/L of AS is prone to secondary effects, as realized by some studies. It was demonstrated that there is a risk of biofouling formation when certain types of antiscalants are dosed in RO and nanofiltration (Vrouwenvelder *et al.*, 2000). The study looked at 14 different antiscalants; and found that some dosages promote biofouling of a biofilm monitor for AS with high content of assimilable organic carbon (AOC). Another study by Zimmer *et al.* (2016) explained that both under-dosing and overdosing cause runaway pressure increase. They explained that large carbonate crystals form when there is nothing or too little AS dosed. According to them, a correct dose guarantees the membrane remains relatively clean of scaling. But, when AS dose was too much, many tiny crystals form and cause runaway pressure increase. RO operators rely on the recommendations of chemical suppliers regarding products and doses. The practice of determining the antiscalant dose is done during commissioning, and rarely, the dose is changed along the operation of a plant. The dose is mainly changed when problems have occurred during operation.

8.8.2 Dosage control and optimization

Since the current practice of dosing antiscalants is to dose at constant dose, the field of dosage control has been limited to proportional flow-control dosing to account for variation of the feed flow. However, this is not that convenient, since feed flows are most of the time constant, and if changed, the change corresponding change in dosage is straightforward and can be manually set-up. One reported attempt of manual optimization was carried out in the Netherlands to achieve high recovery of a system with a free phosphate antiscalant (Jong *et al.*, 2011). The disadvantages of manual optimization are that the system is subject to water quality changes, including content of natural organic matter, operational changes, changes in antiscalant provider, and aging of the antiscalant. Recently, an innovative attempt of optimization of antiscalant dosing has been carried out by the inventors of digital dosing (Grundfos), and has resulted in Smart RO.

Suppliers of anti-scalants use "projection programs". Some are available e.g., Toray, Genesys, Avista to calculate the required dose and type of anti-scalant needed. Some antiscalants can replace acid dose completely to prevent $CaCO_3$ scaling. Remark: Anti-scalants do not increase the solubility of inorganic compounds, but retard the precipitation.

How to determine antiscalant / acid dose and safe recovery?

1. Follow the recommendations of the supplier of antiscalant.
   These recommendations are usually followed and might be on the safe side. Several projection programs recommend dose of antiscalant, even when the concentration of compounds is far below saturation.
   When only acid is required, the pH is usually adapted in the brine to ensure that the SI (saturation index) is around zero. Addition of antiscalant allows reduced acid dose or eliminate the need for any acid dose
2. Optimize and monitor the MTC (Normalized Flux) of the last stage.
3. Optimize and monitor the MTC (Normalized Flux) of the last element of the last stage with Membrane Scale Guard or Integrated Scaling Monitor.

Smart RO by Grundfos is based on previous research, "A smart optimization of antiscalant dosing in water desalination" (Yangali-Quintanilla *et al.*, 2017). Smart RO works by analysing data from standard sensors (pressure, temperature, conductivity) present in an RO system. The sensors monitor the operation and sensors will rapidly react to changes in membrane performance. Data from the sensors can be transmitted to and stored in either the dosing pump or a cloud server, potentially both locations can be used for data storage (local or historical). Smart RO has two main characteristics: 1) Real time data processing and visualization, and 2) Digital intelligence featuring decision making for AS dosing. An upgraded version of a Smart Digital dosing pump is used for the implementation of Smart RO (DDA, advanced digital dosing).

From 2020, Smart RO has become part of Smart Filtration Suite (www.smartfiltrationsuite.com). Smart Filtration Suite is a set of algorithms that communicate directly with the system's PLC to autonomously adjust the membrane filtration system based on feedback from sensors. Through real-time analytics, Smart Filtration Suite optimizes the membrane system to operate in the most cost-efficient and sustainable way, providing savings on energy, water, and chemicals and ensuring reduced downtime. Through dynamic dosing, it optimizes the use of chemicals (incl. antiscalants) in RO systems by always dosing the actual needs of the system, thereby avoiding under- or overdose of chemicals.

8.8.3 Summarizing

Projection programs of antiscalant suppliers are inevitable in determining dose. The membrane scale guard is a strong tool in optimizing anti-scalant dose and recovery. The integrated scaling monitor is currently a cheaper alternative, simpler and more robust and is a strong tool in reducing the risk of scaling in the full-scale plant, optimizing the operating conditions (recovery, acid/scale inhibitor dose, chemical cleaning procedure).

8.9 SCALING IN SEAWATER REVERSE OSMOSIS

Scaling in seawater reverse osmosis is almost exclusively due calcium carbonate and is commonly prevented by dosing acid and more and more with antiscalants. The Stiff &Davis Saturation Index is commonly applied in seawater RO. This application predicts the need for dosing acid. However, several plants and pilot studies show that, there is or might be no need for acid dosing and/or antiscalant.

Two potential reasons may answer this contradiction:

1. The calculations of the pH in the concentrate – as commonly done – are not correct due to:
   a) the fact that the dissociation constants of carbon dioxide CO_2 and hydrogen carbonate HCO_3^- depend on salinity, are not taken into account.
   b) the difficulty to predict the role of CO_2 and CO_3^{2-} in the development of the pH in the RO concentrate.

   The result is that the actual pH is significantly lower that the predicted pH.
2. Precipitation reaction of calcium carbonate in seawater shows long induction times and/or slow crystal growth.

8.9.1 Case study: SWRO pilot plant at the North Sea in the Netherlands

In the pilot plant process scheme shown in Figure 19, the feed water is filtered through a 150 μm strainer before the addition of acid to decrease the pH of the feed water from 8.0 to 6.7. The water is then fed to an UF membrane before being fed to the RO unit. The water recovery of the RO unit is around 40 %. The feed water is considered to be undersaturated according to S&DSI (-0.06) and oversaturated when the saturation is calculated using the SI approach by the PhreeqC program (SI=0.42), based on solubility of calcite. Although the plant is designed to have an antiscalant addition, the antiscalant system was not used. In a normal operational mode, the pre-treated SWRO water feeding the high-pressure pump has a pH of 6.7 after acidification.

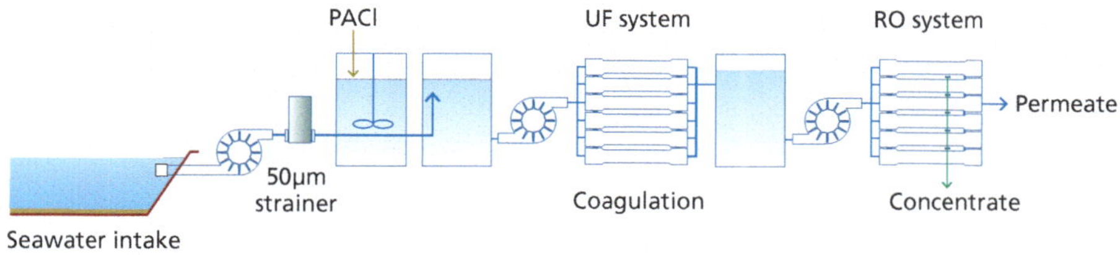

Figure 19 Schematic of the RO pilot plant located in Kamperland (North Sea) in the Netherlands

The outcome of the study of Waly, (2011) indicates that the pH in concentrates of seawater reverse osmosis plants are lower than commonly expected. This opens the opportunity to reduce or even to stop the dose of acid. The seawater reverse osmosis pilot plant run for more than 6 months without any acid dose and showed no indications of precipitation of calcium carbonate (Waly, 2011).

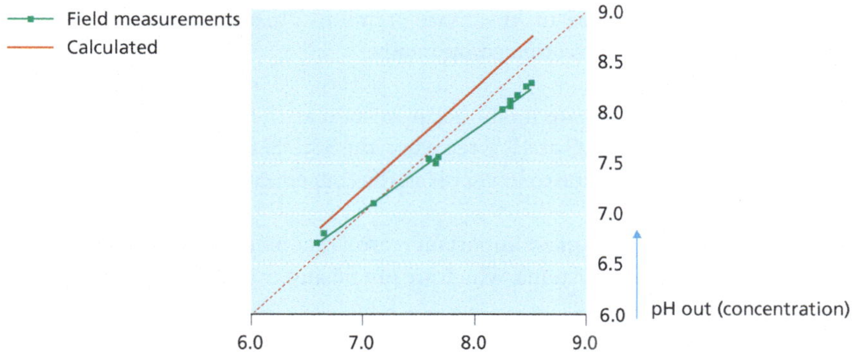

Figure 20 Calculated and measured pH values in concentrate (45% recovery) (Adopted from Waly, 2011)

Laboratory tests showed that magnesium present in seawater is causing much longer induction times than in artificial seawater without magnesium (Waly, 2011). Measured induction times in artificial seawater RO concentrates are surprisingly long (see Figure 20) (Waly, 2011).

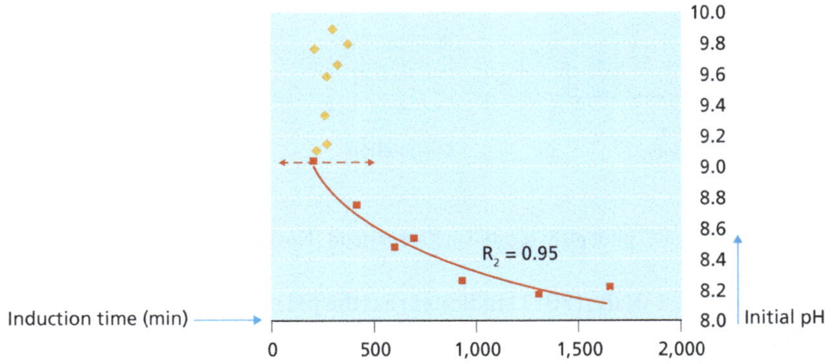

Figure 21 Induction times in artificial seawater RO concentrates at different initial pH levels (50% recovery) (Adopted from Waly, 2011)

In practice. several full-scale Seawater RO plants do not add acid to control calcium carbonate scaling. Some dose antiscalant as an "insurance premium". In addition, antiscalants might reduce membrane fouling due to other components.

Precipitation of calcium carbonate might occur as soon as the concentration exceeds the solubility. In case of seawater RO at 45 % recovery, the S&DSI is positive, but substantially lower than usually calculated, due to lower pH in the concentrate.

Slow kinetics is most likely the most important reason that precipitation (scaling) does not occur in several sea water RO systems, which are not dosing any acid or antiscalant.

8.10 REFERENCES

Abdel-Aal, E. A., Rashad, M. M. & El-Shall, H. 2004. Crystallization of calcium sulfate dihydrate at different supersaturation ratios and different free sulfate concentrations. Crystal Research and Technology, 39, 313-321.

Amjad, Z. 1996. Scale inhibition in desalination applications—an overview. Proceedings of the Corrosion '96, NACE Conference, Paper 230

Amjad, Z. & Koutsoukos, P. 2010. Mineral Scales and Deposits. The Science and Technology of Industrial Water Treatment. CRC Press.

Antony, A., Low, J. H., Gray, S., Childress, A. E., Le-Clech, P. & Leslie, G. 2011. Scale formation and control in high pressure membrane water treatment systems: A review. Journal of Membrane Science, 383, 1-16.

Avista 2016. Technical Bulletin. Scale Inhibitors. Available at https://www.avistatech.com/wp-content/uploads/Avista-TB-Scale-Inhibitors-RO_NF.pdf. [Accessed 25 November 2020].

Boerlage, S. F. E., Kennedy, M. D., Bremere, I., Witkamp, G. J., Van Der Hoek, J. P. & Schippers, J. C. 2000. Stable barium sulphate supersaturation in reverse osmosis. Journal of Membrane Science, 179, 53-68.

Boerlage, S. F. E., Kennedy, M. D., Bremere, I., Witkamp, G. J., Van Der Hoek, J. P. & Schippers, J. C. 2002. The scaling potential of barium sulphate in reverse osmosis systems. Journal of Membrane Science, 197, 251-268.

Brecevic, L. & NIELSEN, A. E. 1989. Solubility of amorphous calcium carbonate. Journal of Crystal Growth, 98, 504-510.

Bremere, I., Kennedy, M., Mhyio, S., Jaljuli, A., Witkamp, G.-J. & Schippers, J. 2000. Prevention of silica scale in membrane systems: removal of monomer and polymer silica. Desalination, 132, 89-100.

Brusilovsky, M., Borden, J. & Hasson, D. 1992. Flux decline due to gypsum precipitation on RO membranes. Desalination, 86, 187-222.

Butt, F. H., Rahman, F. & Baduruthamal, U. 1995. Identification of scale deposits through membrane autopsy. Desalination, 101, 219-230.

Chakraborty, D., Agarwal, V. K., Bhatia, S. K. & Bellare, J. 1994. Steady-State Transitions and Polymorph Transformations in Continuous Precipitation of Calcium Carbonate. Industrial & Engineering Chemistry Research, 33, 2187-2197.

Chen, T., Neville, A. & Yuan, M. 2005. Assessing the effect of on scale formation–bulk precipitation and surface deposition. Journal of Crystal Growth, 275, e1341-e1347.

Chien, W.-C., Lee, C.-C. & Tai, C. Y. 2007. Heterogeneous Nucleation Rate of Calcium Carbonate Derived from Induction Period. American Chemical Society.

Chong, T. H. & Sheikholeslami, R. 2001. Thermodynamics and kinetics for mixed calcium carbonate and calcium sulfate precipitation. Chemical Engineering Science, 56, 5391-5400.

Coleyshaw, E. E., Crump, G. & Griffith, W. P. 2003. Vibrational spectra of the hydrated carbonate minerals ikaite, monohydrocalcite, lansfordite and nesquehonite. Spectrochimica Acta Part A: Molecular and Biomolecular Spectroscopy, 59, 2231-2239.

Crittenden, J. C., Rhodes Trussell, R., Hand, D. W., Howe, K. J. & Tchobanoglous, G. 2012. Water Treatment: Principles and Design / MWH, New Jersey, Montgomery Watson Harza (Firm).

Cubillas, P. & Anderson, M. 2010. Synthesis Mechanism: Crystal Growth and Nucleation.

Dalmolen, J. 2005. Synthesis and application of new chiral amines in Dutch Resolution : family behaviour in nucleation inhibition.

Darton, E. 2000. Membrane chemical research: centuries apart. Desalination, 132, 121-131.

Deckers, J. M., Lash, J. E. & Burns, G. 1984. Heats of Crystallization of CaSO4.2H2O. Bulletin des Sociétés Chimiques Belges, 93, 271-280.

Drak, A., Glucina, K., Busch, M., Hasson, D., Laîne, J.-M. & SEMIAT, R. 2000. Laboratory technique for predicting the scaling propensity of RO feed waters. Desalination, 132, 233-242.

Dydo, P., Turek, M. & Ciba, J. 2003. Scaling analysis of nanofiltration systems fed with saturated calcium sulfate solutions in the presence of carbonate ions. Desalination, 159, 245-251.

Elfil, H. & Roques, H. 2001. Role of hydrate phases of calcium carbonate on the scaling phenomenon. Desalination, 137, 177-186.

Elwell, D. & Scheel, H. J. 1975. Crystal growth from high-temperature solutions, London; New York, Academic Press.

Gabelich, C. J., Chen, W. R., Yun, T. I., Coffey, B. M. & "Mel" Suffet, I. H. 2005. The role of dissolved aluminum in silica chemistry for membrane processes. Desalination, 180, 307-319.

Ghani, S. & Al-Deffeeri, N. S. 2010. Impacts of different antiscalant dosing rates and their thermal performance in Multi Stage Flash (MSF) distiller in Kuwait. Desalination, 250, 463-472.

Gill, J. S. 1999. A novel inhibitor for scale control in water desalination. Desalination, 124, 43-50.

Greenlee, L. F., Testa, F., Lawler, D. F., Freeman, B. D. & Moulin, P. 2010. The effect of antiscalant addition on calcium carbonate precipitation for a simplified synthetic brackish water reverse osmosis concentrate. Water Research, 44, 2957-2969.

Guan, X. 2009. Kinetics Studies of Reactions at Solid-Liquid Interface: Simulation of Calcification, VDM Verlag.

Hatch, G. B. & Rice, O. 1939. Surface-active properties of hexametaphosphate. Industrial & Engineering Chemistry, 31, 51-58.

Jong, R., Duiven, J., Terhorst, G. & Baas, K. 2011. Implementatie onderzoek naar fosfaatvrije antiscalant. H2O.

Kashchiev, D. 2000. Nucleation, Elsevier Science.

Kelle Zeiher, E. H., Ho, B. & Williams, K. D. 2003. Novel antiscalant dosing control. Desalination, 157, 209-216.

Kennedy, M. D., Schippers, J. C. & Salinas-Rodriguez, S. G. 2014. Desalination & Membrane related technology. Lecture note, LN0076/14/1 ,UNESCO-IHE.

Koutsoukos, P. 2010. Calcium Carbonate Scale Control in Industrial Water Systems. The Science and Technology of Industrial Water Treatment. CRC Press.

Kucera, J. 2011. Reverse Osmosis: Design, Processes, and Applications for Engineers, Wiley.

Lauer, G. 1997. Conditioning process and device for producing pure water US Patent applications 5690829.

Le Pevelen, D. D. & Tranter, G. E. 2017. FT-IR and Raman Spectroscopies, Polymorphism Applications. In: LINDON, J. C., TRANTER, G. E. & KOPPENAAL, D. W. (eds.) Encyclopedia of Spectroscopy and Spectrometry (Third Edition). Oxford: Academic Press.

Lee, S., Kim, J. & Lee, C.-H. 1999. Analysis of CaSO4 scale formation mechanism in various nanofiltration modules. Journal of Membrane Science, 163, 63-74.

Lee, S. & Lee, C.-H. 2000. Effect of operating conditions on $CaSO_4$ scale formation mechanism in nanofiltration for water softening. Water Research, 34, 3854-3866.

Macadam, J. & Parsons, S. A. 2004. Calcium carbonate scale formation and control. Re/Views in Environmental Science & Bio/Technology, 3, 159-169.

Mullin, J. W. 2001. 6 - Crystal growth. Crystallization (Fourth Edition). Oxford: Butterworth-Heinemann.

Neofotistou, E. & Demadis, K. D. 2004. Use of antiscalants for mitigation of silica (SiO2) fouling and deposition: fundamentals and applications in desalination systems. Desalination, 167, 257-272.

Nishida, I., Shimada, Y., Saito, T., Okaue, Y. & Yokoyama, T. 2009. Effect of aluminum on the deposition of silica scales in cooling water systems. Journal of Colloid and Interface Science, 335, 18-23.

Oh, H.-J., Choung, Y.-K., Lee, S., Choi, J.-S., Hwang, T.-M. & Kim, J. H. 2009. Scale formation in reverse osmosis desalination: model development. Desalination, 238, 333-346.

Parkhurst, D. L. & Appelo, C. 1999. User's guide to PHREEQC (Version 2): A computer program for speciation, batch-reaction, one-dimensional transport, and inverse geochemical calculations.

Pastero, L., Costa, E., Bruno, M., Rubbo, M., Sgualdino, G. & Aquilano, D. 2004. Morphology of Calcite (CaCO3) Crystals Growing from Aqueous Solutions in the Presence of Li+ Ions. Surface Behavior of the {0001} Form. Crystal Growth & Design, 4, 485-490.

Pervov, A. G. 1991. Scale formation prognosis and cleaning procedure schedules in reverse osmosis systems operation. Desalination, 83, 77-118.

Prisciandaro, M., Lancia, A. & Musmarra, D. 2001. Calcium Sulfate Dihydrate Nucleation in the Presence of Calcium and Sodium Chloride Salts. Industrial & Engineering Chemistry Research, 40, 2335-2339.

Rahman, F. & Amjad, Z. 2010. Scale Formation and Control in Thermal Desalination Systems. The Science and Technology of Industrial Water Treatment. CRC Press.

Sahachaiyunta, P., Koo, T. & Sheikholeslami, R. 2002. Effect of several inorganic species on silica fouling in RO membranes. Desalination, 144, 373-378.

Sawada, K. 1997. The mechanisms of crystallization and transformation of calcium carbonates. Pure and Applied Chemistry, 69, 921-928.

Schausberger, P., Mustafa, G. M., Leslie, G. & Friedl, A. 2009. Scaling prediction based on thermodynamic equilibrium calculation — scopes and limitations. Desalination, 244, 31-47.

Schock, G. & Miquel, A. 1987. Mass transfer and pressure loss in spiral-wound modules. Desalination, 64, 339.

Seewoo, S., Van Hille, R. & Lewis, A. 2004. Aspects of gypsum precipitation in scaling waters. Hydrometallurgy, 75, 135-146.

Sheikholeslami, R. 2003. Mixed salts—scaling limits and propensity. Desalination, 154, 117-127.

Sheikholeslami, R. 2005. Scaling potential index (SPI) for CaCO3 based on Gibbs free energies. AIChE Journal, 51, 1782-1789.

Shih, W.-Y., Gao, J., Rahardianto, A., Glater, J., Cohen, Y. & Gabelich, C. J. 2006. Ranking of antiscalant performance for gypsum scale suppression in the presence of residual aluminum. Desalination, 196, 280-292.

Shih, W.-Y., Rahardianto, A., Lee, R.-W. & Cohen, Y. 2005. Morphometric characterization of calcium sulfate dihydrate (gypsum) scale on reverse osmosis membranes. Journal of Membrane Science, 252, 253-263.

Singh, R. 2014. Membrane Technology and Engineering for Water Purification: Application, Systems Design And Operation, Elsevier Science & Technology.

Söhnel, O. & Mullin, J. W. 1978. A method for the determination of precipitation induction periods. Journal of Crystal Growth, 44, 377-382.

Söhnel, O. & Mullin, J. W. 1988. Interpretation of crystallization induction periods. Journal of Colloid And Interface Science, 123, 43-50.

Stumm, W. & Morgan, J. J. 1981. Aquatic Chemistry: An Introduction Emphasizing Chemical Equilibria in Natural Waters, Wiley.

Synchrony INC 2001. Trends in SCADA for Automated Water Systems.

Troup, D. H. & Richardson, J. A. 1978. Scale nucleation on a heat transfer surface and its prevention. Chemical Engineering Communications, 2, 167-180.

Tzotzi, C., Pahiadaki, T., Yiantsios, S. G., Karabelas, A. J. & Andritsos, N. 2007. A study of $CaCO_3$ scale formation and inhibition in RO and NF membrane processes. Journal of Membrane Science, 296, 171-184.

Van De Lisdonk, C. A. C., Van Paassen, J. A. M. & Schippers, J. C. 2000. Monitoring scaling in Nanofiltration And Reverse Osmosis Membrane Systems. Desalination, 132, 101-108.

van der Leeden, M. C. 1991. The role op polyelectrolytes in barium sulphate precipitation.

Van Engelen, G. & Nolles, R. 2013. A sustainable antiscalant for RO processes. Desalination and Water Treatment, 51, 921-923.

Verdoes, D., Kashchiev, D. & Van Rosmalen, G. M. 1992. Determination of nucleation and growth rates from induction times in seeded and unseeded precipitation of calcium carbonate. Journal of Crystal Growth, 118, 401-413.

Vrouwenvelder, J. S., Manolarakis, S. A., Veenendaal, H. R. & VAN DER KOOIJ, D. 2000. Biofouling potential of chemicals used for scale control in RO and NF membranes. Desalination, 132, 1-10.

Waly, T. 2011. Minimizing the Use of Chemicals to Control Scaling in Sea Water Reverse Osmosis: Improved Prediction of the Scaling Potential of Calcium Carbonate, CRC Press/Balkema. Leiden. Netherlands.

Yangali-Quintanilla, V. A., Dominiak, D. M. & Van De Ven, W. 2017. A smart optimization of antiscalant dosing in water desalination. IDA World Congress Water Reuse and Desalination. Sao Paulo.

Yuchi, A., Gotoh, Y. & Itoh, S. 2007. Potentiometry of effective concentration of polyacrylate as scale inhibitor. Analytica Chimica Acta, 594, 199-203.

Zimmer, K., Hater, W., Jaworski, J., Kruse, N. & Braun, G. 2016. Inhibitors for $CaCO_3$-scaling in reverse-osmosis plants - Influence of suspended matter on membrane clogging. Filtech. Cologne.

Chapter 9

Process design of reverse osmosis systems

Sergio G. Salinas-Rodríguez, Maria D. Kennedy, Jan C. Schippers

The learning objectives of this chapter are the following:

- To apply the principles of membrane filtration in the process design of a seawater reverse osmosis system
- Calculate the number of RO elements, number of pressure vessels, capacity of the high-pressure pump, membrane permeability coefficients for salt and water, permeate flow, feed flow, concentrate flow, permeate of the quality with and without the flux effect, verify the concentration polarization factor, cross-flow velocity of the water inside the membranes, and energy consumption with and without an energy recovery device

9.1 INTRODUCTION

The purpose of the manual calculations described in this chapter is to apply the basic equations governing reverse osmosis (RO) systems described in chapter 2. This process design considers the total dissolved solids (TDS) concentration of the seawater. The focus of this process design is the design of the RO system and not the need for or the type of pre-treatment. The steps described in this chapter are such that they can easily be converted in a calculation sheet with the help of software such as MS Excel. By following a step-wise procedure, the design of RO units does not remain a black box and can help to further understand the design methodology applied by commercial software.
The reader could at the end of the manual design followed in this chapter, compare the results with the ones of computer software for the same design parameters.

Most manufacturers of RO membranes, like Hydranautics (IMS Design), DOW (Wave), Toray, Suez (Winflows) and several others have available design software, enabling to make design projections for their membranes and systems in a quick manner. Figure 1 shows some examples of the available commercial software which in general are free to use.

Editors: Sergio Salinas, Jan Schippers, Gary Amy, In Kim, Maria Kennedy
doi: 10.2166/9781780409863_0243

9.1.1 Basic data

Before we can start with the design of the RO units, we need to know the design capacity of the plant, the type and TDS of the feedwater, and the water temperature. For seawater RO, the recovery ranges in practice between 40 to 50 %. The design information is presented in Table 1.

Table 1 Information for the design of the RO units

Source water	Seawater
Capacity of the plant	$Q_{p\ plant} = 45\ m^3/h$
Salt concentration (feed)	$C_f = 35{,}030\ mg/L$
Total recovery (plant)	R = 40 %
Temperature of the water	T = 25 °C

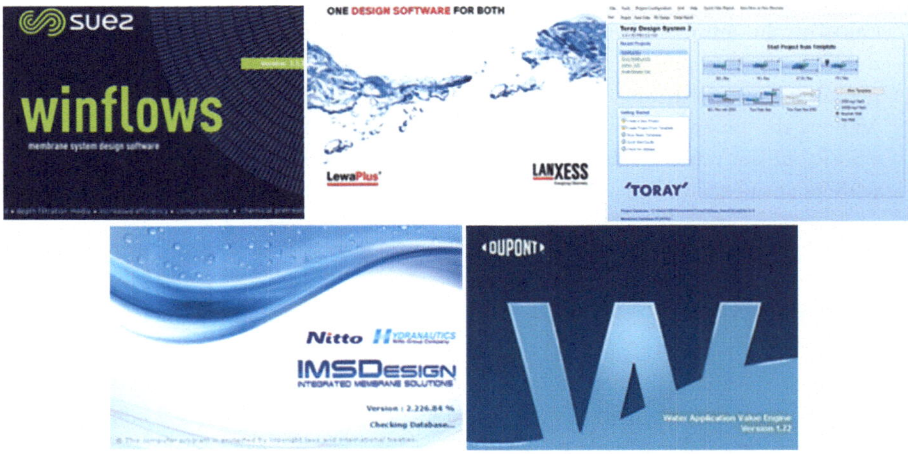

Figure 1 Examples of computer programmes for the design of RO membrane systems

Pre-treatment: Considering that we will make the process design of the RO, there is no need to indicate the type of pretreatment. Pre-treatment is discussed in chapter 3.

Ion composition: Note that we do not specify the ion composition in the feedwater but only the salinity of the seawater, expressed as TDS of the feed water. Commercial softwares allow the input of anions and cations present in the water and they use this information for calculating specific rejections per ion type by the RO membranes.

Recovery: In brackish water RO (BWRO), the scaling potential determines the maximum recovery. In seawater RO (SWRO), the osmotic pressure determines the maximum recovery.

In BWRO, the maximum allowable conversion (in brackish water) in one bank (group of parallel vessels) equipped with spiral wound elements is in practice not more than about 50 %. Reason is that higher conversions result in too low ratio's concentrate to permeate flow per element. As a result too high concentration polarization factor (CPF or β). When the ratio concentrate flow/permeate flow, drops below 5:1 (recovery higher than 18 %, then CPF exceeds 1.2 according the formula 2.57 in chapter 2.

In the last elements in a bank/stage, of BWRO plant, this ratio goes down since:
- concentrate flow drops substantially, at 50 % conversion, this flow drops with a factor 2.
- permeate flow drops as well, but much less than the concentrate flow.

9.1.2 Membrane type

We can select any of the options suggested by the RO manufacturer (Hydranautics). The options for seawater RO desalination are presented in Table 2. Some of the information for selection is the nominal production, the salt rejection, available filtration area, height feed spacer.

Table 2 Seawater reverse osmosis elements recommended by Hydranautics in the IMSDesign software

Model	Nominal production gpd	Nominal production m^3/d	Salt rejection, %	Element type	Size (DxL) (in x in)	Area (ft^2)	Area (m^2)	Spacer (mil)	Spacer (mm)
SWC4 MAX	7,200	27.3	99.8	SWRO Highest rejection	8x40	440	40.9	28	0.71
SWC4-LD	6,500	24.6	99.8	SWRO High rejection	8x40	400	37.2	34	0.86
SWC5 MAX	9,900	37.5	99.8	SWRO High rejection	8x40	440	40.9	28	0.71
SWC5-LD	9,000	34.1	99.8	SWRO High rejection, Low Dp	8x40	400	37.2	34	0.86
SWC6 MAX	6,600	25	99.6	SWRO Highest flow	8x40	440	40.9	28	0.71
SWC6-LD	6,000	22.7	99.6	SWRO High flow Low Dp	8x40	400	37.2	34	0.86

NB 1: The specified test pressure for the first 4 elements is 55 bar and for the last two elements is 41.4 bar.
NB 2: 1 mil = 1/1000 inch.

For elements with similar salt rejection, an element with higher nominal production will produce permeate water with higher salt concentration than an element with lower nominal production. On the other hand, if we consider the same permeate production, then the feed pressure requirement will be lower for the RO element with higher nominal production than for an element with lower nominal production.

Table 3 Basic properties of the selected RO element and selected number of elements per pressure vessel

Manufacturer	Hydranautics
Type	SWC4 MAX
Membrane area per element	$A_e = 40.9\ m^2$
Salt rejection	SR = 99.8%
Elements per vessel	$\#_{elem\,/\,PV} = 6\ [-]$

Each RO element manufacturer provides an element specification sheet (ESS) for each type of membranes they have available. The information in the ESS will be later used to verify maximum flows per element, calculate the permeability coefficient for water and for salt, etc.

The information from the EES that will be used in the following steps is marked by an arrow. The standard testing conditions reported in the EES will be later used to calculated the permeability coefficients for salt and water.

Spiral wound elements are placed in pressure vessels (1 m to 8 m in length). In large plants 6 to 8 elements of 1 m length are placed in one vessel. In small units 1 to 4 elements of 1 m length are placed in vessel, in one vessel.

Total number of elements in one stage follows from: $N_e = A/A_e$, where $A = Q_p / J$.

Where: N_e = number of elements; A = total required membrane area; A_e = membrane area per element (35 m^2); Q_p = permeate flow/capacity; J = Average flux in the stage.

Example 1 – Number of membrane elements

A sea water plant is producing 1,000 m^3/h. Spiral wound elements of 1 m length are placed in vessels of 6 m in length. Surface membrane is 35 m^2. Average flux of the membranes is about 14 L/m^2h

Questions: How many elements are in the plant? How many vessels are installed?

Answers:

Number of elements, N_e = Plant capacity / (Surface area per element · flux)

$N_e = 1{,}000 / (35 \cdot 14 \cdot 10^{-3}) = 2040$ elements

Number of vessels = number of elements / number of elements per vessel

$N_{vessels} = 2040 / 6 = 340$ vessels

SWC4 MAX

Specified Performance*

Permeate Flow:	7,200 gpd (27.3 m³/d)
Salt Rejection:	99.8% (99.7% minimum)

Test Conditions:

Standard 's' test conditions
- 32000 ppm NaCl solution
- 800 psi (5.5 MPa) Applied Pressure
- 77 °F (25 °C) Operating Temperature
- 10% Permeate Recovery
- 6.5 - 7.0 pH Range

*The Specified Performance is based on data taken after approximately 30 minutes of operation. Actual testing of elements may be done at conditions which vary from these exact values; in which case, the performance is normalized back to these standard conditions. Permeate flow for individual elements may vary ±15 percent from the value specified.

General Product Description**

Configuration:	Spiral Wound
Membrane Polymer:	Composite Polyamide
Membrane Active Area**:	440 ft² (40.9 m²)

Packaging: All membrane elements are supplied with a brine seal, interconnector, and O-rings. Elements are enclosed in a sealed polyethylene bag containing less than 1.0% sodium meta-bisulfite solution, and then packaged in a cardboard box.

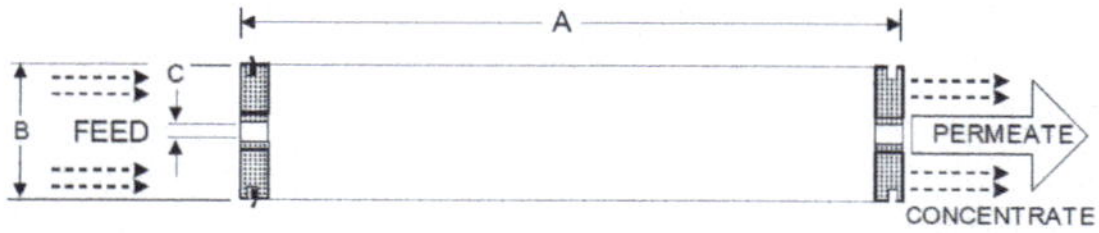

Element Details**

A, inches (mm)	B, inches (mm)	C, inches (mm)	Weight, lbs. (kg)
40.0 (1016)	7.89 (200)	1.125 (28.6)	32.0 ± 2 (14.5 ± 1)

**Values listed are indicative, not specified. For more detailed specifications, see our Technical Service Bulletin documents or contact Hydranautics Technical Department. Element weight values are as shipped from factory. Elements that are used and then gravity drained may still contain at least an additional 2 lbs (1 kg) of liquid.

Product Use and Restrictions^

Maximum Applied Pressure:	1200 psig (8.27 MPa)
Maximum Chlorine Concentration:	< 0.1 ppm
Maximum Operating Temperature:	113 °F (45 °C)
pH Range, Continuous (Cleaning):	2-11 (1-13)
Maximum Feedwater Turbidity:	1.0 NTU
Maximum Feedwater SDI (15 mins):	5.0
Maximum Feed Flow:	75 gpm (17.0 m³/h)
Maximum Pressure Drop for Each Element:	15 psi (0.10 MPa)

^ The limitations shown here are for general use. For specified projects, operation at more conservative values may ensure the best performance and longest life of the membrane. See Hydranautics Technical Bulletins for more details.

Figure 3 Element specification sheet of the SWC4 MAX RO element selected for the design (www.membranes.com [accessed 05 Dec 2018])

9.2 DESIGN GUIDELINES

Most of the RO manufacturers have their own design guidelines considering the many years of experience, the properties of their membranes, etc. In Table 4 a summary of the design recommendations is presented for three membrane manufacturers, namely: DuPont, Toray and Hydranautics.

Table 4 Design guidelines as recommended by DuPont, Hydranautics and Toray (DuPont, 2020, Nitto Hydranautics, 2020, Toray Industries Inc, 2020)

		Dupont			Toray		Hydranautics		
Raw water source		Sea	Sea	Sea	Sea	Sea	Sea	Sea	Sea
Intake type		Well or MF	Open	Open	Well	Open	Well	Surface	Surface
Pre-treatment type		UF	MF/ Conv.	Conv.	Conv.	Conv.	MF/ UF		
SDI @15 minutes		2.5	3	5	3	4	3	4	4
System average flux	L/m^2/h	15-19	14-17	12-17	15-19	12-16	17-20	13.6-17	17-20.4
Lead element flux	L/m^2/h	36	34	32	35	28	42.4	34	42.4
Maximum element recovery	%	15	14	13	13	13			
Flux decline	% year					5	7	5	
Salt passage increase	%/year			7	7	7	10	7	
Beta standard element				1.2	1.2	1.2	1.2	1.2	
Feed flow 8″ (maximum per vessel)	m^3/h	16	15	14	15	13	17	17	17
Reject flow 8″ (min per vessel)	m^3/h	3	3.2	3.4	3.6	3.6	2.7	2.7	2.7
Pressure drop (bar)									
6 m vessel typical	bar			3	2	1.72	1.72	1.72	
6 m vessel max	bar	3.5	3.5	3.5	4	4	3.45	3.45	3.45
Element max	bar	1	1	1	1	1	1.03	1.03	1.03

The maximum feed flow per element (from the data sheet) is 17 m^3/h. The concentration polarization factor $\beta < 1.2$ (β is the concentration polarization factor) or the minimum ratio of concentrate to permeate flow for any element 5:1.

9.3 PROCESS DESIGN STEPS

9.3.1 Step 1 - Simplified calculation of permeate concentration

In the following sections, the subscripts f, c, and p are used for feed, concentrate and permeate, respectively.

Feed flow of the plant	Q_f plant = 112.5 m^3/h	$R = Q_p / Q_f$
Concentrate flow of the plant	Q_c plant = 67.5 m^3/h	$Q_c = Q_{fc} - Q_p$
Concentrate concentration	C_c = 58,336 mg/L	$C_c = C_f\,[1 - R\,(1 - SR)] / (1 - R)$
Avg feed-conc. concentration	C_{fc} = 46,683 mg/L	$C_{fc} = (C_{fc} + C_c) / 2$
Permeate concentrationw	C_p = 93.37 mg/L	$C_p = C_{fc}\,(1 - SR)$

NB. The formula: $C_c = C_f / (1 - R)$ assumes SR = 100 %. Not used in this design. Differences in the final results are not significant.

Summarising the calculations for the whole plant, we have:

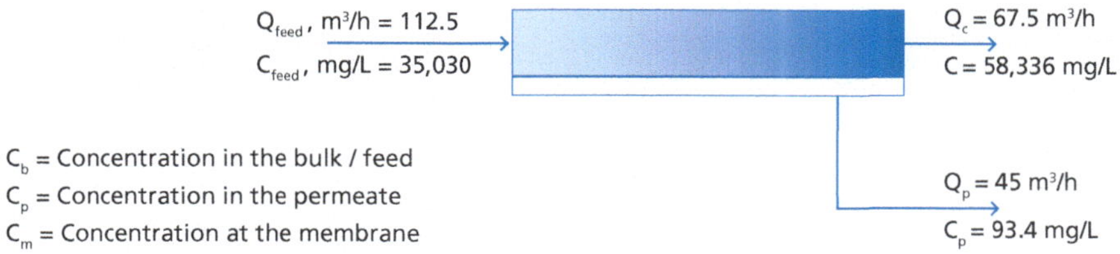

Figure 4 Flows and preliminary calculated concentrations in the RO plant

9.3.2 Step 2 - Calculation number of elements and pressure vessels

We need to select an average design flux. The selection can be based from practice or from design limits suggested by RO membrane manufacturers as presented in Table 4 and Table 5.

The flux which can be achieved in a RO plant is governed by the fouling potential of the feed water. The higher the flux the higher the rate of fouling of the membranes. A high fouling rate results in the need for frequent cleaning of the membranes with chemical cleaning solutions.

Table 5 Recommended RO operating flux ranges as function of water type (Nitto Hydranautics, 2020)

Feed water	J, L/m².h
Sea/ Surface water	14 - 24
Well water	24 - 31
RO permeate	34 - 51

Flux has to be chosen, based on the expected fouling potential feed water. RO plant operators are not in favour of frequent cleaning because:

- cleaning takes time. Long down time e.g., at least 8 hours.
- risk of damaging membranes
- requires careful acting and a lot of attention

Assuming average design flux	J_{avg} = 15 L/m².h	
Flow per element	Q_e = 0.61 m³/h	$Q_e = J_{avg} / A_e$
Nr of elements required in the plant	N_e = 73.5	$\#_{elements} = Q_{plant} / Q_e$
Nr of pressure vessels in the plant	N_{pv} = 12.25	$\#_{PV} = \#_{elem} / \#_{elem/PV}$
	12	Round up digits
Total number of elements	N_e = 72	$\#_{elements} = \#_{PV} \cdot \#_{elem/PV}$
Flux check:	J_{avg} = 15.3 L/m².h	$J_{avg} = Q_{plant} / (\#elements \cdot A_e)$
	OK	$(J_{calc} - J_{assumed}) < 0.5$; "OK"; "not OK"
Flow per element check:	Q_e = 0.63 m³/h	$Q_e = J_{avg} / A_e$ (NB. Considers the new total # of elements, i.e., 72)

Feed, permeate, and concentrate flow per pressure vessel

Feed flow per pressure vessel	$Q_{feed\,PV}$ = 9.38 m³/h	$Q_{feed\,PV} = Q_{feed\,plant} / \#_{PV}$
Permeate flow per pressure vessel	$Q_{perm\,PV}$ = 3.75 m³/h	$Q_{perm\,PV} = R \cdot Q_{feed\,PV}$
Concentrate flow per pressure vessel	$Q_{conc\,PV}$ = 5.63 m³/h	$Q_{conc\,PV} = Q_{feed\,PV} - Q_{perm\,PV}$

Check maximum feed flow in first element

Max. feed flow per element	Q_{max} = 17 m³/h	(from element specification sheet, application data)
	then, OK	$Q_{feed\,PV} < Q_{max}$; "OK" ; "not Ok"

NB. This is to avoid membrane damage

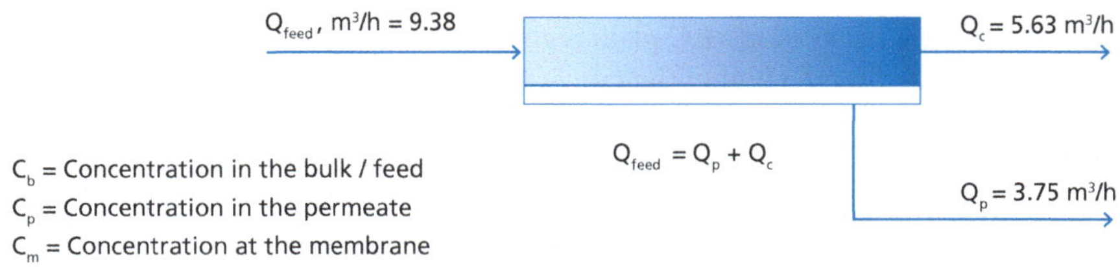

Figure 5 Flows per pressure vessel

9.3.3 Step 3 - Membrane permeability coefficients for water and salt

9.3.3.1 Calculation of membrane permeability coefficient for water (K_w)

K_w is not directly available from manufactures information. So, K_w has to be calculated from test results under standard conditions.

Results at standard conditions (subscript "s" below) are usually made available by membrane manufacturers (see element specification sheet).

We will follow the following steps for the calculation of K_w:

1. $Q = NDP \times K_w \times A$
2. $NDP = P_{feed} - \Delta P/2 - \Delta\pi_{avg} - P_{perm}$

3. P_{feed} from element specification sheet (P_{fs})
4. Pressure loss per element, $\Delta P_e = 0.2$ bar
5. $R_e = 10\,\%$
6. $\Delta_{avg} = (\Delta\pi_{feed} + \Delta\pi_{conc}) / 2$
7. Standard conditions "s"
8. $C_{conc\,s} = C_{feed\,s} / (1 - R_e)$
9. Osmotic pressure: 1,000 mg/L ~ 0.8 bar
10. Nominal capacity (permeate flow) of element = Q_s (element specification sheet)
11. $J_s = Q_s / A_e$
12. Membrane permeability: $K_w = Q_s / (NDP \times A_e)$
13. Membrane productivity = $K_w \cdot A_e$

Ts =	25	°C	(from element specification sheet, test conditions)
P_{fs} =	5.5	MPa	(from element specification sheet, test conditions)
P_{fs} =	55	bar	
ΔP_e =	0.2	bar	Assumed head loss per element
C_{fs} =	32,000	mg/L	(from element specification sheet, test conditions)
R_e =	10 %		(from element specification sheet, test conditions)
C_{cs} =	35,548	mg/L	$C_{cs} = C_{fs} \cdot [1 - R_e\ (1 - SR)] / (1 - R_e)$
1,000 mg/L =	0.8	bar	Equivalence for osmotic pressure, rule of thumb
π_{fs} =	25.6	bar	$\pi_{fs} = C_{fs} \cdot (0.8/1000)$
π_{cs} =	28.4	bar	$\pi_{cs} = C_{cs} \cdot (0.8/1000)$
π_{fcs} =	27.0	bar	$\pi_{fcs} = (\pi_{fs} + \pi_{cs}) / 2$
π_{ps} =	0.270	bar	$\pi_{ps} = 0.01\pi_{fcs}$ [For Brackish water: $0.05\pi_{fc}$]
$\Delta\pi_{avg\,s}$ =	26.7	bar	$\Delta\pi_{avg\,s} = (\pi_{fcs} - \pi_{ps})$
P_{ps} =	0	bar	
NDP_s =	28.2	bar	**$NDP_s = P_{fs} - \Delta P_e/2 - \Delta\pi_{avg\,s} - P_{ps}$**

Nominal capacity:

Q_{ws} =	27.3	m^3/d	(from element specification sheet, performance)
	1.14	m^3/h	

Flux under standard conditions:

J_{ws} =	27.9	$L/m^2 \cdot h$	$J_{ws} = Q_{ws} / A_e$
$K_{w\,25°C}$ =	0.00099	$m^3/m^2 \cdot bar \cdot h$	**$K_w = J_{ws} / NDP_s$**
	0.99	$L/m^2 \cdot bar \cdot h$	

Membrane productivity:

$K_{w\,25}\,A_e$ =	0.040	$m^3/ba \cdot h$

Remark: Temperature has an effect on K_w. The higher the temperature the higher the permeability (It is about 3 % per C).

K_w is linked with the viscosity of water; as a consequence, the higher the temperature, the lower the required pressure to maintain a certain flux (and capacity).

9.3.3.2 Calculation of membrane permeability coefficient for salt (K_s)

We will use the following formula: $C_p = (C_{fc}\ K_s) / J$

Steps for calculating the membrane permeability for salt K_s:

1. Standard conditions "s"
2. J_S from step 2
3. from element specification sheet: C_{fs}, R_e, SR
4. $C_{Conc\ S} = C_{feed\ S} / (1 - R_e)$
5. $C_{fc} = (C_{feed} + C_{conc}) / 2$
6. $C_P = C_{fc} \cdot (1 - SR)$
7. $C_P = (C_{fc} \cdot K_s) / J$
8. $K_s = C_P \times J / C_{fc}$

Remarks:

- Water can pass a membrane; salts as well, however, at a much lower rate. Transport of salts through RO membranes is due to diffusion ($J_s = (C_f - C_p) \cdot K_s$).
- We need K_s to calculate the permeate concentration as function of flux (Step 6b). ($C_p = J_s / J_w = [(C_f - C_p) \cdot K_s] / J_w$, but $C_p << C_f$, then: $C_p \approx (C_f \cdot K_s) / J_w$)
- K_s is not directly available from manufacturers' information. So, K_s has to be calculated from test results under standard conditions.

$T_s =$	25	°C	(from element specification sheet, test conditions)
$C_{fs} =$	32,000	mg/L	(from element specification sheet, test conditions)
$R_e =$	10%		(from element specification sheet, test conditions)
$C_{cs} =$	35,548	mg/L	$C_{cs} = C_{fs} \cdot [1 - R_e(1 - SR)] / (1 - R_e)$
$C_{fcs} =$	33,774	mg/L	$C_{fcs} = (C_{fs} + C_{cs}) / 2$
SR =	99.8%		(from element specification sheet, performance)
$C_{ps} =$	67.5	mg/L	$C_{ps} = C_{fcs} \cdot (1 - SR)$

Nominal flow:

$Q_{ws} =$	27.3	m^3/d	(from element specification sheet, performance)
	1.14	m^3/h	

Flux under standard conditions:

$J_s =$	27.9	$L/m^2 \cdot h$	$J_s = Q_{ws} / A_e$

Now, we can calculate, $K_s = J_s\,(C_{ps} / C_{fcs})$

$K_{s\,25\,°C} =$	0.056	$L/m^2.h$

Comments:

- K_s is independent of applied pressure.
- K_s is dependent of water temperature; the higher the temperature the higher the salt passage.
- Different ions have different K_s values, so, the rejection (SR) is different.
- In general, SR: $Mg^{2+} > Ca^{2+} > Na_+$ and $SO_4^{2-} > Cl^-$. So, calculations should be done for different ions, which makes the whole set of calculations very complicated. Moreover, K_s values for other ions are hardly available.
- High salt rejection (low K_s) combines with low K_w value (due to smaller pores) e.g., Seawater: $K_s = 0.08$ L/m^2.h and $K_w = 1$ L/m^2.h, while for brackish water, $K_s = 1.1$ L/m^2.h and $K_w = 5$ L/m^2.h.
- The larger the pores, the larger the permeability for salt and water.

9.3.4 Step 4 - Preliminary calculation of feed pressure

The required feed pressure (P_f) depends on the average (chosen) flux (J_w), the permeability (K_w) of the RO membrane (selected), the osmotic pressure (π); the pressure loss in the feed/concentrate channel ΔP.

$$J_w = (P_f - \pi) \cdot K_w = \text{net driving pressure} \cdot \text{permeability}$$

or

$$P_f = J_w / K_w + \pi = \text{flux / permeability + osmostic pressure}$$

Salinity, governs together with recovery, the osmotic pressure.

$$C = C_f / (1 - R)$$

The feed pressure, P_f, should overcome all the resistances present in the system to diffuse salts and overcome the osmotic pressure. The net driving pressure, is the effective pressure to push water only.

The chosen flux dictates the required Net driving pressure (NDP) according to the formula: $J = NDP \times K_w$

However, the flux in the RO elements in a system depends on the position of the element inside the pressure vessel, because the NDP depends on pressure losses and osmotic pressure. Even in one element the NDP is not constant, which is clearly shown in the following formula.

$$NDP = P_f - \Delta P - \Delta\pi - P_p$$

Where: NDP = average net driving pressure; P_f = feed pressure; ΔP = head loss across one element (~ 0.2 bar); $\Delta\pi_{avg}$ = average difference osmotic pressure: feed – permeate; P_p = product pressure.

The step to estimate the feed pressure in the RO system, are the following:

1. $J = Q_{perm\ plant}/A_{mem\ plant}$
2. $Q_e = Q_{perm\ plant}/N_{elements}$
3. $NDP = J / K_w$
4. For a pressure vessel:
5. $C_{conc} = C_{feed} / (1\text{-}R)$
6. Osmotic pressure: 1,000 mg/L ~ 0.8 bar
7. $\Delta P = 0.2$ bar
8. $P_{feed} = NDP + \Delta P/2 + \Delta_{avg} + P_{perm}$

Average flux per element	J_{avg} =	15.3	$L/m^2.h$	$J_{avg} = Q_{plant} / (\#_{elements} \cdot A_e)$
Flow per element	Q_e =	0.625	m^3/h	$Q_e = Q_{plant}$ / #elements (or: $Q_e = J_{avg} \cdot A_e$)
	NDP =	**15.5**	bar	$NDP = J_{avg} / K_w$
For a pressure vessel	C_f =	5,030	mg/L	TDS of feed water (data)
	R =	40 %		Plant design recovery
	C_c =	58,336	mg/L	$C_c = C_f \cdot [1 - R \cdot (1 - SR)] / (1 - R)$
		0.8	bar	1,000 mg/L = 0.8 bar. Equivalence for osmotic pressure, rule of thumb.
	π_f =	28.0	bar	$\pi_f = C_f\ (0.8 / 1000)$
	π_c =	46.7	bar	$\pi_c = C_c\ (0.8 / 1000)$
	π_{fc} =	37.3	bar	$\pi_{fc} = (\pi_f + \pi_c) / 2$
	π_p =	0.373	bar	$\pi_p = 0.01\ \pi_{fc}$
	$\Delta\pi_{avg}$ =	37.0	bar	$\Delta\pi_{avg} = (\pi_{fc} - \pi_p)$
Head loss per element	Δp_e =	0.2	bar	Assumed head loss per element
Head loss per pressure vessel	Δp_{PV} =	1.2	bar	$\Delta p_{PV} = \Delta p_e \cdot \#_{elements}$
Pressure in permeate	P_p =	0.0	bar	Negligible
Estimated feed pressure	P_f =	53.0	bar	$P_f = NDP + \Delta P_{PV}/2 + \Delta\pi_{avg} + P_p$
	say:	**54**	bar	round up

9.3.5 Step 5 - Calculations of flows, recovery, and concentration polarization factor for each element

We will calculate the flow and recovery for each element in a pressure vessel, as illustrated in Figure 6.

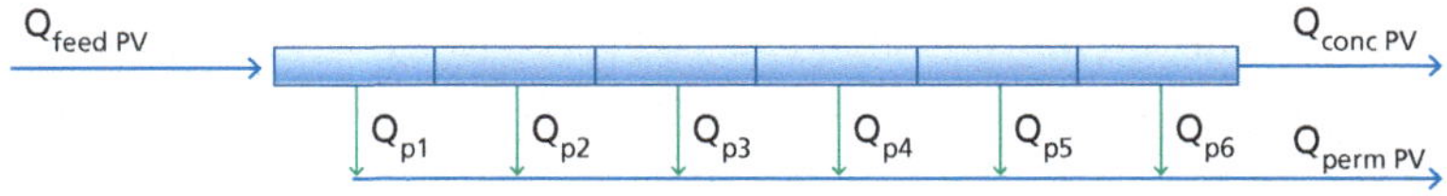

Figure 6 Schematic of the flow streams inside a RO pressure vessel with 6 elements in series

This is to verify: the feed pressure, the recovery, and check the concentration polarization factor β for each element (and/or the ratio of concentrate to permeate flow per element and the permeate quality per element).

$$NDP_i = P_{fi} - \Delta P_{ei}/2 - \Delta\pi_{avg\,i} - P_{pi}$$

Where: P_{fi} = feed pressure for element "*i*", ΔP_{ei} = pressure headloss for element "*i*", $\Delta\pi_{avg_i}$ = average osmotic pressure for element "*i*", $P_{perm\,i}$ = pressure in the permeate side of element "*i*".

The pressure headloss per pressure vessel in in the range 1.2 – 2 bar. In our design we will consider that the headloss is the same per element and equal to 0.2 bar.

Another consideration is that we will calculate the values at the middle of the element, thus we will take the feed-concentrate concentration $[C_{fc} = (C_f + C_c)/2]$.

The procedure to follow will be the following:

1. We start with the *Element 1*. (From feed to rear in the pressure vessel)
2. P_{f1} = Feed pressure calculated in Step 4.
3. $Q_{f1} = Q_{Perm\,plant} / N_{PV}$ (This value was already calculated in Step 2)
4. C_{f1} (From initial data for the design)
5. Assume R_{e1} to calculate osmotic pressure. (We will assume a value and later on verify if the assumption was correct).
6. $C_{c1} = C_{f1} / (1 - R_{e1\,assumed})$
7. Osmotic pressure: 1,000 mg/L ~ 0.8 bar (Rule of thumb)
8. $NDP_1 = P_{f1} - \Delta P/2 - \Delta\pi_{avg} - P_{perm}$
9. $Q_{p1} = ND_{p1} \cdot K_w \cdot A_e$ (This is the main formula)
10. $R = Q_{p1} / Q_{f1}$
11. Is $R_{e1\,assumed} = R_{e1\,calculated}$?
    No? Then repeat procedure
    Yes? Then continue to next element
12. $Q_{c1} = Q_{f1} - Q_{p1}$

For step 5 and 6 we can calculate:

1. $C_{fc1} = (C_{f1} + C_{c1}) / 2$
2. $Q_{fc1} = (Q_{f1} + Q_{c1}) / 2$

And then we can proceed for *Element 2*, considering the information from element 1.

1. Element 2
2. $P_{f2} = P_{f1} - \Delta P$
3. $Q_{f2} = Q_{c1}$
4. $C_{f2} = C_{c1}$
5. Assume R_{e2} to calculate osmotic pressure

We can repeat the procedure for the following elements in the pressure vessel.

	1	2	3	4	5	6	Unit	
P_{fi} =	54	53.8	53.6	53.4	53.2	53.0	bar	
ΔP_{ei} =	0.2	0.2	0.2	0.2	0.2	0.2	bar	headloss per element
Q_{fi} =	9.38	8.38	7.54	6.84	6.29	5.87	m^3/h	$Q_{f1} = Q_{f\,PV}$, $Q_{f2} = Q_{c1}$, etc
C_{fi} =	35,030	39,163	43,556	47,982	52,181	55,868	mg/L	$C_{f1} = C_{f\,PV}$, $C_{f\,i+1} = C_{c\,i}$
Iter. 1 R_{ei} =	10.5	10.0	9.3	8.0	6.7	5.2	%	Assume R_{ei} to start iteration
C_{ci} =	39,131	43,506	48,013	52,146	55,921	58,926	mg/L	$C_{ci} = C_{fi} \cdot [1 - R_{ei} \cdot (1 - SR)] / (1 - R_{ei})$
π_{fi} =	28.02	31.33	34.84	38.39	41.74	44.69	bar	$\pi_{fi} = C_{fi}\ (0.8/1000)$
π_{ci} =	31.30	34.80	38.41	41.72	44.74	47.14	bar	$\pi_{ci} = C_{ci}\ (0.8/1000)$
$\pi_{fc\,i}$ =	29.66	33.07	36.63	40.05	43.24	45.92	bar	$\pi f_{c\,i} = (\pi_{fi} + \pi_{ci}) / 2$
$\pi_{p\,i}$ =	0.297	0.331	0.366	0.401	0.432	0.459	bar	$\pi_{p\,i} = 0.01 \cdot \pi_{fc\,i}$
$\Delta\pi_{avg\,i}$ =	29.37	32.74	36.26	39.65	42.81	45.46	bar	$\Delta\pi_{avg\,i} = (\pi_{fc\,i} - \pi_{pi})$
NDP_i =	24.53	20.96	17.24	13.65	10.29	7.44	bar	$NDP_i = P_{fi} - \Delta P_{ei}/2 - \Delta\pi_{avg\,i} - P_{pi}$
Q_{pi} =	0.99	0.85	0.70	0.55	0.42	0.30	m^3/h	$Q_{pi} = NDP_i\ K_w\ A_e$
R_{ei} =	10.57	10.10	9.24	8.06	6.61	5.12	%	$R_{ei} = Q_{pi} / Q_{fi}$
Iter. 2 R_{ei} =	10.57	10.10	9.24	8.06	6.61	5.12	%	
C_{ci} =	39,163	43,556	47,982	52,181	55,868	58,876	mg/L	$C_{ci} = C_{fi}\ [1 - R_{ei} \cdot (1 - SR)] / (1 - R_{ei})$
	OK	OK	OK	OK	OK	OK		$(C_{ci\,iter2} - C_{ci\,iter1}) < 100$? ; "OK" ; "not OK"
π_{fi} =	28.02	31.33	34.84	38.39	41.74	44.69	bar	$\pi_{fi} = C_{fi} \cdot 0.8/1000$
π_{ci} =	31.33	34.84	38.39	41.74	44.69	47.10	bar	$\pi_{ci} = C_{ci} \cdot 0.8/1000$
	OK	OK	OK	OK	OK	OK		$(\pi_{ci\,iter2} - \pi_{ci\,iter1}) < 0.5$? ; "OK" ; "not OK"
$\pi_{fc\,i}$ =	29.68	33.09	36.62	40.07	43.22	45.90	bar	$\pi_{fc\,i} = (\pi_{fi} + \pi_{ci}) / 2$
$\pi_{p\,i}$ =	0.297	0.331	0.366	0.401	0.432	0.459	bar	$\pi_{p\,i} = 0.01 \cdot \pi_{fc\,i}$
$\Delta\pi_{avg\,i}$ =	29.38	32.76	36.25	39.66	42.79	45.44	bar	$\Delta\pi_{avg\,i} = (\pi_{fc\,i} - \pi_{p\,i})$
NDP_i =	24.52	20.94	17.25	13.64	10.31	7.46	bar	$NDP_i = P_{fi} - \Delta P_{ei}/2 - \Delta\pi_{avg_i} - P_{pi}$
Q_{pi} =	0.99	0.85	0.70	0.55	0.42	0.30	m^3/h	$Q_{pi} = NDP_i \cdot K_w \cdot A_e$
	OK	OK	OK	OK	OK	OK		$(Q_{pi\,iter2} - Q_{pi\,iter1}) < 0.25$? ; "OK" ; "not OK"
R_{ei} =	10.57	10.09	9.25	8.05	6.62	5.13	%	$R_{ei} = Q_{pi} / Q_{fi}$
Q_{ci} =	8.38	7.54	6.84	6.29	5.87	5.57	m^3/h	$Q_{ci} = (Q_{fi} - Q_{pi})$
C_{fci} =	37,097	41,360	45,769	50,081	54,024	57,372	mg/L	$C_{fci} = (C_{fi} + C_{ci}) / 2$
P_{ci} =	53.8	53.6	53.4	53.2	53.0	52.8	bar	$P_{ci} = P_{fi} - \Delta P_{ei}$

The results so far of the previous design steps can be displayed as follow:

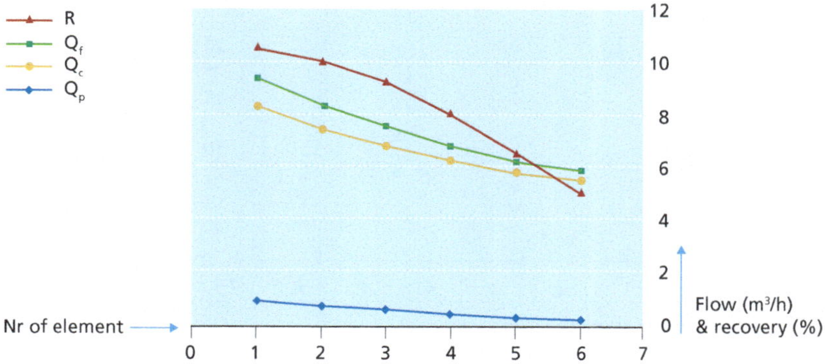

Figure 7 Flow and recovery per element along the pressure vessel

As can be observed in Figure 7, the recovery per element and permeate flow per element are not uniform in a pressure vessel. The front element produces more than three times that the last element in the pressure vessel. The recovery of the first element is 10.6 % while the last element has a recovery of 5.6 %.

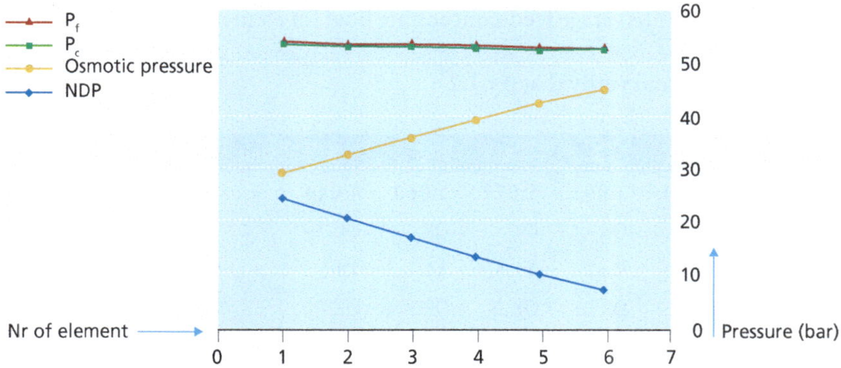

Figure 8 Pressure and osmotic pressure per element along the pressure vessel

The net driving pressure decreases along the pressure vessel as illustrated in Figure 8. The net driving pressure is calculated from the feed pressure, minus the head losses per element, minus the increasing osmotic pressure (due to the salt rejection by the RO membranes). The first element has an NDP of 24.5 bar while the last element has an NDP of 7.5 bar which results in a flux of 24.3 $L/m^2/h$ for the first element and 7.4 $L/m^2/h$ for the last element (see Figure 9).

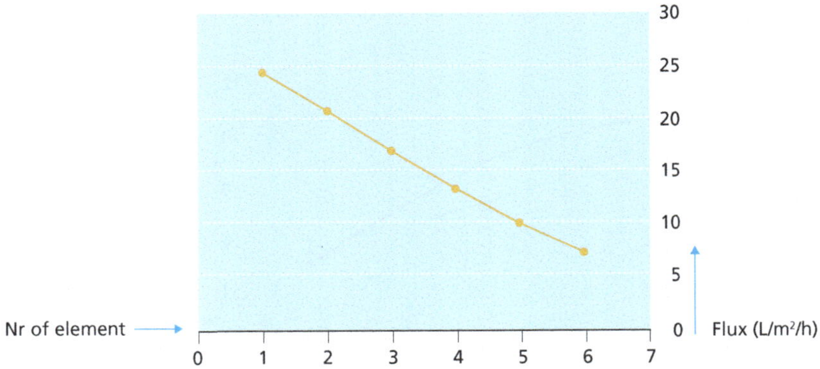

Figure 9 Flux per element along the pressure vessel

9.3.5.1 Calculation of the concentration polarization factor

With the results obtained per element, we can calculate the concentration polarization factor (CPF or β) for each element, with the following formula:

$$\beta_1 = K_p \cdot e^{\left(\frac{Q_{p1}}{Q_{fc1}}\right)}$$

Where:
Q_{p1} = Permeate flow of element 1
$Q_{fc1} = (Q_{f1} + Q_{c1}) / 2$ (Average feed-concentrate flow for element 1)
$K_p = 0.99$ (Hydranautics)
repeat for other elements and verify that $\beta < 1.2$.

	1	2	3	4	5	6	Unit	
$\beta_i =$	1.107	1.101	1.091	1.077	1.060	1.044	-	$\beta_i = K_p \cdot Exp^{(Qpi / Qfci)}$
	OK	OK	OK	OK	OK	OK		$\beta < 1.2$?
$Q_{ci} / Q_{pi} =$	8.5	8.9	9.8	11.4	14.1	18.5	-	
	OK	OK	OK	OK	OK	OK		$Q_{ci} / Q_{pi} > 5$?

NB. We have used the formula applied by Hydranautics. DOW has a different formula. Dupont Filmtec applies for their elements the formula: CPF = $\exp^{(0.7R)}$. Where: CPF = Concentration Polarization Factor, R = Recovery.

The recommended recovery (by Dow) varies with the quality of the feed water e.g. Seawater (10-12 %); filtered treated domestic waste water 10-12 %; pre-treated surface water 15-18 %; Softened well water 19-25 %.
Remark: In practice commonly $\beta < 1.2$ is used as a guideline (Hydranautics) as a maximum for CPF, to avoid operational problems e.g., scaling and fouling. When the ratio concentrate flow/permeate flow, drops below 5:1 (recovery higher than 18 %), then β exceeds 1.2 according the formula.

In sea water RO systems, the concentration polarization will decrease with increasing recovery. Reason is that the flux is dropping dramatically with increasing recovery.

9.3.6 Step 6 - Calculations of permeate quality

9.3.6.1 Assuming a constant salt rejection (no flux effect)

We will calculate the permeate quality for each element and per pressure vessel.
To simplify the calculations, it is assumed that salt rejection is constant, namely 99.7 % (standard conditions).

The step will be the following:

1. $C_{fc1} = (C_{f1} + C_{c1}) / 2$ (calculated in step 5, iteration 2)
2. $C_{p1} = CF_{fc1} \times (1 - SR)$
3. Repeat same procedure for elements 2, 3, 4, 5 and 6.
4. $Q_{p1}, Q_{p2}, Q_{p3}, Q_{p4}, Q_{p5}, Q_{p6}$ from step 5.
5.

$$C_{product} = \frac{C_{p1} \cdot Q_{p1} + C_{p2} \cdot Q_{p2} + C_{p3} \cdot Q_{p3} + C_{p4} \cdot Q_{p4} + C_{p5} \cdot Q_{p5} + C_{p6} \cdot Q_{p6}}{Q_{p1} + Q_{p2} + Q_{p3} + Q_{p4} + Q_{p5} + Q_{p6}}$$

6. Compare with Step 2.

	1	2	3	4	5	6	Unit	
C_{fci} =	37,097	41,360	45,769	50,081	54,024	57,372	mg/L	from step 5, iteration 2
C_{pi} =	74.2	82.7	91.5	100.2	108.0	114.7	mg/L	$C_{pi} = C_{fci}(1 - SR)$

Now, we can calculate the permeate concentration per pressure vessel with the mentioned formula:

$$C_{product} = \frac{74.2 \cdot 0.99 + 82.7 \cdot 0.85 + 91.5 \cdot 0.70 + 100.2 \cdot 0.55 + 108.0 \cdot 0.42 + 114.7 \cdot 0.30}{0.99 + 0.85 + 0.70 + 0.55 + 0.42 + 0.30}$$

$C_{product} = 89.96$ mg/L

9.3.6.2 Salt rejection depends on the flux

We will calculate the salt rejection and permeate quality per element, taking into account the effect of flux.

Salinity of the feed water determines the permeate salinity, together with membrane performance (K_s), flux and recovery.
The steps to calculate the permeate quality with effect of flux are the following:

1. $C_{fc1} = (C_{f1} + C_{c1}) / 2$
2. $J_1 = Q_{p1} / A_e$
3. $C_{P1} = (C_{fc1} \times K_s) / J_1$
4. Repeat same procedure for elements 2, 3, 4, 5, and 6.

5.

$$C_{product} = \frac{C_{p1} \cdot J_1 + C_{p2} \cdot J_2 + C_{p3} \cdot J_3 + C_{p4} \cdot J_4 + C_{p5} \cdot J_5 + C_{p6} \cdot J_6}{J_1 + J_2 + J_3 + J_4 + J_5 + J_6}$$

6. Compare with Step 2 and with calculation without the flux effect.

	1	2	3	4	5	6	Unit	
C_{fci} =	37,097	41,360	45,769	50,081	54,024	57,372	mg/L	from step 5, iteration 2
J_i =	24.28	20.74	17.09	13.50	10.21	7.39	L/m².h	$J_i = Q_{pi} / A_e$
C_{pi} =	85.2	111.2	149.4	206.8	294.9	433	mg/L	$C_{pi} = C_{fci} \cdot K_s / J_i$

Now, we can calculate the permeate concentration per pressure vessel with the mentioned formula:

$$C_{product} = \frac{85.2 \cdot 24.28 + 111.2 \cdot 20.74 + 149.4 \cdot 17.09 + 206.8 \cdot 13.50 + 249.9 \cdot 10.21 + 433 \cdot 7.39}{24.28 + 20.74 + 17.09 + 13.50 + 10.21 + 7.39}$$

$C_{product}$ = 170.9 mg/L

What is the effect of temperature on salinity of product water? The higher the temperature, the higher the K_s [$K_s = J\ (C_p / C_{fc})$ and $J = \Delta P / (\eta \cdot R)$]. As a consequence the salinity in the product will increase (~3 % increase per °C).

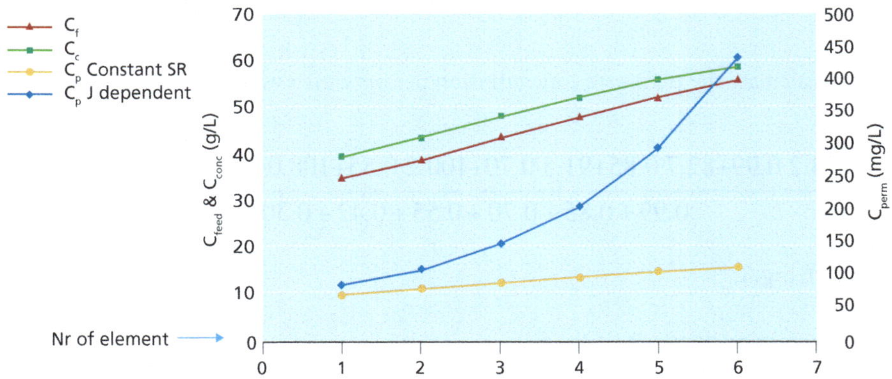

Figure 10 Feed concentration, concentrate concentration, and permeate concentration (considering salt rejection is constant & considering the effect of flux on permeate concentration) per element along the pressure vessel

Why is the salinity of the product water (permeate) of the elements increasing with the position (1, 2, 3, ...6)? This is due to two reasons: *i*) the salinity in the feed/concentrate stream is increasing with increasing recovery (element 1 to element 6); *ii*) the flux is decreasing with increasing recovery.

9.3.7 Step 7 - Cross-flow velocity calculation

SWC4 MAX	Type of membrane element	
A_e = 40.8 m^2	Area of the membrane element	
L_e = 1.0 m	Length of the membrane element	
h = 0.00071 m	Height of the feed space	
ε = 085	Porosity of the feed spacer	0.8-0.85 (Vrouwenvelder, 2009)

The steps to follow are the next ones:
Q_{fc} = flow in the feed-concentrate stream
w = total spacer width

	1	2	3	4	5	6	Unit	
$Q_{fc\,i}$ =	8.88	7.96	7.19	6.57	6.08	5.72	m^3/h	$Q_{fc\,i} = (Q_{f\,i} + Q_{c\,i}) / 2$
w =	20.4	20.4	20.4	20.4	20.4	20.4	m	$w = (A_e / L_e) / 2$
A_{eff} =	0.0123	0.0123	0.0123	0.0123	0.0123	0.0123	m^2	$A_{effective} = \varepsilon \cdot h \cdot w$
$v_{fc\,i}$ =	0.20	0.18	0.16	0.15	0.14	0.13	m/s	$v_{fc\,i} = Q_{fc\,i} / A_{effective}$

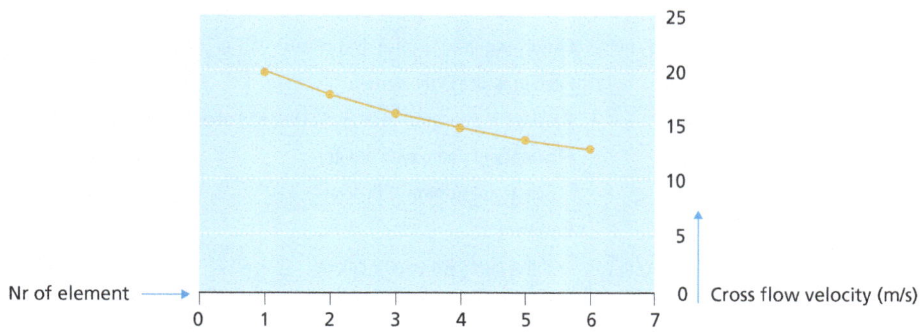

Figure 11 Cross flow velocity per element along the pressure vessel

9.3.8 Step 8 - Energy consumption

9.3.8.1 Energy to raise the pressure of 1 m^3 to 1 bar

By definition: Work = Energy = force × displacement = N × m = Joule
1 Joule = 1 N × m = 1 kg·m/s^2 × m = 1 kg·m^2/s^2

To bring water at a higher level of e.g., h metres (1 bar = 10 m), the work will be per m^3:

Force $= F = m \cdot g = (\rho \cdot V) \cdot g$ or force per $m^3 = (\rho \cdot g)$

Work $=$ Energy $=$ force $\times$ distance $= (\rho \cdot g) \cdot h$

$= (1{,}000\ kg/m^3 \cdot 9.8\ m/s^2) \cdot 10\ m = 98{,}000$ Joule

Note: Joule / s = Watt or Joule = W·s = Ws

Then, we have:

= 98,000 W·s = 98,000 / 3,600 W·h

= 0.0275 kW·h per m^3.

9.3.8.2 Without energy recovery device (ERD)

$E = (0.0275 \cdot P_{feed}) / (N_{pump} \cdot R)$

	0.0275	kWh/m^3	Energy required to raise the pressure of 1 m^3 water to 1 bar
P_{feed} =	54	bar	Feed pressure in the system
N_{pump} =	0.77		Efficiency of pump with driver
	5,030	mg/L	TDS of feed water (data)
R =	40%		Recovery
E =	4.82	kWh/m^3	without energy recovery device

9.3.8.3 With energy recovery device (ERD)

$E = [(0.0275 \cdot P_{feed}) / (N_{pump} \cdot R)] - [0.0275 \cdot (1 - R) \cdot P_{conc} \cdot N_{turbine}] / R$

	0.0275	kWh/m^3	Energy required to raise the pressure of 1 m^3 water to 1 bar
P_{feed} =	54	bar	Feed pressure in the system
P_{conc} =	52.8	bar	Pressure in concentrate stream at the end of pressure vessel
N_{pump} =	0.77		Efficiency of pump with driver
$N_{turbine}$ =	0.8		Efficiency of turbine with driver
R =	40%		Recovery
E =	3.08	kWh/m^3	without energy recovery device

9.3.9 Step 9 – Summary

Nr of pressure vessels in the plant =	12		#PV
Total number of elements =	72		#elements
R =	40.6	%	$R = Q_{p\,PV} / Q_{f\,PV}$, where $Q_{p\,PV} = \Sigma Q_{p\,i}$
J_{avg} =	15.5	L/m^2.h	$J_{avg} = \Sigma J_i / \#_{elements}$

$Q_{P\,Plant}$ =	45.6	m^3/h	$Q_{P\,Plant} = \#_{PV} \times Q_{P\,PV}$
E =	4.8	kWh/m^3	without energy recovery device
E =	3.1	kWh/m^3	with energy recovery device

Summary of the process design per RO pressure vessel:

							C_p, mg/L	
Element	**R, %**	**Q_f, m^3/h**	**Q_c, m^3/h**	**Q_p, m^3/h**	**C_f, mg/L**	**C_c, mg/L**	**Const SR**	**J dependent**
1	10.6	**9.38**	8.38	0.99	35,030	39,163	74	85
2	10.1	8.38	7.54	0.85	39,163	43,556	83	111
3	9.2	7.54	6.84	0.70	43,556	47,982	92	149
4	8.1	6.84	6.29	0.55	47,982	52,181	100	207
5	6.6	6.29	5.87	0.42	52,181	55,868	108	295
6	5.1	5.87	5.57	0.30	55,868	58,876	115	433
Total			$Q_{p\,PV}$ =	**3.80**			**90**	**171**

The total recovery of the designed RO unit can be calculated by dividing the permeate flow over the feed flow. This is 3.8 m^3/h / 9.38 m^3/h equals 0.405 or 40.5 % which corresponds with out initial target for the RO recovery.

The total flow of the plant is equal to 3.8 m^3/h multiplied by the number of pressure vessels, equal to 45.6 m^3/h which also matches our designed capacity. Note that the pre-treatment units need to be designed for the RO feed flow (45.6 m^3/h divided by 40 % recovery equals to 112 m^3/h).

Element	**P_f, bar**	**P_c, bar**	**NDP, bar**	**J, $L/m^2.h$**	**Q_{fc}, m^3/h**	**Beta**	**Q_c:Q_P**	**v_{fc}, m/s**
1	54	53.8	24.52	24.3	8.88	1.107	8.5	0.20
2	53.8	53.6	20.94	20.7	7.96	1.101	8.9	0.18
3	53.6	53.4	17.25	17.1	7.19	1.091	9.8	0.16
4	53.4	53.2	13.64	13.5	6.57	1.077	11.4	0.15
5	53.2	53.0	10.31	10.2	6.08	1.060	14.1	0.14
6	53.0	52.8	7.46	7.4	5.72	1.044	18.5	0.13
Total			J_{avg} =	**15.5**				

The average flux also matches our initially selected design flux of 15 $L/m^2/h$. The front elements in the RO unit operate at higher flux rate than the rear elements. In our design, the rear element operated at 3.5 times less flux than the front element.

In all the elements, the concentration polarization factor was below 1.2, as required by the RO membrane manufacturers.

The cross flow velocity ranged between 0.1 and 0.2 m/s which is considered normal in RO installations. In order to keep the production constant, the feed pressure will increase to overcome the extra resistance due to membrane fouling.

Comment:

- For drinking water, 500 mg/L is usually the guideline for TDS (WHO, 2011).
- For industrial waters much lower guidelines are often adopted, e.g., 10 to 50 mg/L.
- Usually a second pass is installed when lower salinity is required.

9.4 REFERENCES

DuPont (2020) FILMTEC™ Reverse Osmosis Membranes Technical Manual. In: Water solutions (ed), pp. 207.

Nitto Hydranautics (2020) Integrated membrane solutions design (IMS Design), 2.226.1907.23.84 edn.

Toray Industries Inc (2020) Toray Design System 2, 2.1.10.179 edn.

Vrouwenvelder JS (2009) Biofouling of spiral wound membrane systems Ipskamp Drukkers, Enschede, Netherlands

WHO (2011) Guidelines for drinking-water quality, Fourth edn WHO Press

Chapter 10

Recent advances in SWRO and emerging membrane-based processes for seawater desalination

Gary L. Amy, Zhenyu Li, Lijo Francis and Noreddine Ghaffour

The main learning objectives of this chapter are the following:

- Have an overview of the recent progress in seawater reverse osmosis
- Know the various emerging membrane-based processes for seawater desalination
- Have an overview of innovations and trends in SWRO pre- and post-treatment

10.1 INTRODUCTION AND BACKGROUND

Seawater reverse osmosis (SWRO), a membrane-based process, has largely supplanted thermal processes as the conventional seawater desalination technology globally (Amy *et al.*, 2017), although some multi-effect distillation (MED) facilities have recently been constructed in Gulf Cooperation Council (GCC) region proximate to the Gulf of Arabia and the Red Sea. The SWRO technology itself is dynamically evolving through improvements in desalting performance (overall salinity rejection as well as problematical constituents (e.g., boron (B)), unit cost ($\$/m^3$), specific energy consumption (kWh/m^3), permeability (LMH/bar), and fouling resistance. There is considerable ongoing material-science work on improving SWRO performance through development of high-permeability RO (HP-RO) membranes, permitting less membrane area for a given operating pressure, and anti-fouling RO (AF-RO) membranes, enabling longer operational cycles between cleaning-in-place (CIP) events with less chemical wastes (spent cleaning solutions). In addition, novel process configurations such as closed circuit RO (CC-RO) and flow reversal RO (FR-RO) are being developed, using standard SWRO spiral wound elements, to promote higher water recovery with lower scaling.

Editors: Sergio Salinas, Jan Schippers, Gary Amy, In Kim, Maria Kennedy
doi: 10.2166/9781780409863_0265

In contrast to SWRO as a pressure-driven membrane process, the seawater-desalination technology landscape includes other emerging membrane processes driven by osmotic, temperature, and electrical gradients: forward osmosis (FO), membrane distillation (MD), and electrodialysis (ED), respectively. Given the energy intensity of SWRO systems (a specific energy consumption (SEC) approaching about 2.5 kWh/m^3 for the RO step itself), there is interest in harvesting salinity gradient energy (SGE) associated with SWRO brine as an energy offset through integration of SGE membranes processes such as pressure retarded osmosis (PRO) or reverse electrodialysis (RED). Our discussion below will focus on membrane-related processes since non-membrane desalting processes such as capacitive deionization (CDI) are currently limited to brackish (lower salinity) applications although adsorption desalination (AD) has recently been tested for seawater desalination at demonstration-scale. Moreover, while MD and FO are presently being marketed as brine concentration processes, their associated volume reduction increases recovery, providing desalinated water; moreover, with techno-economic improvements, they may have the ability to function as primary desalination processes in certain cases.

A deterrent to deployment of SWRO in the GCC region along the Gulf of Arabia and Red Sea has been challenging water quality conditions in terms of high salinity levels (about 40,000 – 45,000 ppm) , high temperatures (30 – 35 °C), and occurrence of harmful algal blooms (HABs), releasing algal organic matter (AOM) as an RO membrane foulant as well as algal toxins. Consequently, more HAB-resilient SWRO pretreatment processes such as dissolved air flotation (DAF) and/or ultrafiltration (UF) are increasingly being implemented, with growing interest in subsurface intakes that provide pretreatment through biofiltration (Missimer *et al.*, 2013). The recent development of hollow-fiber softening nanofiltration (NF) membranes, capable of being hydraulically-washed, may play a role in pretreatment for SWRO, providing scaling control with higher overall water recovery.

10.2 SEAWATER REVERSE OSMOSIS

10.2.1 Recent trends in seawater reverse osmosis (SWRO)

The present technology choice for new seawater desalination facilities clearly points to selection of SWRO, with 92% of new plants in 2018 being SWRO facilities (Voutchkov, 2019). The SWRO energy footprint has decreased to levels approaching a specific energy consumption of 2.5 and 3.5 kWh/m^3, respectively, for the RO step alone and the overall system. There is a thermodynamic limit to the energy required for desalting seawater with a practical limit of about 1 kWh/m^3, which is well below present practice. Present system units costs are in the general vicinity of \$1.00/m^3 (Voutchkov, 2018), depending on size and financing, roughly double that for drinking water from freshwater sources (notably, intakes and outfalls can account for up to 20 % of capital costs).

Some recent SWRO trends include: (*i*) a standard practice of integrating pressure recovery devices (retrospectively, a truly *disruptive* technology developed over 20 years ago); (*ii*) larger capacity facilities (e.g., Sorek, Israel, 624 MLD (commissioned in 2013)); (*iii*) larger elements (16-inch); (*iv*) new pretreatment approaches for challenging feed waters (e.g., ultrafiltration (UF) and dissolved air flotation (DAF); (*v*) improved operations (fouling control and sensors); and (vi) integration of renewable energy to drive desalination with associated reduction in greenhouse gas (GHG) emissions. Further SWRO process improvements are now focusing on improved SWRO membranes, higher permeability and lower fouling, and new process configurations, e.g., semi-batch and flow-reversal modes of operation.

Given environmental concerns about SWRO (Elimelech and Philip, 2011), there has been a movement toward the *greening* of SWRO (Lattemann *et al.*, 2010), focusing on intakes (minimization of impingement and entrainment), chemical minimization (chemical-free pretreatment), waste generation (less frequent cleaning-in-place (CIP_0), energy footprint minimization (integration of salinity gradient energy), GHG reduction (integration of renewable energy), and outfalls (minimization of impacts on marine ecosystem through a multi-port diffuser).

10.2.2 High permeability reverse osmosis (HR-RO) membranes

The fabrication of HP-RO membranes can provide a lower energy consumption and added operational flexibility in achieving higher flux at typical SWRO operating pressures or comparable flux at lower pressures, although for the former, scaling must be controlled. Material science innovations have opened up this area through fabrication of mixed-matrix inorganic-organic nanocomposite membranes (e.g. impregnation of zeolites, metal oxide frameworks (MOFs), or biocides like Ag nanoparticles), biomimetic membranes (aquaporins, synthetic water and ion channels) and graphene oxide (GO) membranes (stacked GO sheets) (Pendergast and Hoek, 2011). However, there is a limit to lowering SWRO pressure because one cannot escape the inherent osmotic pressure penalty (30 bar for 35,000 ppm seawater). Another challenge is to ensure that there is no sacrifice in salt or salt-constituent rejection while increasing water permeability; from the opposite perspective, boron selective membranes (Rahmawati *et al.*, 2012) have been developed without a sacrifice in permeability and are commercially available. Thus far, only inorganic-organic nanocomposite and aquaporin RO membranes have been commercialized for desalination while other HP-RO membranes are still under development (Subramani and Jacangelo, 2015).

10.2.3 Anti-fouling reverse osmosis (AF-RO) membranes

A general approach to fabrication of AF-RO membranes has been surface modification of conventional RO membranes by physical and chemical methods (Choudhury *et al.*, 2018; Shahkaramipour *et al.*, 2017), including creation of surface patterns, as well as development of organic/inorganic composite RO membranes with good fouling resistance. Materials with antifouling properties may be coated or grafted onto the membrane surface, including hydrophilic materials (e.g., zwitterions). Work has also been done on evaluating Ag-impregnated polymeric membrane for biocidal properties to inhibit biofouling. There are a number of commercial products available but their surface chemistries and compositions are largely proprietary. Besides promoting membrane surface properties to minimize fouling, work has also been done on modifying membrane spacers and their geometry to promote near-surface hydrodynamic conditions that minimize fouling.

10.2.4 Closed circuit reverse osmosis (CC-RO)

A semi-batch mode of RO operation, designated as closed circuit RO (CC-RO), provides an opportunity for further reduction in SWRO energy consumption and water recovery (Efrati, 2011). In CC-RO, standard RO elements are incorporated in the CC-RO system. CC-RO is operated as a semi-batch process (versus steady-state conventional RO) and runs at up to 98% recovery in a single stage. The mode of operation does not produce concentrate for a period, then goes through a flush cycle, and then resumes closed circuit operation. Cross-flow is provided by a circulation pump to limit fouling/scaling. CC-RO is projected to attain a 15 – 20 percent reduction in specific energy consumption, suggesting that it could approach 2.0 kWh/m^3 in seawater desalination although one should also consider the complexity and additional costs of the CC-RO system. Compared to continuous RO, model simulations of CCRO showed up to 37% energy savings for *brackish* water desalination at a high water recovery (Warsinger *et al.*, 2016). Thus far, CC-RO has only been deployed at smaller scale for seawater desalination.

10.2.5 Flow reversal reverse osmosis (FR-RO)

The operational mode of FR-RO involves a feed-flow reversal in an RO element, achieved by switching the connections of feed and concentrate before a supersaturated solution can precipitate from the concentrate onto the membrane. The timing is determined by knowledge of feed composition and operating conditions. Scaling is prevented by changing conditions to under-saturation before a super-saturated solution can precipitate as determined by its induction time. Besides minimizing or eliminating anti-scalant chemical addition to control scaling, an overall increase in water recovery and decrease in residual brine is also realized as well as less-frequent cleaning-in-place (CIP). Thus far, FR-RO has only been deployed for brackish water desalination at smaller scale.

10.3 OTHER MEMBRANE-BASED SEAWATER DESALINATION PROCESSES

10.3.1 Forward osmosis (FO) desalination

While forward osmosis (FO) can be adapted to other applications (e.g., brine concentration), the focus here is on its potential role in desalination. Osmotic pressure is employed as the driving force in the FO process, created by the osmotic gradient between a draw solution (DS) with high concentration (high osmotic pressure) and a feed solution (FS) with low concentration (low osmotic pressure). The DS flows along one side of the FO membrane while the FS along the other side. Water is extracted from the lower osmotic pressure FS into the higher osmotic pressure DS while the FS constituents are rejected by FO membrane as a barrier. Since water movement is driven by the osmotic pressure, almost no external hydraulic pressure is required in the FO process. FO is generally hybridized with other processes to recover product water from the diluted DS and/or reuse DS components. An attribute of the FO process is that water flux decline due to FO fouling is lower than conventional pressure-driven RO processes for desalination and most FO fouling is reversible (Holloway *et al.*, 2007; Siddiqui *et al.*, 2018).

Opportunities for improvement in the forward osmosis (FO) process can be discussed within the context of membranes, draw solutions, process configurations, and hybrids. The present FO commercialization landscape includes a number of companies who operate as suppliers of FO membranes/modules, draw solutions, and/or processes/systems. FO membranes include commercially-available flat sheet membranes (polymeric and biomimetic) and under-development hollow fiber membranes that are in the translational research arena. Membrane advances have come through achieving support-layer properties to minimize internal concentration polarization; open support layer, low thickness, high porosity, and low tortuosity; and separation layer properties to achieve high flux and low salt leakage. FO modules include plate-and-frame and spiral wound for flat sheet membranes and under-development hollow fiber membrane modules. FO draw solutions include simple salts (e.g., NaCl), volatile salts influenced by temperature (e.g., NH_4HCO_3/NH_3-CO_2), hydrogels (temperature or pH sensitive), and switchable polarity solvents (temperature or pH sensitive). Temperature-dependent draw solutions are deemed as thermolytic. Attributes of an ideal draw solution include high osmotic pressure, low reverse draw solute flux, easy regeneration, non-toxic, and inexpensive. Presently, only NaCl and thermolytic draw solutions (volatile salts and hydrogels) have found their way into practice (for brine control). Given significant recent improvements in FO membranes, effective draw solutions have arguably become the limiting-factor component in further advancing the FO technology, including potential seawater desalination.

There are two alternatives for forward osmosis desalination (FOD) (Valladares-Linares *et al.*, 2014): (*i*) *direct* FOD with a saline *feed* water and a synthetic draw solution with a subsequent draw solution recovery step to recirculate draw-solution components and produce product water or (*ii*) *indirect* FOD with an impaired-quality feed water (e.g., a municipal wastewater effluent) and a saline *draw* solution (i.e., seawater) with the FO step extracting water from the feed, followed by a low pressure RO (LPRO) step to desalt the diluted draw solution, i.e., an FO-LPRO hybrid, to thus reduce the total cost of the desalination process (Devin *et al.*, 2015). The estimated specific energy consumption for FO-LPRO has been estimated to be 2.5 kWh/m^3 (0.7 for FO & 1.8 for LPRO) without pretreatment (Valladares-Linares *et al.*, 2016), comparing favorably to SWRO alone. Moreover, the FO step can provide pretreatment for the LPRO step. Unless low cost renewable energy or waste heat is available, direct FO desalination cannot reduce the energy consumption required for desalinations, regardless of the type of DS used (McGovern *et al.*, 2014; Elimelech *et al.*, 2011). In the case of FO alone as a stand-alone process, it only provides osmotic dilution at a low-energy investment, necessitating another step: e.g., low-pressure RO or a draw solution recovery and recirculation process, with an added energy investment, to truly achieve desalting.

10.3.2 Membrane distillation (MD) desalination

As with FO, membrane distillation (MD) has received more attention as a brine concentration process, but the discussion here will be limited to it as a desalination process. The driving force for the MD process is the partial vapor pressure difference between the two sides of a hydrophobic microporous membrane. MD yields high quality fresh water by desalting seawater with 99.99% rejection of salt or non-volatile particles (Camacho *et al.*, 2013; Alkhudhiri *et al.*, 2012). The MD process operates at low pressures (< 2 bars) and at low temperatures (30-80 °C), making it suitable for the use of low grade energy sources such as waste heat, geothermal or solar energy. Furthermore MD is a compact process and hence uses less space (foot print) and requires less maintenance (Francis *et al.*, 2014; Francis, *et al.*, 2013).

The present MD commercialization landscape encompasses several companies mostly using air gap membrane distillation (AGMD) with flat sheet membranes and, to a lesser extent, an extension of vacuum membrane distillation (VMD) deemed as vacuum multi-effect MD (VMEMD). In AGMD, a stagnant air gap is maintained between the membrane and a condensation surface on the permeate side. In VMD, a vacuum is exerted on the permeate side so that the vapor passing across the membrane from the feed side condenses outside the MD module. Energetically, AGMD is more favorable than direct contact membrane distillation (DCMD) which is often used as a standard testing protocol of new membranes. However, VMD is also more energetically attractive than DCMD. Considering AGMD versus VMD, AGMD is constrained to the use of flat sheet membranes because of the need for a condensation surface whereas VMD can operate using either flat sheet or hollow fiber membranes, with a more favorable footprint. AGMD provides a lower flux but seawater can be used as a coolant and it is thermally efficient; conversely, VMD provide a higher flux but requires a vacuum and it is vulnerable to membrane pore wetting.

There are commercially-available MD membranes in a flat sheet configuration as well as hollow fiber configuration but many of the hollow-fiber MD membranes are considered to be *dual-purpose (re-purposed) microfiltration (MF) membranes* with generally hydrophobic properties. Given the present dominant status of flat sheet membranes, further development of hollow fiber MD membranes may provide more favorable physical and energy footprints. High-performance hollow fiber MD membranes are now under development in Singapore. Compared to SWRO desalting, MD can accommodate a much higher feed stream (e.g., 200,000 ppm salinity), making it attractive for higher salinity seawater (e.g., Gulf of Arabia), and exhibits different scaling and fouling issues.

MD desalination involves high thermal (e.g., 100 kWh/m^3) but low electrical (e.g., pumping) energy investments. Given that thermal energy requirements have been a major deterrent to adopting the technology, there is great interest in integration of MD with (low grade) waste heat through waste heat recovery and utilization. The electrical energy requirement can be low, e.g., 1 kWh/m^3 versus about 2.5 kWh/m^3 for SWRO, excluding pretreatment. Thus, there is an interest in implementing MD in large industrial complexes where various industries have both waste heat availability and desalted water needs. Another area of interest is MD integrated with, and driven by, solar energy (Solar-MD) with the main focus being on solar-thermal energy for direct heat utilization. Attributes of solar-thermal MD include reduced GHG emissions and potential deployment in off-the-grid remote/rural locations).

10.3.3 Electrodialysis (ED) desalination

While ED has traditionally been a brackish-water desalination process, it has recently been applied to seawater desalination through a demonstration project funded by the Singapore Public Utilities Board in support of Evoqua. The Evoqua approach has been to design a process consisting of ED modules in series with each step operated with a unique ion exchange membrane having a specific resistance and thickness fabricated for different salinity conditions. The resultant electro-deionization (EDI) technology, designated as NexED® by Evoqua, has been able to desalt seawater with a specific energy consumption of 2.2 kWh/m^3 at demonstration-scale (3,800 m^3/day feed capacity), with a longer-term target of achieving 1.5kWh/m^3. EDI is similar to ED but also includes mixed-bed ion-exchange resins between anion and cation membranes (o facilitate ion removals.

10.4 MEMBRANE BASED SALINITY GRADIENT ENERGY PROCESSES

10.4.1 Pressure retarded osmosis (PRO)

PRO is an osmotically-driven membrane process that is similar to FO process, but there is an applied hydraulic pressure (e.g. a piston) at the DS side. The volume expansion in the DS by extracting fresh water from the low salinity side using osmotic pressure is restricted and increases the hydraulic pressure of the DS reservoir. The pressurized flow of DS is then driven through a hydro turbine to generate power (Tufa *et al.*, 2015). Salinity gradient energy (SGE) is production from a higher-salinity water (draw solution; e.g., a desalination brine) used in combination with a low-value/-salinity water (feed solution; e.g., a municipal wastewater effluent).

There are commercially-available PRO membranes: a flat sheet membrane/spiral wound module and a hollow *fine* fiber membrane/module; high-performance (i.e., high-power density (W/m^2) hollow fiber PRO membranes are under development [Chung *et al.*, 2015]. A key distinction between PRO versus FO membrane properties is added mechanical strength in the former because of the need to also withstand a maximum operating pressure (bar).

SWRO-PRO is a potential PRO hybrid (Achilli *et al.*, 2014), representing an opportunity for retrofit of SWRO facilities for potential PRO energy offset of up to about 0.5 kWh/m^3. Moreover, environmental impact mitigation is provided by harnessing energy without GHG emissions, and brine dilution is provided before marine disposal. The seawater-river water combination with PRO was earlier shown to not be viable in a demonstration project (i.e., the former Statkraft Project in Norway) due to insufficient salinity gradient as well as both seawater and river water fouling issues; however, these limitations may be overcome in the future by higher power-density/lower-fouling PRO membranes. Moreover, Straub *et al.* (2016) showed that PRO is particularly proficient at extracting salinity energy from large concentration differences, e.g., SWRO brine. The Singapore Public Utilities Board (PUB) has recently been exploring the use of a wastewater RO brine, a low value FS, coupled with SWRO brine (Wan and Chung, 2015).

There have recently ben two major PRO demonstration projects in Japan and Korea. The Japanese Mega-ton Water System project constructed a PRO pilot plant at Fukuoka to use RO brine and treated wastewater for power generation (Kurihawa, 2015). A maximum PRO power density of 13.3 W/m^2 was achieved (He *et al.*, 2015). The Korean National Research Project, Global MVP (Membrane Distillation, Valuable Source Recovery, and PRO), harvested osmotic pressure by using RO brine as the DS and wastewater as the FS. Coupled with a high efficiency isobaric pressure exchanger, the harvested osmotic pressure was directly applied to pre-pressurize the seawater before RO and reduce the overall energy consumption of desalination process (Anastasio *et al.*, 2015).

10.4.2 Reverse electrodialysis (RED)

RED is another membrane-related salinity energy gradient process, building upon the electrodialysis (ED) process. Like ED, RED employs an array (stack) of cation (CEM) and anion (AEM) exchange membranes which are separated by spacer channels that allow for water flow along the membranes to transport ions (not water), with electrodes capturing electrical current associated with the flow of ions (Mei and Tang, 2018). Both AEMs and CEMs allow only the passage of counter-ions and the co-ions are rejected. Like PRO, both a saline brine and low-salinity water are required to provide a salinity gradient.

Key issues for further process development include improved (lower resistance) CEM and AEM membranes, optimal stack configuration, and applications involving a higher osmotic-pressure gradient. Work by Vermaas *et al.* (2013) showed the dependence of RED energy efficiency on the ratio between the magnitudes of flow, flow directions (counter- versus co-) of higher salinity water (e.g., seawater) and lower salinity water (e.g., river water flow), and the number of electrode segments. As with PRO membranes, a targeted attribute of RED membranes is power density (W/m^2). An existing demonstration project in the Netherlands (http://www.redstack.nl) involves the use of North Sea water and IJssel Lake (derived from the Rhine River) water, separated by a dike. An advantage of RED versus PRO is that electricity is generated directly from a salinity gradient whereas a turbine would be required by PRO (Logan and Elimelech, 2012).

10.5 RENEWABLE ENERGY-DRIVEN DESALINATION

When energy for seawater desalination is provided by fossil fuels, there is also a significant carbon footprint of desalting seawater, e.g., up to about 1 kg CO_2^-eq/kWh, depending on fossil-fuel mix. This has promoted an interest in renewable energy (RE) (Ghaffour *et al.*, 2015; Ghaffour *et al.*, 2014), especially solar, to drive SWRO as well as ED, MD, and thermolytic FO. RE options include solar-thermal (solar heat collectors or concentrated solar power (CSP)) and solar electric (photovoltaic) as well as wind and geothermal. For solar, there is a tradeoff between capital costs (e.g., solar-electric PV panel or solar-thermal collector investments) and operational (energy) costs. A key constraint for solar is its intermittency, necessitating storage or augmentation by the electrical grid; an innovative hybrid approach being considered is combining solar and geothermal energy using an alternating 12-hour cycle. RE electricity is required for SWRO and ED while MD and thermolytic FO are thermally-driven processes.

The world's largest solar-electric SWRO (60,000 m^3/day) is located in Al Khafji, Saudi Arabia next to a neighboring solar power plant with a capacity of 20 mW, but is also connected to the electrical grid. The Perth Australia SWRO facility is the largest plant (capacity of 140,000 m^3/d and energy demand: (3.5 kWh/m^3) driven by wind energy, indirectly. The plant's total energy consumption is offset by energy production from a wind farm 260 km from the plant with 67 turbines producing 132 MW versus 82 MW needed by plant. The Perth RE approach is one of energy compensation with grid connections. Another related opportunity would be to integrate waste heat recovery and utilization to substitute for solar-thermal to drive MD and thermolytic-FO, both needing only low-grade waste heat (60 – 80 °C) or solar-thermal.

10.6 INNOVATIONS AND TRENDS IN SWRO PRE- AND POST-TREATMENT

10.6.1 Innovations in SWRO pre-treatment

Conventional seawater pretreatment typically consists of dual-media filtration (DMF) followed by cartridge filtration. However, more robust pretreatment schemes are now being implemented in response to challenging water quality conditions associated with HABs, including ultrafiltration (UF) and/or dissolved air flotation (DAF). UF has been shown to be HAB-resilience through changes in operating conditions, e.g., coagulant addition, lower flux, and more frequent backwashing. DAF has also shown an ability to remove oil-and-grease associated with shipping-channel impacts on seawater quality. There is increasing interest in subsurface impacts such as seabed galleries given their ability to provide pretreatment for SWRO through biodegradation of organic- and bio-foulants.

There is a growing consensus that biopolymers and transparent exo-polymeric particles (TEP) are principal SWRO organic foulants, with TEP serving as a precursor to SWRO membrane biofouling promoted by assimilable organic carbon (AOC) (Qasim., 2019). Removal of these components by pretreatment minimizes the need for pre-chlorination and de-chlorination before the RO step. Calcium carbonate ($CaCO_3$) and sulfate ($CaSO_4$) remain as the principal SWRO scalants, with the former controlled by acid addition and the latter by an anti-scalant. Emerging strategies for SWRO scaling control include managing first-stage SWRO recovery to permit acid addition only; when necessary, use a biodegradable anti-scalant; consider NF pretreatment for Ca^{2+} and Mg^{2+} removal; and/or consider new RO process configurations like FR-RO to control scalant formation through induction time.

10.6.2 SWRO post-treatment trends

Given that seawater, on average, contains 65 mg/L of bromide (Br^-), past work has shown that final disinfection after SWRO using chlorine produces significant amounts of brominated disinfection by-products (DBPs), even though there is fairly effective Br-retention by SWRO membranes. However, more recent work has shown the formation of significant levels of iodinated DBPs from iodide (I^-) present, having a higher public health concern; moreover, it has been shown that about half of the total iodine in seawater (50 - 60 ug/L) exists as iodate (IO_3^-), itself considered as a potentially harmful DBP. Our understanding of DBPs after SWRO lags behind that for freshwater drinking-water sources.

10.7 REFERENCES

Achilli, A., Prante, J. L., Hancock, N. T., Maxwell, E. B., Childress, Amy, Experimental Results from RO-PRO: A Next Generation System for Low-Energy Desalination, Environ. Sci. Technol., 48 (11), 6437–6443 (2014).

Alkhudhiri, A., Darwish, N., Hilal, N., Membrane distillation: a comprehensive review, Desalination 287:2-18 (2012).

Amy, G., Ghaffour, N., Li, Z., Francis, L., Valladares Linares, R., Missimer, T., Lattemann, S., Membrane-based seawater desalination: Present and future prospects, Desalination, 401:16–21 (2017).

Anastasio, D., Arena, J., Cole, E., McCutcheon, J., Impact of temperature on power density in closed-loop pressure retarded osmosis for grid storage, J. Membr. Sci. 479:240-245 (2015).

Camacho, L. M., Dumée, L., Zhang, J., Li, J., Duke, M., Gomez, J., Gray, S., Advances in Membrane Distillation for Water Desalination and Purification Applications, Water, 5, 94-196; doi:10.3390/w5010094 (2013).

Choudhury, R., Gohil, J., Mohanty, S., Nayak, S., Antifouling, fouling release and antimicrobial materials for surface modification of reverse osmosis and nanofiltration membranes, Journal of Materials Chemistry, 2:313-333 (2018).

Chung, T-S, Luo, L., Wan, C. F., Cui, Y., and Amy, G., What is next for forward osmosis (FO) and pressure retarded osmosis (PRO), Separation and Purification Technology 156, 856–860 (2015).

Devin, L., Shaffer, D.L., Werber, J.R., Jaramillo, H., Lin, S.H., Elimelech, M., Forward osmosis: Where are we now? Desalination 356, 271-284 (2015).

Elimelech, M., and Philip, W., The Future of Seawater Desalination: Energy, Technology, and the Environment, Science 333, 712 (2011).

Francis, L., Ghaffour, N., Alsaadi, A., Amy, G., Material gap membrane distillation: a new design for water vapor flux enhancement, J. Membr. Sci. 448:240-247 (2013).

Francis, L., Ghaffour, N., AlSaadi, A., Amy, G., PVDF Hollow Fiber and Nanofiber Membranes for Fresh Water Reclamation using Membrane Distillation, Journal of Material Science 49:2045-2053 (2014).

Efrati, A., Closed circuit desalination series no-4: high recovery low energy desalination of brackish water by a new single stage method without any loss of brine energy, Desalination and Water Treatment, 31, 95-101 (2011).

Elimelech, M., Phillip, W., The future of seawater desalination: Energy, technology, and environment. Science , 333:712-717 (2011).

Ghaffour, N., Bundschuh, J., Mahmoudi, H., Goosen, M., Renewable energy-driven desalination technologies: A comprehensive review on challenges and potential applications of integrated systems, Desalination 356, 94-114 (2015).

Ghaffour, N., Lattemann, S., Missimer, T., Sinha, S., Amy, G., Renewable energy-driven innovative energy-efficient desalination technologies, Applied Energy, 136, 1155-1165 (2014).

Holloway, R., Childress, A., Dennett, K., Cath, T., Forward osmosis for concentration of anaerobic digester centrate, Water Res. 41:4005-4014 (2007).

Kurihara, M., Role of Pressure Retarded Osmosis (PRO) in the MEGATON Project, http://www.desaltech2015.com/assets/presenters/Kurihara_Masaru.pdf (2015).

Lattemann, S., Kennedy, M.D., Amy, G., Seawater desalination - A green technology?, Journal of Water Supply: Research and Technology - AQUA 59 (2-3), 134-151 (2010).

Logan, B. E., and Elimelech, M., Membrane-based processes for sustainable power generation using water, Nature, 488, 313 (2012).

McGovern, R., Lienhard, J., On the potential of forward osmosis to energetically outperform reverse osmosis desalination, J. Membr. Sci. 469:245-250 (2014).

Mei, Y., Tang, C., Recent developments and future perspectives of reverse electrodialysis technology: A review, Desalination, 425:156–174 (2018).

Missimer, T.M., Ghaffour, N., Dehwah, A.H.A., Maliva, R.G., Amy, G., Subsurface intakes for seawater reverse osmosis facilities: Capacity limitation, water quality improvement, and economics, Desalination, 322, 37-51 (2013).

Qasim, M., Mohamed Badrelzaman, M., Noora N. Darwish, Noora, Darwish, Naif, Hilal, N., Reverse osmosis desalination: A state-of-the-art review, Desalination, 459:59–104 (2019).

Rahmawati, K., Ghaffour, N., Aubry, C., Amy, G.L., Boron removal efficiency from Red Sea water using different SWRO/BWRO membranes, Journal of Membrane Science 423-424, 522-529 (2012).

Shahkaramipour, N., Tran, T., Ramanan, S., Lin, H., Membranes with Surface-Enhanced Antifouling Properties for Water Purification, Membranes 7, 13 (2017). doi:10.3390/membranes7010013

Siddiqui, F., She, Q., Fane, A., Field, R., Exploring the differences between forward osmosis and reverse osmosis fouling, J. Membr. Sci. 565:241-253 (2018).

Straub, A., Deshmukh, , Elimelech, M., Pressure-retarded osmosis for power generation from salinity gradients: is it viable?, Energy Environ. Sci. 9:31-48 (2016).

Subramani, A., Jacangelo, J. G., Emerging desalination technologies for water treatment: A critical review, Water Research 75, 164 -187 (2015).

Valladares Linares, R., Li, Z., Yangali-Quintanilla, V., Ghaffour, N., Amy, G., Leiknes, T., Vrouwenvelder, H., Life cycle cost of a hybrid forward osmosis – low pressure reverse osmosis (FO–LPRO) system for seawater desalination and wastewater recovery, Water Research, 88, 225 – 234 (2016).

Valladares-Linares, R., Li, Z., Sarp, S., Bucs, S., Amy, G., Vrouwenvelder, J., Forward osmosis niches in seawater desalination and wastewater reuse, Water Res. 66:122-139 (2014).

Vermaas, D. A., Veerman, J., Yip, N. Y., Elimelech, M., Saakes, M., Nijmeijer, K., High Efficiency in Energy Generation from Salinity Gradients with Reverse Electrodialysis, ACS Sustainable Chem. Eng. 1, 1295–1302 (2013).

Voutchkov, N., Seawater Desalination Current Status and Trends, Texas Desalination Coastal Issues Forum (2019)

Voutchkov, N., Energy use for membrane seawater desalination – current status and trends, Desalination, 431:2–14 (2018).

Wan, C. F., and Chung, T. S., Osmotic power generation by pressure retarded osmosis using seawater brine as the draw solution and wastewater retentate as the feed

Journal of Membrane Science, 479, 148–158 (2015).

Warsinger, D., Tow, E., Nayar, K., Maswadeh, L., Lienhard, J., Energy efficiency of batch and semi-batch (CCRO) reverse osmosis desalination, Water Research, 106:272-282 (2016).

IWA Publishing's authorised EU representative for General Product Safety Regulations is Diane D'Arras, 15 rue Duret, 75116 Paris, France, e-mail: safety@iwap.co.uk.

Printed and bound by CPI Group (UK) Ltd, Croydon, CR0 4YY
12/05/2026
02108884-0001